An Introductory Course in Computational Neuroscience

Computational Neuroscience
Terrence J. Sejnowski and Tomaso. A Poggio, editors

For a complete list of books in this series, see the back of the book and https://mitpress.mit.edu/books/series/computational-neuroscience

An Introductory Course in Computational Neuroscience

Paul Miller

The MIT Press
Cambridge, Massachusetts
London, England

This book was set in Times by Toppan Best-set Premedia Limited Printed and bound in the United States of America.

Library of Congress Cataloging-in-Publication Data

Names: Miller, Paul, 1969- author.
Title: An introductory course in computational neuroscience / Paul Miller.
Description: Cambridge, MA : The MIT Press, 2018. | Series: Computational
  neuroscience series | Includes bibliographical references and index.
Identifiers: LCCN 2018003118 | ISBN 9780262038256 (hardcover : alk. paper)
Subjects: LCSH: Computational neuroscience--Textbooks. |
  Neurosciences--Mathematics.
Classification: LCC QP357.5 .M55 2018 | DDC 612.8/233--dc23 LC record available at https://lccn.loc.
gov/2018003118

10 9 8 7 6 5 4 3

Contents

Series Foreword

Computational neuroscience is an approach to understanding the development and function of nervous systems at many different structural scales, including the biophysical, the circuit, and the systems levels. Methods include theoretical analysis and modeling of neurons, networks, and brain systems and are complementary to empirical techniques in neuroscience. Areas and topics of particular interest to this book series include computational mechanisms in neurons, analysis of signal processing in neural circuits, representation of sensory information, systems models of sensorimotor integration, computational approaches to biological motor control, and models of learning and memory. Further topics of interest include the intersection of computational neuroscience with engineering, from representation and dynamics, to observation and control.

Terrence J. Sejnowski

Tomaso Poggio

Acknowledgments

I am grateful to the following people for their constructive comments and suggestions, which helped improve this book: Jonathan Cannon, Irv Epstein, John Ksander, Stephen Lovatt, Eve Marder, Alexandra Miller, Candace Miller, Ray Morin, Narendra Muckerjee, Alireza Soltani, Stephen Van Hooser, Ryan Young; and the following members of the Brandeis University Computational Neuroscience Classes (Spring 2017 and Spring 2018): Taniz Abid, Rabia Anjum, Apoorva Arora, Sam Aviles, Remi Boros, Brian Cary, Kieran Cooper, Ron Gadot, Sophie Grutzner, Noah Guzman, Lily He, Dahlia Kushinsky, Jasmine Quynh Le, Andrew Lipnick, Cherubin Manokaran, Sigal Sax, Nathan Schneider, Daniel Shin, Elizabeth Tilden, David Tresner-Kirsch, Nick Trojanowski, Vardges Tserunyan, and Jeffrey Zhu.

Several tutorials in this book evolved from course materials produced by Larry Abbott, Tim Vogels, and Xiao-Jing Wang, to whom I am grateful for introducing me to neuroscience.

I am particularly thankful to Candace Miller for her encouragement during this enterprise and to Brandeis University for its support.

Preface

I designed this book to help beginning students access the exciting and blossoming field of computational neuroscience and lead them to the point where they can understand, simulate, and analyze the quite complex behaviors of individual neurons and brain circuits. I was motivated to write the book when progressing to the "flipped" or "inverted" classroom approach to teaching, in which much of the time in the classroom is spent assisting students with the computer tutorials while the majority of information-delivery is via students reading the material outside of class. To facilitate this process, I assume less mathematical background of the reader than is required for many similar texts (I confine calculus-based proofs to appendices) and intersperse the text with computer tutorials that can be used in (or outside of) class. Many of the topics are discussed in more depth in the book *Theoretical Neuroscience* by Peter Dayan and Larry Abbott, the book I used to learn theoretical neuroscience and which I recommend for students with a strong mathematical background.

The majority of figures, as well as the tutorials, have associated computer codes available online, at github, https://github.com/primon23/Intro-Comp-Neuro, at my website, http://people .brandeis.edu/~pmiller, and at the website of MIT Press, https://mitpress.mit.edu/computation alneuroscience. I hope these codes may be a useful resource for anyone teaching or wishing to further their understanding of neural systems.

1

Preliminary Material

When using this book for a course without prerequisites in calculus or computer coding, the first two weeks of the course (at a minimum) should be spent covering the preliminary material found in chapter 1. The contents of the different sections of this chapter are introduced here.

1.1 Introduction

1.1.1 The Cell, the Circuit, and the Brain

In my experience, many students who enjoy solving mathematical or computational problems take a course such as computational neuroscience as their first introduction to neuroscience, or even as their first university-level course in the life sciences. For such students, section 1.2 offers a very basic summary of the meaning and relevance of biological and neurological terms that are used but not introduced elsewhere in the book. The newcomer to neuroscience should read section 1.2 before commencing the course.

1.1.2 Physics of Electrical Circuits

The ability of neurons to convey and process information depends on their electrical properties, in particular the spatial and temporal characteristics of the potential difference across the neuron's membrane—its *membrane potential*. Nearly all of single neuron modeling revolves around calculating the causes and effects of changes in the membrane potential. To understand fully the relevant chapters in this book, it is first necessary for the reader to appreciate some of the underlying physics, so a background is provided in section 1.3 of this chapter.

Membrane potential, V_m   The potential difference across the membrane of a cell, which is highly variable in neurons, ranging over a scale of tens of millivolts.

1.1.3 Mathematical Preliminaries

The universe runs on differential equations, thanks to the continuity of space and time. The same applies to the brain, so at the heart of this computational modeling course is the requirement to write computer codes that solve differential equations. This may sound daunting, but it is in fact a lot easier than solving the same differential equations by the analytical methods one might find in a mathematics course. As a preliminary to delving into the various specific ordinary differential equations that we will find in this course, it is first important to understand what an ordinary differential equation is and what it means.

Variable   A property of the system that changes with time.

Parameter   A property of the system that is fixed during an experiment or simulation.

Differential equation   An equation describing the rate of change of a variable in terms of all quantities—which may include the current value of the variable—that impact its rate of change.

Differential equations describe rates of change in terms of the current properties of a system. Those properties that change are *variables* of the system. A complete description of the system requires one differential equation for each variable. Typically, in neuroscience, the rate of change of one variable depends on other variables, in which case the description of the system comprises coupled (i.e., connected) ordinary differential equations.

Leak potential, E_L   The value that the membrane potential a neuron returns to after any temporary charge imbalance leaks away.

Leak conductance, G_L   The ability of charge to leak in and out of the cell through its membrane, the inverse of the membrane resistance. G_L increases with the surface area of the cell.

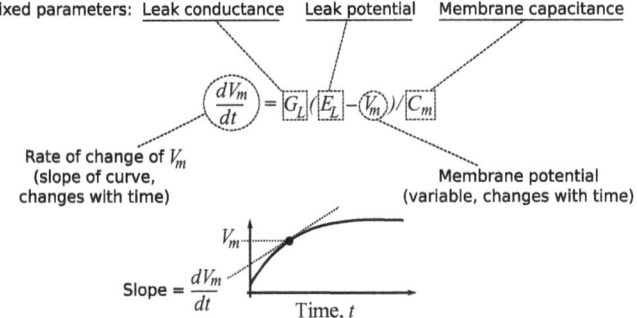

Figure 1.1
Annotation of a differential equation. The equation $dV_m/dt = G_L(E_L - V_m)/C_m$ is evaluated at each point in time (each point on lower curve), to determine the slope of the curve, which is the rate of change of the variable, V_m. By knowing the rate of change, the complete curve of V_m as a function of time can be determined. The parameters G_L, E_L, and C_m are fixed electrical properties of the neuron, which would be provided in any question, or must be measured in any experiment.

Membrane capacitance, C_m The capacity of the cell's membrane to store electrical charge, a quantity that increases with the surface area of the cell.

For now, we will focus on a single variable changing in time. For example, the most common form of equation we will come across in this course can be written (see figure 1.1):

$$\frac{dV_m}{dt} = G_L(E_L - V_m)/C_m \tag{1.1}$$

or equivalently

$$\dot{V}_m = G_L(E_L - V_m)/C_m \tag{1.2}$$

(the first form with dV_m/dt is due to Leibniz, the second with $\dot{V}_m$ is due to Newton—these were the two feuding originators of calculus). Note that a single differential equation has just a single variable, in this case the membrane potential, V_m, whose rate of change at any point in time is given by evaluating the quantity on the right-hand side. In the absence of other information, we must assume the other quantities: in this case leak conductance, G_L, leak potential, E_L, and membrane capacitance, C_m, are all fixed *parameters*. The equation is not simple to solve, because as the membrane potential changes, so too does the term on the right-hand side, which tells us its rate of change. A little calculus will tell us the solution, but if we are writing a computer code we can proceed without calculus.

Section 1.4 provides more mathematical background for this process. Section 1.6 provides methods for solving such equations computationally.

1.1.4 Writing Computer Code

A major aim of this book is to enable students with limited or no prior coding background to solve interesting scientific problems by computer simulation or to analyze large datasets with sophisticated methods. For those to whom MATLAB or writing computer code is a new enterprise, it is essential to spend as much time as necessary, at least several hours, getting comfortable with the coding language before attempting tutorial 2.1. To aid in this process, section 1.5 provides an introductory MATLAB tutorial (there are many others available online) that focuses on the essential skills needed for this course. Once the tutorial in section 1.5 is completed, the new programmer will find section 1.6 a helpful first step in solving many of the problems found in the tutorials of this book.

1.2 The Neuron, the Circuit, and the Brain

This section is designed for students who have not taken any prior courses in neuroscience and perhaps have little background in cell biology.

1.2.1 The Cellular Level

Neurons are cells found in animals, so they possess all of the standard components of animal cells: a nucleus containing chromosomes of DNA where transcription takes place (production of RNA in the first step of protein synthesis); the nucleus is surrounded by a nuclear envelope outside of which is the cytoplasm containing various specialized subcompartments called organelles, such as mitochondria, the "power packs" of the cell; the cytoplasm itself is surrounded by a cell membrane marking the boundary of the cell. The cell membrane, also called the plasma membrane, consists of a lipid bilayer and is an electrical insulator. Within the membrane are situated transmembrane proteins, the most important of which for neural function are those that bind together to form ion channels. Ion channels allow specific ions to flow into and/or out of the cell.

Soma   The main body of a cell, containing the nucleus.

Process   A part of a cell that is elongated, extending far from the soma.

Dendrites   Branched processes of a neuron that usually receive input from other cells or receptors and transfer the input to the soma.

Axons   Branched processes of a neuron, usually thinner so less visible than dendrites, which usually convey neural activity generated in the soma or in the part of the axon adjacent to the soma (its initial segment, also called the "axon hillock"), to other neurons.

Neurons are unusual in their spatial structure, which allows them to connect with specificity to other neurons that may not be close neighbors. While the nucleus resides in the cell's body (its soma) branching structures (processes) that connect to other cells are contiguous with the soma (figure 1.2). Dendrites are such processes, named after their treelike structure, that extend over hundreds of microns to receive inputs from other cells. Another branching process, the axon, can extend for over a meter in some animals.

Neurons are also unusual in their active electrical properties, the most prominent of which is their ability to produce and transmit a large pulse of deviation in the electrical potential difference across the cell membrane. Such a pulse, called an action potential or simply "spike," is carried away from the cell body down the axon. The voltage spike is caused by active ion channels, which open or close in response to changes in the potential difference across the cell membrane. In chapter 4 we will study and simulate the cause and properties of such a spike.

Vesicle   A small spherical container of biochemical molecules.

At the terminals of the axon, the spike in the membrane potential causes voltage-gated channels to open and admit a brief pulse of calcium ions. These calcium ions activate proteins that lead to release of one or more vesicles—spheres surrounded in lipid and containing a cargo. In this case, the cargo is a neurotransmitter, which upon release of a vesicle diffuses rapidly to bind to receptors of a neighboring neuron. The binding to the neighboring neuron initiates the opening of ion channels in that neuron, completing the passing of a message that allows the changes in the membrane potential of one neuron to impact the membrane potential of any other neuron to which it is connected.

Glial cells   The general name for many cells in the brain that are not neurons, with multiple roles, mostly thought to support the activity of neurons, but also with their own electrical properties.

Neurons are by no means the only cells in the brain. We deliberately focus on them in this book, because their electrical properties have a clear, direct role in mental processing and they

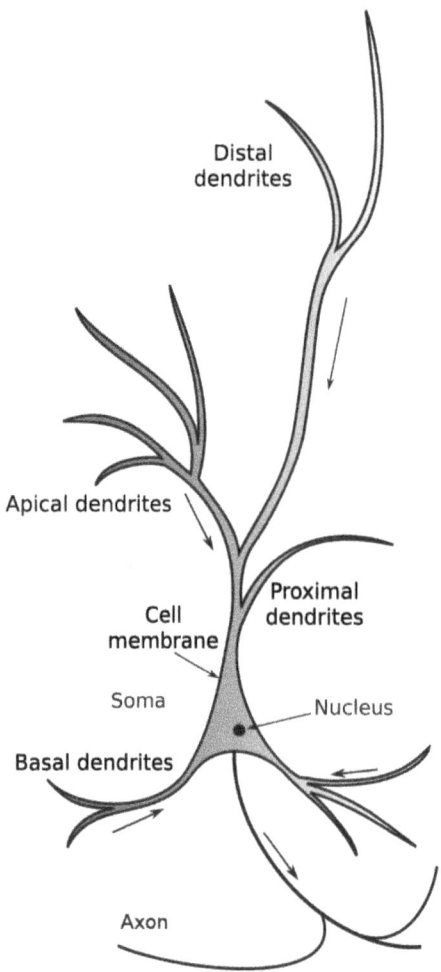

Figure 1.2
Structure of a neuron. Arrows on dendrites and axon depict the standard direction of information flow: Inputs are received on the dendrites and, once sufficient electrical charge accumulates on the surface of the soma, a spike of voltage propagates away from the soma down the axon. Reverse (retrograde) propagation is also possible.

are the best studied; voltage spikes are easily detectable by electrodes within brains, and the electrical fields generated by currents flowing into and out of neurons produce the strongest signals detectable outside the brain. However, it should be noted that the most significant difference between Einstein's brain and the average person's is an increased abundance of glial cells, which are normally thought of as support cells for neurons. Recent experiments have shown that glia produce their own electrical signals and communicate with each other. It is reasonable to assume that glia respond to overall activity in a local circuit and to some extent modulate this activity, but a great deal more remains to be uncovered. I hope that after completing this book you will have gained sufficient confidence in computational modeling to be ready to simulate more novel models testing how glia contribute to information processing in the brain.

1.2.2 The Circuit Level

The amount of branching—or ramification—of the processes of a neuron is strongly correlated with the number of cells that connect with it. Pyramidal cells (figure 1.3A), the most common excitatory cell in the cortex and hippocampus of mammalian brains, receive on the order of 10,000 inputs. Purkinje cells (figure 1.3B), the major output cells of the cerebellum, possess the most highly branched of dendrites: Each cell receives on the order of 100,000 inputs. Its axon, on the other hand, is relatively unbranched: After accumulating so much input, it sends its output to just a few cells in the cerebellar nuclei.

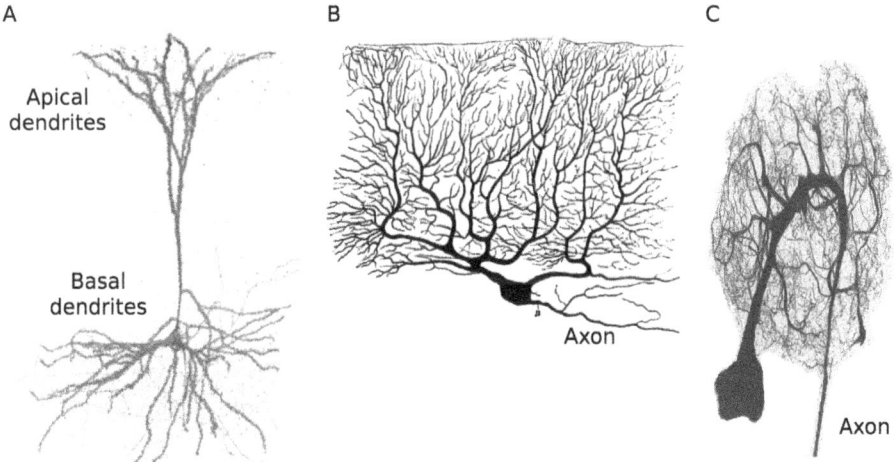

Figure 1.3
Structures of some distinct cells. (A) Layer 3 pyramidal cell from the frontal cortex of a mouse (courtesy of the Luebke Laboratory, Boston University). Cells like these are the predominant excitatory cells of the cerebral cortex. (B) A human Purkinje cell, the inhibitory output cell of the cerebellum, as drawn by Ramón y Cajal (from *Popular Science Monthly* 71, 1907). (C) A neuron from the stomatogastric ganglion of the crab. In this case, the axon branches away from the dendrites instead of being attached directly to the cell body (courtesy of the Marder laboratory, Brandeis University).

Connectome   The complete circuit diagram, in terms of neurons and their connections, of a
particular region of a central nervous system, or the entire neural circuitry of an animal.

Central nervous system   The brain, spinal cord, and other areas with significant neural
processing, such as the retina, of an animal.

Neurons do not just connect with other cells at random. Many invertebrates have a fixed
number of neurons with stereotypical connections. In a small number of species, the entire
circuit—called the connectome—is known and is reproducible across different members of the
species. The first such circuit to be mapped was for *Caenorhabditis elegans*, a small worm with
302 neurons. Small motor circuits in marine creatures such as lamprey and lobsters have been
fully described and, more important, matched with their function. A full description of the connectome of *Drosophila melanogaster*, a fruit fly with more than 80,000 neurons, is close to being
complete.

In mammals, the description of the connectome is more of a statistical process. In each small
circuit the numbers and locations of each different type of cell can be evaluated, in combination with the average numbers of connections between the different cell types. Thus, one can
build up stereotypical circuits in different regions of the central nervous system, with complete
descriptions now available for isolated areas such as the retina. Given the 80 billion (8×10^{10})
neurons and the 100 trillion (10^{14}) connections of a human brain, it is currently infeasible to
produce a circuit at the neural level for humans. Moreover, even the statistical approach is not
straightforward: Different individuals of a mammalian species have different connectomes, with
the existence of a connection and its strength in any particular animal correlated with the properties of a number of different cells. Therefore, it is quite likely that a connectome acquired by
averaging over many individuals, missing some of these crucial correlations, would produce a
nonfunctioning brain.

1.2.3 The Regional Level

Cerebral cortex   The outer region of mammalian brains that is highly folded in humans.

At the grossest scale, brains can be divided into regions according to their structure and function.
These regions can be further subdivided as dictated by differences in the relative abundances of

cell types or differences in the neurons' functional properties. For example, the most recent evolutionary addition to mammals' brains is the cerebral cortex, which forms the outer, upper surface of our brains (figure 1.4A). It contains gray matter mostly comprising neurons and is supported by underlying white matter comprising other cell types such as glia. In primates, it is highly folded with ridges (gyri) and crevasses (sulci) as a consequence of its large surface area. These bumps and dips, combined with location inside the skull, allow anatomists to identify equivalent areas in different brains. The areas themselves can be named by their anatomical location or by their function, which reliably maps to location across individuals. For example, motor planning areas are found in front of the central sulcus, whereas sensory processing areas for visual, auditory, and touch stimuli are found behind the central sulcus.

Perhaps the most studied area in mammals is the visual cortex, which is right at the back of the brain (part of the occipital cortex). The visual cortex is further subdivided (primary, V1; secondary, V2; etc.) because: (1) these subregions are structurally distinct; (2) a complete representation of the visual scene is found in each subregion; and (3) neurons have distinct types of response in the different subregions (cells in V1 respond primarily to contrast boundaries, whereas cells in V2 respond better to more complex combinations of stimuli). Furthermore, a single subregion such as V1 can be subdivided in some animals according to which eye the neurons in a location respond to more, which orientation direction they respond to, whenever such features are clustered by spatial location. This latter level of spatial subdivision is called a topographic map (figure 1.4B).

Topographic map   An area of the brain where the neurons are positioned in an organized arrangement that is directly related to the arrangement of the different stimulus features to which they respond.

The divisions of cortex into subregions described above are based on the different properties of different areas of the cortical sheet. The folded cortical sheet also has a thickness of 1–2 mm within which distinct layers of cells can be identified. Six layers are enumerated, but layer 1 lacks cell bodies (soma); layers 2 and 3 are typically combined, with cells therein having similar properties; and layer 4 can be expanded into sublayers in some sensory areas (such as V1), whereas it is absent from other areas (such as prefrontal cortex). While cells in different layers of a column may respond to related stimulus features (e.g., an edge in the same location of the visual scene), there are clear differences in the dominant input cells (from other parts of the brain, or other areas of cortex, or from a higher or lower layer in the same location) and output cells of the neurons in these different layers. Thus, the layer-specific connections in a region of cortex form a stereotypical circuit, whose functional properties are an ongoing subject of research.

A

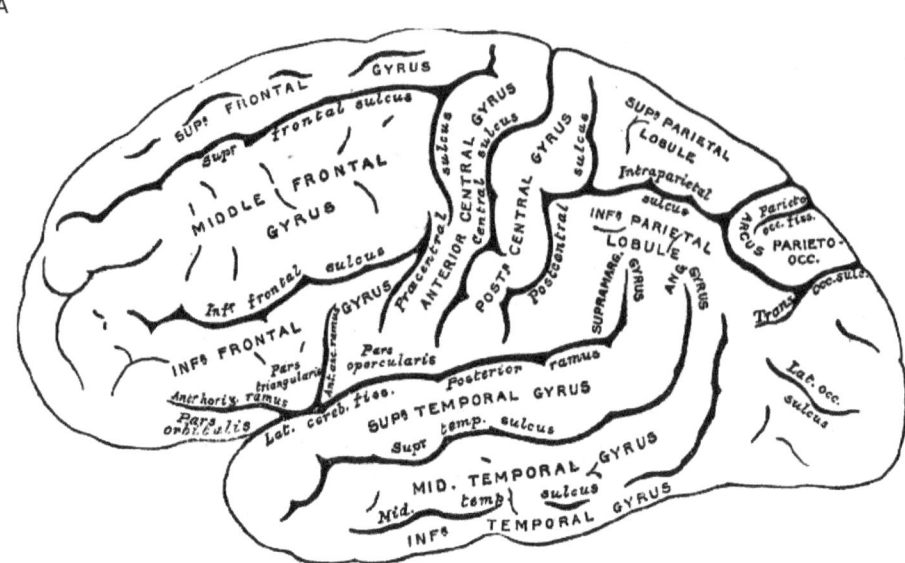

B

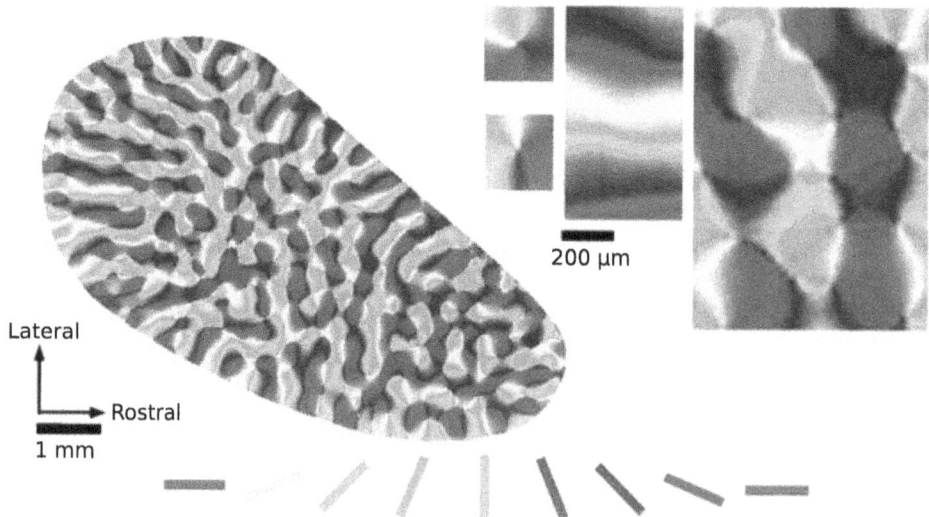

Figure 1.4
Regions and topographic maps. (A) Cerebral cortex with gyri, from H. Gray, *Anatomy of the Human Body* (1918), illustrated by H. V. Carter. (B) The shading of an orientation selectivity map indicates that neurons in primary visual cortex are clustered according to the orientation of a bar to which they most strongly respond (see the mapping from orientation to shading below). Insets show pinwheel centers (see section 6.9) and boundaries. This figure panel was originally published as part B of figure 1 in W. H. Bosking, Y. Zhang, B. Schofield, and D. Fitzpatrick (1997), "Orientation selectivity and the arrangement of horizontal connections in tree shrew striate cortex," *Journal of Neuroscience, 17*(6), 2212–2221. Reprinted with permission.

1.3   Physics of Electrical Circuits

We usually think of electrical circuits as requiring wires and connections made of metal. However, wherever electrical charge is moved or is able to move from one place to another, an electrical circuit can be produced. Whereas in metals the movement of electrical charge is due to free electrons, in solutions—such as the intracellular and extracellular fluids of living organisms—the charge carriers are ions. This leads to some important distinctions between the operation of neurons and of common electrical circuits. Before focusing on the biologically important distinctions, first we review the more general features and terminology of electrical circuits.

1.3.1   Terms and Properties

Charges   Many chemicals are soluble and dissociate (split up) into complementary positively and negatively charged ions in solution. Oppositely charged ions attract each other, while similarly charged ions repel each other. The motion of the charges is mediated by an electric field, which is the spatial gradient of an electric potential. Positive charges move to a more negative potential, while negative charges move to a more positive potential. Each charge produces its own electric field such that approach to a positive charge is to a more positive potential and approach to a negative charge is to a more negative potential.

Currents   An electrical current is a flow of electrical charge from one place to another. Of particular interest in neuroscience is the electrical current through a particular channel in the cell's membrane or the total current across a patch of membrane. In this case the inward membrane current, I_m, is equal to the rate of change of charge, Q, on the inside of the cell's membrane. As an equation, this reads:

$$I_m = \frac{dQ}{dt}. \tag{1.3}$$

Potential Difference (Voltage)   A potential difference generates the drive for charges to move (hence it is also known as electromotive force, or EMF). Significant potential differences arise between the inner and outer surfaces of the cell membrane of neurons. Within the cell, smaller differences in potential arise because charge can more easily flow to balance out such differences.

Capacitance (Capacity to Store Charge)   The capacitance, C, of a surface relates the amount of charge, Q, stored on a surface to the potential difference, V, produced across that surface via:

$$Q = CV. \tag{1.4}$$

Conductance (Ability to Let Current Flow)   The conductance, G, between two regions determines how easily current flows from one region to the other to equalize any potential difference between the regions. A high conductance means that current is high when there is a potential difference between the two regions, such that charge flows rapidly in order to counteract any potential difference between the two regions. The inverse of conductance is resistance, R, so the two equations:

$$V = IR \quad \text{and} \quad I = GV \tag{1.5}$$

are equivalent to each other.

1.3.2   Pumps, Reservoirs, and Pipes

It can be helpful to make the analogy of water flowing through pipes when thinking about electrical currents (see table 1.1 and figure 1.5). Many of the words used, such as pumps, current flow, channel, and potential, have equivalent meanings in the two settings. Water flows downhill from a high potential energy to a low potential energy. If two water containers are connected with a pipe, or channel, the water will flow until the water level is the same in both containers, or at the same potential. If a pump is continuously operating, water can be maintained at a higher level in one container than another. If the channel (pipe) is closed, the higher level remains in the absence of further pumping. The closed pipe is equivalent to a high resistance that prevents current flow. If the pipe is opened, then its resistance drops and current flows in the direction that will reduce the potential difference (i.e., water flows downhill).

The main limitation of the analogy is that charge can be either positive or negative. When one considers negative charges, the physical movement of the actual ions is opposite to the net electric current produced. However, if one sticks with positive charges then the analogy of water flowing through pipes between containers can be helpful. Opening holes in the cell membrane

Table 1.1
Analogy between flows of electrical charges and fluids

Electrical		Water Analogy	
Term	Description	Term	Description
Charge	Fundamental electrical property due to excess or lack of electrons	Water	A quantifiable substance that can move or flow
Current	Flow of charge	Current	Flow of water
Potential difference	Produces driving force for flow of charge	Height difference	Produces driving force for flow of water
Resistance	Inhibits flow of charge	Obstruction	Reduces flow of water
Conductance	Allows flow of charge	Opening	Permits flow of water
Pump	Moves charge to locations it would not naturally go	Pump	Moves water to locations it would not naturally go
Open channel	Channel in a high conductance state	Open channel	Channel with an opening
Closed channel	Channel in a low conductance state	Closed channel	Channel with an obstruction

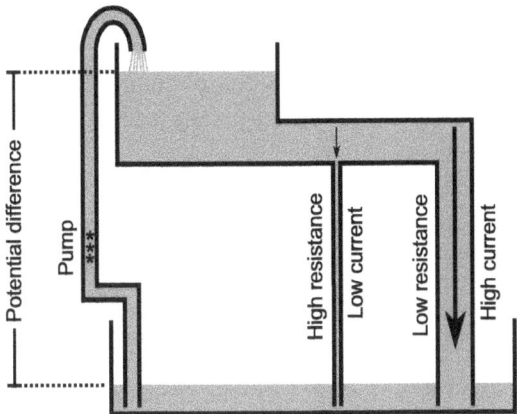

Figure 1.5
Pump and reservoir analogy to electrical circuits. Open channels have a low resistance, permitting high current flow across a potential difference from high potential to low potential. Pumps (effectively batteries) can maintain the potential difference, but without them the passive current flow would tend to erase any potential difference.

(ion channels) allows current to flow. This corresponds to a decrease in membrane resistance or, equivalently, an increase in membrane conductance. Pumps in cells can cause systematic deviations from equilibrium, moving ions to regions of higher chemical or electrical potential energy (respectively to regions of higher concentration or to a more positive electrical potential for a positive charge), just as they can move water to a higher potential energy (uphill).

1.3.3 Some Peculiarities of the Electrical Properties of Neurons

Charge flow in neurons is due to movement of ions—usually atoms, but also molecules, that have gained or lost one or more electrons. The necessity of physical movement of these ions allows for the cell to control charge flow and in so doing, to control the potential difference between the inside and the outside of the cell. Such control is achieved via proteins in the cell membrane, which can interact with specific types of ions in a manner that permits movement of some ions and not others.

1. *Ion pumps* preferentially pump into or out of the cell specific ions, producing a concentration difference across the cell membrane that is particular to each type of ion. Functionally, the process is equivalent to making a series of mini-batteries—whose strength and direction depends on ion type—along the cell membrane.

2. *Ion channels* permit flow of a subset of ions, often only one type of ion. As these channels open and close, the effect is to functionally connect or disconnect the corresponding mini-batteries. Active channels are those that open or close in response to biophysical properties, such as local potential difference, concentration of a particular chemical, or temperature.

Aqueous solutions like the cell's cytoplasm are a million or more times worse than good conductors at transferring charge (for example, the electrical resistivity of sea water is 0.2 Ωm —in the range of semiconductors—whereas the electrical resistivity of copper is $2 \times 10^{-8} \Omega$m). Therefore, charge that enters a cell can leak out before spreading out equally within the cell, causing charge imbalances within the cell, so the potential difference across the cell membrane in its distal dendrites (figure 1.2) can be quite different from that in the soma. Many models of neurons focus solely on the dynamical properties of the membrane potential and ignore this spatial feature, treating the neuron as a pointlike object with a single membrane potential. Multicompartment models (section 4.7) are spatially extensive and take into account intracellular variations in membrane potential.

A Note on Time Constants   Strictly, a time constant is defined by the amount of time needed for a system to get closer to equilibrium by a factor of $1/e$ (where e is Euler's constant, used in exponential notation). Even when a system is not approaching equilibrium, one can often calculate "instantaneous time constants" that tell us how quickly things are changing (compared to the state the system is trying to reach) at that point in time. When we are thinking about current flowing across the membrane of a cell, the two key quantities that set the time constant are the amount of charge that is to flow and the rate of flow of that charge. Amount divided by rate gives the timescale. Both of these are proportional to the potential difference, while the amount of charge is also proportional to the capacitance (equation 1.4) and the rate of flow is also proportional to the conductance (equation 1.5). Therefore, the timescale is given by capacitance divided by the conductance, or equivalently, the capacitance multiplied by the resistance. This makes sense in terms of the water analogy—the time to empty a bath goes up if the bath is bigger, because it stores more water, but goes down if the drain is wider, because water flows out more quickly. A larger bathtub has greater capacity to store water so is like a larger cell with greater capacitance. A larger drain, which allows water to flow out more quickly, is like a larger conductance. If the area of the drainage were scaled with the area of the bathtub, the time to empty would be independent of size of bathtub. Similarly, for a neuron, if the number of open channels in a cell membrane scales up in proportion to the total surface area where the charge is stored, then the time constant does not change.

Note that the time constant of a neuron is not typically static. The more channels that are open at any time, the more rapidly charge can flow into or out of the cell, so the shorter its effective time constant.

1.4   Mathematical Background

This section is intended to supply those readers who have not taken a college-level calculus course with the intuition and fundamental results needed to complete an introductory course in computational neuroscience.

1.4.1 Ordinary Differential Equations

A hallmark of all living things is that we change over time—that is, we are dynamical systems. Differential equations are used to describe dynamical systems. The structure of a differential equation reflects the continuous nature of space and time, combined with the fact that for a system to get from one state to another it must pass through a continuous path of intervening states. That is, in spatial terms, there is no such thing as teleportation. The position of an object a tiny amount of time in the future can be calculated from its current position and current velocity. Differential equations quantify this procedure.

In this book, we will only deal with ordinary differential equations (ODEs), which describe how individual properties of a system change with time, or along a single spatial dimension, but not with both together (the combination of time and space requires partial differential equations). We treat spatial variation by coupling together ordinary differential equations that represent the dynamics at different points in space—such as in different neurons or different compartments of a single neuron—and include their effects on each other.

We will find that even though using mathematics to solve ODEs can become inordinately difficult, writing computer codes to solve them is straightforward.

Example 1: A Falling Stone Gravity produces a constant acceleration, so that if we neglect air resistance, a dropped stone's speed is proportional to the time for which it has been falling. We will define the downward velocity as negative (since the stone's height decreases with time) and can ask how the position of the stone changes with time. In doing this we will carry out integration twice—once to obtain velocity from acceleration, then again to obtain position from velocity. That is:

$$\frac{dv}{dt} = a \tag{1.6}$$

(in words, the rate of change of velocity is equal to the acceleration) and

$$\frac{dy}{dt} = v \tag{1.7}$$

(in words, the rate of change of y-position is equal to the velocity).

Table 1.2 indicates how velocity and position accumulate over time, if we approximate the effect of gravity with $a = -10$ ms^{-2}. In the third row, the velocity is given by the accumulation over time of acceleration, which is a fixed quantity. The result of such accumulation is a linear dependence: Velocity is simply acceleration times time, $v = at$. In the last row, position is obtained by accumulating the average velocity across each timestep. That is, the value of position is equal to its previous value plus the mean of the current value of velocity and the prior value of velocity multiplied by the change in time. Such a procedure can be carried out by computers extremely rapidly and with very small timesteps. Indeed, by the end of chapter 1 you should

Table 1.2
Dynamics of a falling stone

Time (s)	0	1	2	3	4
Acceleration (ms^{-2})	−10	−10	−10	−10	−10
Velocity (ms^{-1})	0	−10	−20	−30	−40
Position (m)	0	−5	−20	−45	−80

Note: As time increases (top row) the velocity (third row) accumulates the mean of successive values of acceleration (second row) and the position (bottom row) accumulates the mean of successive values of velocity.

A

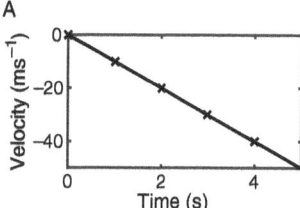

B

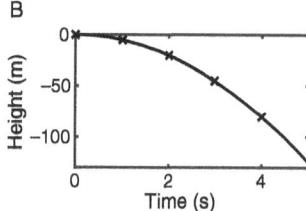

Figure 1.6
Dynamics of a falling stone. (A) Velocity, $v(t)$, decreases linearly with time. (B) As a result, height, $y(t)$, decreases quadratically with time. Crosses are data points from table 1.2. This figure was produced by the available online code `falling_stone.m`.

understand how to program a computer to produce tables with a similar form to table 1.2 and to plot them as graphs such as figure 1.6.

Example 2: The Sprinter Let's next consider the example of a sprinter in a 100-meter race. Initially the sprinter is stationary and so has zero velocity. We can assume the sprinter accelerates up to a maximal velocity and maintains it until finishing the event. We might want to know how long the sprinter takes to complete 100 meters, which we can do if we know the velocity as a function of time. Perhaps the velocity is provided by the following equation:

$$v(t) = v_{max}\left(1 - e^{-t/\tau}\right), \tag{1.8}$$

where v_{max} is the maximum velocity (e.g., $v_{max} = 10$ m/s) and τ indicates how long it takes to approach maximum velocity (e.g., $\tau = 1$ s).

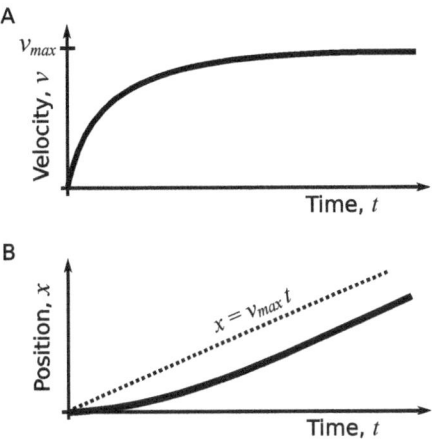

Figure 1.7
Time-dependent velocity. (A) Velocity of a hypothetical sprinter increases from zero toward a maximum, v_{max}, at which it remains. (B) The velocity corresponds to the gradient of the position versus time curve, which starts instantaneously flat, then steepens to lie parallel, but below the dotted line $x = v_{max}t$.

The shape of the curve in equation 1.8 is shown in figure 1.7A. The ordinary differential equation to describe the position, x, of the sprinter along the track is then:

$$\frac{dx}{dt} = v(t) = v_{max}\left(1 - e^{-t/\tau}\right). \tag{1.9}$$

Equation 1.9 states that the differential of position, x, with respect to time, t, is the velocity at that point in time. To put it another way, the rate of change of position is the velocity. Throughout this course, we will be using the fact that the rate of change of a variable can be described in terms of its current value and/or the current values of other variables.

For this course, you do not need to know how to use mathematics to solve the equation for position, $x(t)$. Instead, section 1.6 explains how a computer code can be written to solve the equation. Rather, it is important to know what various functions look like, so that you can see if the behavior makes sense. For example, the exponential function, $e^{-t/\tau}$, has a value of 1 for $t = 0$, and decays to become negligible for $t \gg \tau$. That means the function $\left(1 - e^{-t/\tau}\right)$ has the value of zero if $t = 0$ and the value of 1 if $t \gg \tau$, as necessary for $v(t)$ to commence at zero and reach a maximum value, v_{max}, in equations 1.8 and 1.9.

In this example, $v(t)$ is a simple enough function that I can write down the solution:

$$x(t) = v_{max}t - v_{max}\tau\left(1 - e^{-t/\tau}\right). \tag{1.10}$$

This means that the position is a little less than $v_{max}t$ (see figure 1.7B). $v_{max}t$ is the position reached if the sprinter were able to run at maximum velocity from the starting line (as may happen in a relay). If $t \gg \tau$ the distance "lost" by acceleration is $v_{max}\tau$ (or 10 meters) in the current example. This implies the time to complete 100 meters would be approximately 11 s.

In section 1.6, we will see how to solve equations like equation 1.9, using a computer simulation. When an analytic (i.e., mathematical) solution like equation 1.10 is also known, the exact analytic solution can be compared with simulations to test the accuracy of various simulation methods.

Example 3. Cycling over a Hill   In the first two examples, the velocity was determined purely by the time elapsed since the start of the process. More generally, the velocity may have a known dependence on the variable being calculated, such as position. For example, when a child cycles over a hill she may ride five times faster on the downhill than the uphill. In this case, the rate of change of position will depend directly on the current position as opposed to the current time.

We will consider a hill of total distance 2 miles, shaped steepest at the bottom, so that the cyclist's velocity increases with position along the hill as $v(x) = v_0 \left(1 + x^2\right)$ (see figure 1.8). Then the rate of change of position is written as:

$$\frac{dx}{dt} = v(x,t) = v_0 \left[1 + x(t)^2\right]. \tag{1.11}$$

Notice that writing $v(x,t)$ in equation 1.11, rather than just $v(x)$, is done simply to remind the reader that the velocity does depend on time, even though it is explicitly written only in terms of position, x. Similarly, $x(t)$ is written to remind the reader of the time-dependence of position, x.

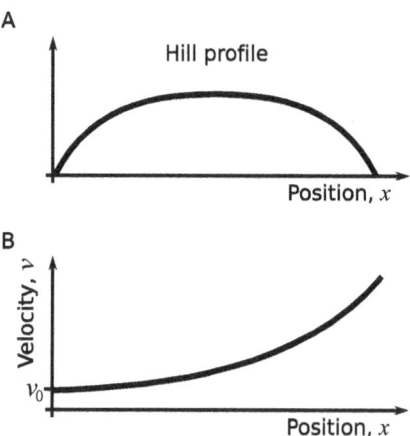

Figure 1.8
Position-dependent velocity. (A) Profile of a hill so that with increasing distance the terrain progresses from steep uphill to shallow uphill to flat to shallow downhill to steep downhill. (B) The velocity of a child on a bicycle is slowest at the steepest uphill and increases monotonically with position, to become fastest at the steep downhill. The velocity as a function of position might be approximated as $v(x) = v_0 \left(1 + x^2\right)$.

A first course in calculus will teach you how to find the analytic expression for the solution of equation 1.11 as $x(t) = \tan(v_0 t)$, which then tells us the time taken to complete the 2-mile journey is $(\tan^{-1}(2))/v_0$. While the analytic method for solving such an equation is significantly trickier than for solving examples 1 and 2, the computational method for solving example 3 requires only a trivial alteration from example 1. Again, for this course it is much more important to understand what the equations mean and how their solutions make sense than worrying about finding exact formulas. The one formula for the solution of an ordinary differential equation that you should get to know is provided in example 4.

Example 4. Exponential Decay to a Point of Equilibrium—Cooling a House   Any process with an equilibrium point and a rate of change that is directly proportional to the difference between the system's current state and its equilibrium point follows an exponential decay. Such processes are very common in chemistry and physics, so we will consider a couple of examples.

For this example, we consider the change in temperature of a house on a hot day when air conditioners are first switched on. The key to the change in temperature is that while air conditioners extract heat from the house at a constant rate, the high temperature outside of the house causes heat to enter the house at a rate proportional to the difference between inside and outside temperatures. So, as the house cools, the flow of heat from outside to inside increases until the air conditioners can no longer keep up and an equilibrium is reached (we assume that if there is a thermostat it is set to a lower temperature than this eventual equilibrium, so the air conditioners never switch off). The equilibrium is dynamic—as it is for most biological and chemical equilibria—based on a balance of outward heat flow matching the inward heat flow.

Mathematically we can write the rate of change of temperature, T, as

$$K \frac{dT}{dt} = -A + B(T_{ext} - T) \tag{1.12}$$

where K is the heat capacity of the house (greater for a bigger house), A is the rate of heat extraction (proportional to the power of the air conditioner), T_{ext} is the outside air temperature, and B is a measure of how easily heat flows through the walls of the house, or "leakiness" (better insulation reduces B). Note that we are using upper case "T" to represent temperature in equation (1.12), while the lower case "t" always represents time in this book. Therefore dT/dt means rate of change of temperature, so would have standard units of degrees Celsius per second, though more manageable units for a building would be degrees per hour.

Equation (1.12) is both an ODE and linear in the variable of interest. That is, if we plotted dT/dt as a function of T, we would get a straight line (see figure 1.9A). The full name of this type of equation is therefore a linear first-order ordinary differential equation (ODE). It is worth knowing the general solution of such equations, since they are so common.

The most important point in understanding the behavior of equation 1.12 is to realize that while the rate of change of temperature is a linear function of temperature, if we plot the result-

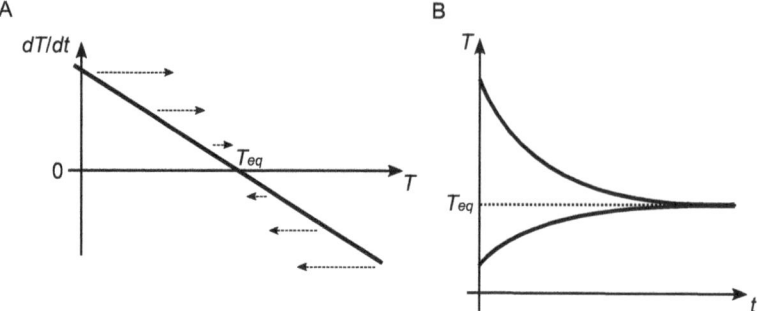

Figure 1.9
Rate of change of temperature as a linear function of temperature. (A) The diagonal straight line indicates how the rate of change of temperature depends linearly on temperature in a simplified temperature-regulation system. Horizontal dotted lines indicate the direction and magnitude of the rate of temperature change. They point toward the equilibrium temperature, T_{eq}, where $dT/dt = 0$. (B) Since dT/dt does not have a fixed value, a plot of temperature, T, against time, t, does not have a fixed gradient and so is nonlinear.

ing temperature as a function of time, we get a nonlinear function—as the room heats or cools its temperature changes, and thus the gradient of temperature as a function of time changes, until it reaches a point where rate of change of temperature is zero ($dT/dt = 0$) when temperature stops changing.

In all such equations, we can calculate the equilibrium point—in this case, the temperature, T_{eq}, as the value of T at which $dT/dt = 0$. Following equation 1.12, we set

$$0 = -A + B(T_{ext} - T), \tag{1.13}$$

when $T = T_{eq}$, which leads to

$$T_{eq} = T_{ext} - A/B. \tag{1.14}$$

This result means that the equilibrium temperature is below the external temperature by an amount that is proportional to the power of the air conditioner (A) and inversely proportional to the "leakiness" of the house (B). Of course, this makes sense—stronger air conditioners or better insulation will allow the house to be cooler on hot days.

Just as any straight line can be written in terms of two parameters—its gradient and its intercept on the y-axis—so too can any linear first-order ODE be written in terms of two parameters. The important parameters are the equilibrium point and a time constant. The equilibrium point found above is the intersection on the x-axis (figure 1.9A), whereas the time constant is the inverse of the gradient. In our example, we can rewrite the straight line in equation 1.12 representing dT/dt as

$$\frac{dT}{dt} = \frac{T_{eq} - T}{\tau}, \tag{1.15}$$

so we see the gradient is $-1/\tau$ and the y-intercept is T_{eq}/τ. In our example, we have already found T_{eq} and only a little algebra is required to show that $\tau = K/B$. (We use a very similar look-ing, but distinct symbol, τ, the Greek "tau," for the inverse-gradient, because it becomes the time constant of the process, and time constants are traditionally given the symbol "τ.")

The final property to notice is that the equilibrium point is stable. The requirement for stability is that the temperature returns to its equilibrium following small deviations. In fact, for a linear system it does not matter how far from equilibrium the system is; if the equilibrium is stable, the system will return to equilibrium. In our example, the key property is that the gradient of the dT/dt versus T line is negative (figure 1.9A), so that if T is greater than its equilibrium value it will decrease over time (negative dT/dt), whereas if T is lower than its equilibrium value it will increase over time (positive dT/dt), as shown in figure 1.9B.

Once we know a linear system has a stable equilibrium point, we know it will reach that equi-librium as an exponential decay (proof to follow). If the initial temperature is T_0 at the time $t = 0$ when the air conditioners are first switched on, then the temperature will change as a function of time according to:

$$T(t) = T_{eq} + (T_0 - T_{eq})\exp(-t/\tau). \tag{1.16}$$

We will see this equation many times in this book. You should convince yourself it behaves as expected, so that by setting $t = 0$, we find $T(0) = T_0$, and if $t \gg \tau$ then $T(t) \approx T_{eq}$.

Finally, for our example, we can assume the initial temperature of the house is the external temperature, so $T_0 = T_{ext}$, in which case $T_0 - T_{eq} = A/B$ and the solution can be written as:

$$T(t) = T_{ext} - (A/B)[1 - \exp(-Bt/K)]. \tag{1.17}$$

Proof of Exponential Solution* For those with a little calculus background, we will use the for-mula for the integral of $1/(a-x)$ with respect to x as $-\ln(a-x)$ in order to prove the exponential solution of a linear first-order ODE. This proof, like all of those in the appendices of this book, can be skipped without loss of content, but should be followed by those who want to deepen their mathematical understanding and have sufficient background in calculus.

In general, if an equation can be written in the form of equation 1.15 as:

$$\frac{dx}{dt} = \frac{x_{eq} - x}{\tau}, \tag{1.18}$$

we can rearrange it in a format suitable for integration as

$$\frac{dx}{x_{eq} - x} = \frac{dt}{\tau}. \tag{1.19}$$

After integrating we find

$$\left[-\ln(x_{eq} - x) \right]_{x(0)}^{x(t)} = t/\tau, \tag{1.20}$$

which becomes

$$\ln\left[x_{eq} - x(t)\right] - \ln\left[x_{eq} - x(0)\right] = \ln\left[\frac{x_{eq} - x(t)}{x_{eq} - x(0)}\right] = -t / \tau. \tag{1.21}$$

Equation 1.21 can be solved to find $x(t)$ by taking the exponential of both sides:

$$\frac{x_{eq} - x(t)}{x_{eq} - x(0)} = \exp(-t / \tau). \tag{1.22}$$

Rearrangement of equation 1.22 gives the solution,

$$x(t) = x_{eq} + \left[x(0) - x_{eq}\right] \cdot \exp(-t / \tau). \tag{1.23}$$

The solution proves that τ is the time constant and x_{eq} is the equilibrium point reached after a time long compared to the time constant. Therefore, knowing this solution, all that must be found when solving any linear first-order differential equation is the equilibrium point and the time constant. We shall use such a shortcut in example 5 below.

Example 5. Dynamics of Gating Variables   The term "gating variable" is used to represent the fraction of a particular type of ion channel in a state that allows current to flow. We assume the channel consists of a protein with two states, which we will call "closed" and "open." The key parameters are the rates for closed channels to open and the rates for open channels to close. These rate constants are α and β respectively and have units of inverse time (s^{-1}). The chemical equation is simply

$$C \underset{\beta}{\overset{\alpha}{\rightleftharpoons}} O. \tag{1.24}$$

If we write N_O as the number of open channels and N_C as the number of closed channels then equation 1.24 tells us that the rate of closed channels opening is αN_C while the rate of open channels closing is βN_O. If we write the total number of channels as N_T then the number of closed channels is given by $N_C = N_T - N_O$, so rate of closed channels opening is $\alpha(N_T - N_O)$. Therefore, we can write the rate of change of the number of open channels as the difference between opening and closing rates as

$$\frac{dN_O}{dt} = \alpha(N_T - N_O) - \beta N_O. \tag{1.25}$$

Notice that if all channels are open, then the first term on the right of equation 1.25 is zero (no more channels can open and N_O cannot increase), whereas if all channels are closed the second term on the right is zero (no more channels can close and N_O cannot decrease).

We define the gating variable, s, as the fraction of channels in the open state, $s = N_O / N_T$, such that the dynamics of s is found after dividing equation 1.25 by N_T to give:

$$\frac{ds}{dt} = \alpha(1-s) - \beta s. \tag{1.26}$$

Basic algebra allows us to rewrite equation 1.26 as

$$\frac{ds}{dt} = \frac{s_{eq} - s}{\tau_s} \tag{1.27}$$

so long as we identify

$$s_{eq} = \frac{\alpha}{\alpha + \beta} \tag{1.28}$$

and

$$\tau_s = \frac{1}{\alpha + \beta}. \tag{1.29}$$

Finally, we can write down the solution of the equation 1.27 (following the methods of example 4) to give:

$$s(t) = s_{eq} + [s(0) - s_{eq}] \cdot \exp(-t/\tau_s). \tag{1.30}$$

It is worth noting that the following properties of the solution (equations 1.28–1.30) match our expectations or intuition:

1. The equilibrium value, $s_{eq} = \alpha/(\alpha + \beta)$, can only vary between 0 and 1 (since α and β are positive rate constants);

2. s_{eq} approaches 1 if $\alpha \gg \beta$ (if opening rate is much greater than closing rate, nearly all channels are open);

3. s_{eq} approaches 0 if $\beta \gg \alpha$ (if closing rate is much greater than opening rate, nearly all channels are closed);

4. The time constant, $\tau_s = 1/(\alpha + \beta)$, is dominated by the largest rate constant, while remaining inversely proportional to rates.

Equation 1.30 fully describes the behavior of a two-state system, for which the channel can be in only either a closed or open state. However, the solution can be applied to systems with multiple pairs of such states, so long as the transitions within one pair of states do not impact the transition rates within other pairs.

For example, in chapter 4, when we focus on conductance-based models, we will consider ion channels with separate activation and inactivation variables. These variables can be related to the states of individual proteins that comprise the (mostly) independent subunits of a channel. In most models, each of the variables has two states, one of which allows the channel to open and the other of which keeps it closed. Multiplication of these distinct gating variables together to evaluate the total probability of a channel being open—proportional to the expected instanta-neous conductance of many such channels—is valid if each variable is independent. Indeed, for the sodium conductance, the total conductance is proportional to a term of m^3h, which assumes three independent activation variables, m, and a single independent inactivation variable, h, each of which follow the dynamics of equation 1.30.

1.4.2 Vectors, Matrices, and Their Basic Operations

When dealing with a large quantity of variables—say, the value of the firing rate of many neu-rons, each at many different points in time—it becomes essential to store them together in vectors or matrices. Computer codes written in MATLAB can perform calculations on all of the variables very efficiently when they are combined together in a matrix (the "Mat" in MATLAB stands for Matrix). Therefore, for this course it is important for you to learn how matrices and vectors are combined by addition or multiplication.

A vector is a list of numbers, with each number labeled according to its position on the list. A matrix is a rectangular grid of numbers, with each number labeled according to its position in the grid, in particular via its row index then column index. It is worth noting that each row of a matrix is a vector, as is each column—vectors can be rows or columns. A matrix can be thought of as a vector of vectors—either a row vector of column vectors, or a column vector of row vectors.

In this book, a single line under the variable's name indicates it is a vector. For example, the vector $\underline{r}$ could represent a list of the firing rates of many cells. The underline is not present when a particular entry of the vector is used; for example, $r(3)$ or r_3 would mean the firing rate of the third cell in the list and that firing rate is a single value, not itself a vector.

Similarly, two lines under a variable's name indicate that it is a matrix. For example, $\underline{\underline{W}}$ could represent the set of connection strengths between all pairs of cells in a group, each row labeling the presynaptic cell and each column labeling the postsynaptic cell. Again, no underlines are used when writing a particular entry in the matrix, so $W(2,3)$ or $W_{2,3}$ would be the strength of con-nection from cell 2 to cell 3, which is a single value.

Examples of Matrices and Vectors

a. If $\underline{r} = \begin{pmatrix} 10 \\ 2 \\ 15 \\ 9 \end{pmatrix}$ then $r(3) = 15$, or alternatively $r_3 = 15$.

b. If $\underline{W} = \begin{pmatrix} 2 & 0 & 5 \\ 0 & 4 & 1 \\ 0 & 3 & 1 \end{pmatrix}$ (a 3×3 matrix) then $W(1,3) = 5$ and $W(3,2) = 3$, or alternatively

$W_{13} = 5$ and $W_{32} = 3$.

Matrix and Vector Addition   Matrices or vectors of the same size can be added together to produce a matrix or vector of the same size. Each entry in one matrix is added to the corresponding entry in the other matrix to produce that entry in the sum.

Examples of Addition

a. If $\underline{a} = \begin{pmatrix} 1 \\ 3 \\ 2 \\ 9 \end{pmatrix}$, $\underline{b} = \begin{pmatrix} -3 \\ 4 \\ 1 \\ 2 \end{pmatrix}$, and $\underline{c} = \underline{a} + \underline{b}$, then $\underline{c} = \begin{pmatrix} -2 \\ 7 \\ 3 \\ 11 \end{pmatrix}$.

b. If $\underline{\underline{A}} = \begin{pmatrix} 1 & 3 & 0 \\ 1 & 4 & 2 \end{pmatrix}$, $\underline{\underline{B}} = \begin{pmatrix} 2 & 0 & 2 \\ -3 & -2 & 3 \end{pmatrix}$, and $\underline{\underline{C}} = \underline{\underline{A}} + \underline{\underline{B}}$, then $\underline{\underline{C}} = \begin{pmatrix} 3 & 3 & 2 \\ -2 & 2 & 5 \end{pmatrix}$.

Matrix and Vector Subtraction   Matrices or vectors of the same size can be subtracted from each other to produce a matrix or vector of the same size. Each entry in one matrix is subtracted from the corresponding entry in the other matrix to produce that entry in the difference.

a. If $\underline{a} = \begin{pmatrix} 1 \\ 3 \\ 2 \\ 9 \end{pmatrix}$, $\underline{b} = \begin{pmatrix} -3 \\ 4 \\ 1 \\ 2 \end{pmatrix}$, and $\underline{d} = \underline{a} - \underline{b}$, then $\underline{d} = \begin{pmatrix} 4 \\ -1 \\ 1 \\ 7 \end{pmatrix}$. Note that if $\underline{e} = \underline{b} - \underline{a}$ then $\underline{e} = -\underline{d}$.

b. If $\underline{\underline{A}} = \begin{pmatrix} 1 & 3 & 0 \\ 1 & 4 & 2 \end{pmatrix}$, $\underline{\underline{B}} = \begin{pmatrix} 2 & 0 & 2 \\ -3 & -2 & 3 \end{pmatrix}$, and $\underline{\underline{C}} = \underline{\underline{A}} - \underline{\underline{B}}$, then $\underline{\underline{C}} = \begin{pmatrix} -1 & 3 & -2 \\ 4 & 6 & -1 \end{pmatrix}$.

Matrix and Vector Multiplication   Multiplication of two matrices or of a matrix and a vector is possible only if the number of columns in the first term is equal to the number of rows of the second term. The product will be a new matrix or vector with the number of rows of the first term and number of columns of the second term. For example, a row vector multiplied by a column vector produces a single value (a scalar).

> **Scalar product**   The product of two vectors with the same number of elements to produce a scalar (a single number) by multiplying together the corresponding pairs of elements in the two vectors and summing the results of each multiplication.

We will establish the rules of multiplication by first evaluating the product of a row vector by a column vector. The scalar product is obtained by summing the products of corresponding entries in the two vectors.

Example

If $\underline{x} = \begin{pmatrix} 1 \\ 3 \\ 2 \\ 9 \end{pmatrix}$ and $\underline{y} = (2 \quad 0 \quad -1 \quad 3)$, and $\underline{z} = \underline{x} \times \underline{y}$, then

$$\underline{z} = (1 \times 2) + (3 \times 0) + (2 \times (-1)) + (9 \times 3) = 27.$$

It is clear that this formula can only work if the number of rows of the column vector, $\underline{x}$, matches the number of columns of the row vector, $\underline{y}$, such that each element in $\underline{x}$ has a corresponding element in $\underline{y}$ to be multiplied by.

Rule for Matrix Multiplication   If $\underline{\underline{C}} = \underline{\underline{A}} \times \underline{\underline{B}}$ then the value of the entry in row-i and column-j of $\underline{\underline{C}}$ is the scalar product of row-i of matrix $\underline{\underline{A}}$ with column-j of matrix $\underline{\underline{B}}$.

Examples

Let $\underline{\underline{A}} = \begin{pmatrix} 1 & 3 & 0 \\ 1 & 4 & 2 \end{pmatrix}$, $\underline{\underline{B}} = \begin{pmatrix} 2 & 1 \\ 5 & 0 \\ 4 & 3 \end{pmatrix}$, $\underline{\underline{C}} = \underline{\underline{A}} \times \underline{\underline{B}}$ and $\underline{\underline{D}} = \underline{\underline{B}} \times \underline{\underline{A}}$.

Then $\underline{\underline{C}}$ is formed from a 2×3 matrix multiplied by a 3×2 matrix, so it is a 2×2 matrix.

$C(1,1)$ is the scalar product of the first row of $\underline{\underline{A}}$ with the first column of $\underline{\underline{B}}$ so
$C(1,1) = (1 \times 2) + (3 \times 5) + (0 \times 4) = 17$.

$C(2,1)$ is the scalar product of the second row of $\underline{\underline{A}}$ with the first column of $\underline{\underline{B}}$ so
$C(2,1) = (1 \times 2) + (4 \times 5) + (2 \times 4) = 30$.

Proceeding in this manner, one can show that

$$\underline{\underline{C}} = \begin{pmatrix} 17 & 1 \\ 30 & 7 \end{pmatrix}.$$

$\underline{\underline{D}}$ is formed from a 3×2 matrix multiplied by a 2×3 matrix, so it is a 3×3 matrix.

$D(1,1)$ is the scalar product of the first row of $\underline{\underline{B}}$ with the first column of $\underline{\underline{A}}$ so
$D(1,1) = (2 \times 1) + (1 \times 1) = 3$.

$D(2,1)$ is the scalar product of the second row of $\underline{\underline{B}}$ with the first column of $\underline{\underline{A}}$ so
$D(2,1) = (5 \times 1) + (0 \times 1) = 5$.

Proceeding in this manner, one can show that $\underline{\underline{D}} = \begin{pmatrix} 3 & 10 & 2 \\ 5 & 15 & 0 \\ 7 & 24 & 6 \end{pmatrix}$.

Transpose of a Matrix or Vector From time to time we may want to switch rows and columns of a matrix or vector. This operation is called taking the transpose. The transpose of a vector, $\underline{x}$, is indicated by a superscript as $\underline{x}^T$ or simply $\underline{x}'$.

For example, if we want to multiply a vector by itself we cannot do so, but if we take the transpose of one copy of the vector then we can multiply the row vector by the column vector to get a scalar product.

If $\underline{x} = \begin{pmatrix} 1 \\ 3 \\ 2 \\ 9 \end{pmatrix}$ then $\underline{x}^T = (1 \quad 3 \quad 2 \quad 9)$ and $\underline{x}^T \cdot \underline{x} = (1 \times 1) + (3 \times 3) + (2 \times 2) + (9 \times 9) = 95$ is the sum

of the squares of each entry.

Multiplication of a Vector/Matrix by a Square Matrix Produces a Product the Same Size as the Initial Vector/Matrix Let

$$\underline{\underline{W}} = \begin{pmatrix} 2 & 0 & 5 \\ 0 & 4 & 1 \\ 0 & 3 & 1 \end{pmatrix}$$

be a matrix of connection strengths such that $W(i, j)$ is the strength of connection from neuron-i to neuron-j and

$$\underline{r} = \begin{pmatrix} 10 \\ 5 \\ 20 \end{pmatrix}$$

represent the firing rates of each neuron. We will assume the total inputs to each of the three neurons can be calculated as the sum of individual inputs, where the individual input from neuron-i to neuron-j is the product of the connection strength and the firing rate of neuron-i. The total input, $\underline{I}$, is a vector of the same size as $\underline{r}$ and can be calculated as

$$\underline{I} = \underline{\underline{W}}^T \underline{r}.$$

To see this, we notice that input to the first neuron,

$$I(1) = W(1,1)r(1) + W(2,1)r(2) + W(3,1)r(3)$$

is the first column of $\underline{\underline{W}}$ multiplied term by term by the column-vector, $\underline{r}$. The first column of $\underline{\underline{W}}$ is the same as the first row of $\underline{\underline{W}}^T$, that is

$$I(1) = W^T(1,1)r(1) + W^T(1,2)r(2) + W^T(1,3)r(3).$$

Proceeding for the other rows of $\underline{I}$ using the other rows of W^T leads to the complete set of total inputs as

$$\underline{I} = \underline{\underline{W}}^T \underline{r} = \begin{pmatrix} 20 \\ 80 \\ 75 \end{pmatrix}.$$

Satisfy yourself that a similar result can be reached by calculating the inputs and the rates as row vectors using

$$\underline{I}^T = \underline{r}^T \underline{\underline{W}}.$$

1.4.3 Probability and Bayes' Theorem

We should all be familiar with the fact that the probability of an event can range from zero (impossible) to one (certain) such that a probability near zero means highly unlikely and a probability near one means almost certain. Fifty-fifty means a probability of one-half, which is the probability of tossing a head on a fair coin. In general, if there are N equally likely alternatives that are mutually exclusive (meaning they cannot both happen, e.g., a coin cannot land on both faces at the same time), then the probability of each alternative is $1/N$. This result arises because if all possible and distinct outcomes are accounted for, then the probabilities of each distinct outcome must sum to one. This is another way of saying that one of the distinct alternatives is certain to happen.

Rule 1a: *If A and B are distinct, mutually exclusive alternatives, then the probability of either A or B is equal to the sum of the probability of A and the probability of B.*

More generally, when A and B may not be mutually exclusive:

Rule 1b: *The probability of either A or B occurring is equal to the sum of the probability of A and the probability of B, minus the probability of them both happening.*

In mathematical form this statement is written as

$$P(A \cup B) = P(A) + P(B) - P(A \cap B). \tag{1.31}$$

The Venn diagram in figure 1.10 demonstrates this point—the intersection, $P(A \cap B)$ represents the probability of both A and B occurring (the intersection is within both circles). Since the sum of the areas of the two circles would include the intersection twice, the total shaded area is the sum of the two circles minus the area of the intersection.

In examples such as coin tosses and dice rolls, where only one outcome is possible, then $P(A \cap B) = 0$, where A and B refer to any of the distinct results.

For example, if we want to know the probability of rolling a 5 or higher with a standard six-faced die, we should add together the probability of rolling a 5 and the probability of rolling a 6. That is, if n is the number rolled, then

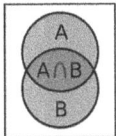

Figure 1.10
In the Venn diagram, left, the entire top circle represents $P(A)$, the entire bottom circle represents $P(B)$, and the overlap is $P(A \cap B)$. The total shaded region is $P(A \cup B)$ whose area is the sum of the areas of each circle minus the area of intersection, which is otherwise counted twice.

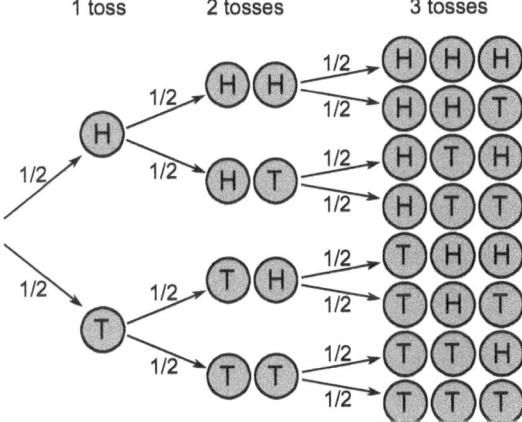

Figure 1.11
Probability of multiple independent events. If a coin is tossed multiple times, the number of possible outcomes for the sequence of tosses doubles following each toss, so there are eight outcomes after three tosses. The probability of heads or tails is 1/2 for every toss, independent of the number of prior heads or tails. Each of the final sequences (HHT, HHT, etc.) is equally probable, with a probability of $1/2^N$, where N is the number of coin tosses.

$$P(n \geq 5) = P(n = 5) + P(n = 6) = \frac{1}{6} + \frac{1}{6} = \frac{1}{3}.$$

Our next example will lead us to rule 2. Suppose we want to know the likelihood of tossing heads three times in a row. A key point in allowing us to solve the problem is the notion of independence—the probability of tossing heads on any one toss does not depend on the results of the other tosses. In this case, since each toss has two equally likely outcomes, there are $2 \times 2 \times 2 = 8$ possible outcomes: HHH, HHT, HTH, HTT, THH, THT, TTH, TTT (where H means heads tossed and T means tails tossed—see figure 1.11). So, the probability of three heads in a row is $1/8 = 1/2 \times 1/2 \times 1/2$. Rule 2 expresses this result more generally.

Rule 2: *The probability of two or more independent events all occurring is the product of the probabilities of each event occurring individually.*

Mathematically, independence means the probability of one event is unchanged by the occurrence of the other event. More precisely, two events, A and B, are independent if

$$P(A \mid B) = P(A),$$

where $P(A \mid B)$ means the probability of A if B occurs, which is said as "the probability of A given B."

Independent events   Two events are independent of each other if the outcome of one event does not alter the probabilities of any outcome of the other event.

Note on Independent Events   It is not always clear when events are independent, and intuitively we can make mistakes by forgetting independence. For example, if you had just tossed tails three times in a row, you might think that the next toss is more likely to be heads, simply because tossing four tails in a row is very unlikely. However, it is not possible for previous coin tosses to affect a later one, so the probability of heads is still 1/2. The sequence of results TTTH is exactly as likely as the sequence TTTT. The reason that getting one head is more likely than none when a coin is tossed four times is because there are four ways of getting a single head (HTTT, THTT, TTHT, and TTTH). Once we know the first three tosses are all tails, three of the four ways of getting a single heads are no longer available, and

$$P(H \mid TTT) = P(H) = 1/2.$$

Also,

$$P(TTTH) = P(TTT)P(H) = (1/8)(1/2) = 1/16 = P(TTTT).$$

In other cases, we might think that two events are independent and proceed as if they are, in which case our calculations would be incorrect if the events were not independent, meaning they were correlated. For example, the probability of a hot and sunny day is not simply the product of the two separate probabilities, i.e.,

$$P(Hot, Sunny) \neq P(Hot) \times P(Sunny),$$

since a day is likelier to be hot if it is sunny, which means the two are positively correlated.

Positively correlated   Events are positively correlated if the occurrence of one event means the other event is more likely than otherwise. Mathematically, two events, A and B, are positively correlated if $P(A|B) > P(A)$, which also implies $P(B|A) > P(B)$ and $P(A \cap B) > P(A)P(B)$.

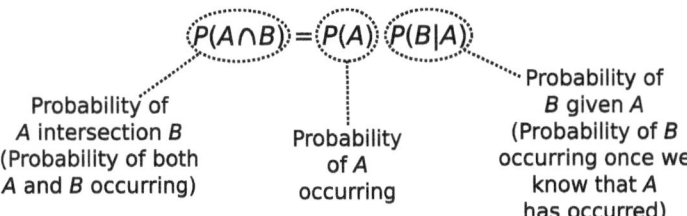

Figure 1.12
Annotation of rule 3.

Rule 3: *The probability of any two events occurring is the product of the probability of one event occurring and the probability of the other once we know the first has occurred, or mathematically,* $P(A \cap B) = P(A)P(B|A)$.

This equation is stated as "the probability of both A and B (or A intersection B) is equal to the probability of A multiplied by the probability of B given A" (figure 1.12).

For example, suppose a goalkeeper in a soccer game (figure 1.13) is facing a penalty kicker who shoots left 2/3 the time and right 1/3 of the time. When the kicker shoots left, the ball goes to the top-left corner 1/4 of the time and the bottom-left corner 3/4 of the time. When the kicker shoots right, the ball goes to the top-right corner 1/3 of the time and the bottom-right corner 2/3 of the time. The goalkeeper has to choose where to dive and hope to be correct in order to save the penalty. The goalkeeper dives to the bottom left. What is the probability the choice is correct?

Answer: From rule 3, the probability of a bottom-left kick is the probability the kicker aims leftward, multiplied by the probability the ball goes to the bottom-left when it is aimed leftward. That is, $P(B \cap L) = P(L)P(B|L) = 2/3 \times 3/4 = 1/2$, so the answer is 1/2.

You should convince yourself that all four probabilities shown in figure 1.13 can be generated in a similar manner.

Rule 3 can be stated as: The probability of both A and B occurring is the probability of A *and* the probability of B occurring if A occurs. This can be rewritten by symmetry as $P(A \cap B) = P(B)P(A|B)$. The second way of writing the equation states the probability of both occurring is the probability of B *and* the probability of A occurring if B occurs. Both of these must be true, and both methods of calculating the result must lead to the same answer.

For example, in figure 1.13, the probability of the penalty-taker aiming low is $(1/2 + 2/9)$ = 13/18. The probability of the penalty-taker aiming left if we know the aim is low, is $(1/2)/(13/18) = 9/13$. We can combine these to say the probability the shot is to the bottom left is the probability the shot is low multiplied by the probability the shot is left when it is low. That is,

$$P(B \cap L) = P(L)P(B|L) = \frac{13}{18} \times \frac{9}{13} = \frac{1}{2}.$$

Figure 1.13
Probabilities for each of four choices of target by a soccer player taking a penalty. In this example, the probability that the target is bottom-left is $P(B \cap L) = 1/2$. This can be calculated as the probability the player aims left, $P(L) = 1/2 + 1/6 = 2/3$, multiplied by the probability the player shoots low when aiming left, $P(B|L) = \dfrac{P(B \cap L)}{P(L)} = \dfrac{1/2}{2/3} = 3/4$. Illustration by Jaleel Draws of Pencil on Paper (POP).

Bayes' Theorem We have just seen that rule 3 can be written in two ways, both as $P(A \cap B) = P(A)P(B|A)$ and, by symmetry, $P(A \cap B) = P(B)P(A|B)$. Bayes' theorem simply identifies these two alternative equations as $P(A)P(B|A) = P(B)P(A|B)$. This is usually rearranged to give:

Rule 4 (Bayes' Theorem)

$$P(B|A) = \frac{P(B)P(A|B)}{P(A)}. \tag{1.32}$$

If A and B are independent then knowing that Event A has occurred has no impact on the probability of Event B occurring, so $P(B|A) = P(B)$. This shows that *rule 2* is a special case of *rule 3*, since $P(A \cap B) = P(A)P(B|A) = P(A)P(B)$ if A and B are independent.

The test for independence need only be carried out in one direction: If the probability of event A is not affected by event B, then the probability of event B is not affected by event A. This

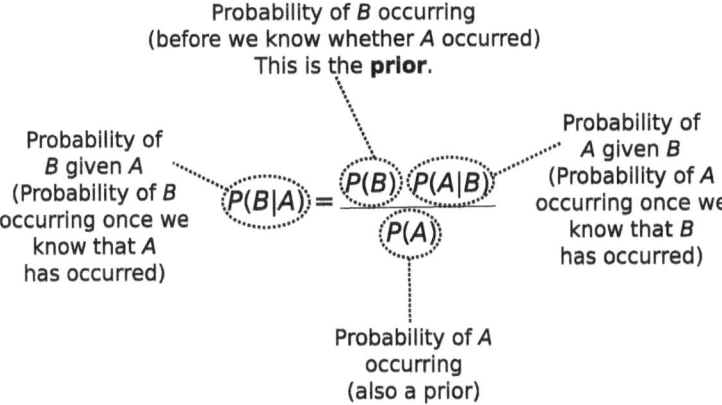

Figure 1.14
Annotation of Bayes' theorem, rule 4. The formula is typically used once we know that event A has occurred. The terms $P(B)$ and $P(A)$ are priors, since these are respectively the probabilities of events B and A occurring, before we know whether A occurred.

can be seen from Bayes' theorem (equation 1.32, figure 1.14), since if $P(A \mid B) = P(A)$ then $P(B \mid A) = P(B)$.

Examples of Bayes' Theorem

(i) Biased coin Suppose you have five coins, one of which is biased and produces heads with a probability of 3/4, while the others are fair so produce heads with a probability of 1/2 (see figure 1.15). Otherwise the coins are identical. You choose one of the coins at random, flip it three times, and get heads each time. What is the probability it is the biased coin?

For this problem, we let event B be "Biased coin is randomly selected" and event A be "Three Heads in a row" are tossed. We use the nomenclature $\tilde{B}$ to represent the event "Not B," meaning a fair coin was randomly selected.

In this case, we know or can calculate the following from the question, combined with the preceding four rules:

$P(B) = 1/5$ (1 out of 5 coins is biased, so probability it is chosen is 1/5).

$P(A \mid B) = \dfrac{3}{4} \times \dfrac{3}{4} \times \dfrac{3}{4} = \dfrac{27}{64}$ (if the coin is biased, this is the probability of tossing three heads in a row from *rule 2*).

Now $P(A)$ is the probability of tossing three heads in a row, irrespective of what coin is used. This is the sum of the probability of choosing the biased coin and then tossing three heads *plus* the probability of choosing a fair coin and then tossing three heads. Using *rule 3*, this can be written as:

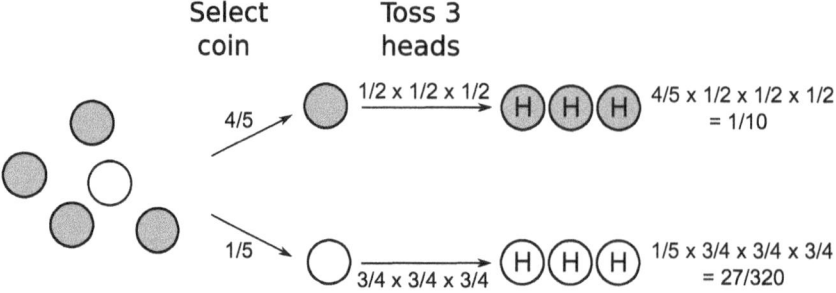

Figure 1.15
Flipping a biased or unbiased coin (example i). The probability of selecting a biased coin, $P(B)$, is 1/5, and the probability of selecting an unbiased coin, $P(\tilde{B})$, is 4/5. The probability of tossing three successive heads when the biased coin is selected, $P(A\mid B)=\dfrac{3}{4}\times\dfrac{3}{4}\times\dfrac{3}{4}=\dfrac{27}{64}$, while the probability of tossing three successive heads when the unbiased coin is selected, $P\left(A\mid\tilde{B}\right)=\dfrac{1}{2}\times\dfrac{1}{2}\times\dfrac{1}{2}=\dfrac{1}{8}$. Combining these yields the probability of selecting the biased coin and tossing three heads as $P(A\cap B)=P(B)P(A\mid B)=\dfrac{1}{5}\times\dfrac{27}{64}=\dfrac{27}{320}$, while the probability of selecting the unbiased coin and tossing three successive heads is $P\left(A\cap\tilde{B}\right)=P\left(\tilde{B}\right)P\left(A\mid\tilde{B}\right)=\dfrac{4}{5}\times\dfrac{1}{8}=\dfrac{1}{10}$.

$$P(A)=P(A\mid B)P(B)+P\left(A\mid\tilde{B}\right)P\left(\tilde{B}\right)=\left(\frac{1}{5}\right)\left(\frac{27}{64}\right)+\left(\frac{4}{5}\right)\left(\frac{1}{8}\right)=\frac{59}{320}.$$

Finally, we can use Bayes' theorem (*rule 4*) to evaluate the probability that the coin chosen was biased, given that you tossed three heads in a row, $P(B\mid A)$, as

$$P(B\mid A)=\frac{P(B)P(A\mid B)}{P(A)}=\frac{\left(\dfrac{1}{5}\right)\left(\dfrac{27}{64}\right)}{\dfrac{59}{320}}=\frac{27}{59}=0.458.$$

You should verify you get the same answer if you use an alternative form (see figure 1.15):

$$P(B\mid A)=\frac{P(A\cap B)}{P(A\cap B)+P\left(A\cap\tilde{B}\right)}.$$

The answer is less than 1/2, which tells us that even though the biased coin is much more likely to produce three heads in a row than a fair coin, the coin that produced three heads in a row is more likely to be a fair coin simply because there are more fair coins to choose from.

In this example, the *prior* probability of the biased coin was 1/5. The experimental result of tossing three heads produced a *posterior* probability of 0.458, which is higher than the prior. If we continued with a new independent test of the coin, we could begin the new experiment assuming the prior probability is 0.458. That is, in a sequence of experiments, the posterior probability following one experiment can serve as the prior probability of the next one.

There is evidence that within our brains the experiences of a lifetime act as a prior for interpretation of any new experience. When a stimulus induces a pattern of activity among neurons, connections change between the neurons such that the particular activity pattern is more likely to be induced by future stimuli. Thus, our brains may have evolved such that we naturally use Bayes' theorem when responding to the outside world (see tutorial 8.4).

(ii) Medical condition   When doctors examine a patient, they must use Bayes' theorem, either explicitly or implicitly, to determine the probability of a particular illness given the symptoms. From prior cases, the doctor can know the probability of any symptom when a patient has a certain illness. Similarly, the doctor can know how common an illness is and perhaps form an appropriate prior—i.e., the probability the patient has an illness before carrying out any examination or test—by taking into account the patient's age, gender, and family history. Moreover, if the patient arrived in the middle of an outbreak of one disease, then the prior probability of that disease would increase. Finally, although precise probabilities are not known in the way they might be in games with coins or cards, the doctor should be able to use historical data to estimate the probabilities.

For a concrete example, we will consider a patient with a rash that matches a symptom of a rare disease affecting only 1 in 10 million individuals, but the symptom is always present in patients with that disease. However, the doctor knows that such a rash appears in 10 percent of patients with a more common condition affecting 1 in 50,000 individuals. What is the likelihood the patient has the rare disease?

We write $P(R) = 10^{-7}$ as the prior probability of the rare disease, $P(C) = 2 \times 10^{-5}$ as the prior probability of the more common illness, and $P(S)$ as the probability of the symptoms. From the question, we know $P(S|R) = 1$ (probability of the symptoms in someone with the rare disease is 1) and $P(S|C) = 0.1$ (probability of the symptoms in someone with the common condition is 0.1).

If the rare and common diseases are the only ways to get the symptoms, then $P(S) = P(S|R)P(R) + P(S|C)P(C)$ (using *rule 3*).

Then Bayes' theorem (rule 4, equation 1.32) tells us the probability of the rare disease given the symptoms is:

$$P(R|S) = \frac{P(S|R)P(R)}{P(S)} = \frac{P(S|R)P(R)}{P(S|R)P(R) + P(S|C)P(C)}$$

$$= \frac{1 \times 10^{-7}}{1 \times 10^{-7} + 0.1 \times 2 \times 10^{-5}} = 0.0476$$

That is, the patient is twenty times more likely to have the common disease even though the symptoms are ten times more likely for a patient with the rare disease than for a patient with the common disease.

(iii) A lie-detector test Suppose that 90 percent of criminals fail a lie-detector test when responding "No" when asked if they committed their crime, but 1 in 100 innocent parties fail the same test because of nervousness in the interrogation room. In a city of 100,000, a person is randomly selected and fails the lie detector test when asked about a crime that could have been committed only by a resident. What is the probability that the person did in fact commit the crime?

In this example, we write $P(C) = 1/100,000$ as the prior probability of a randomly selected person being the criminal, $P(I) = P(\tilde{C}) = 1 - P(C) = 99,999/100,000$ as the prior probability a randomly selected person is innocent, and $P(F)$ as the probability that someone fails the lie detector test. From the question, we know $P(F|C) = 0.9$ and $P(F|I) = 0.01$.

We can evaluate $P(F) = P(F|C)P(C) + P(F|I)P(I)$ (from *rule* 3) and calculate the probability that someone who fails the test is a criminal via Bayes' theorem (rule 4, equation 1.32):

$$P(C|F) = \frac{P(F|C)P(C)}{P(F)} = \frac{P(F|C)P(C)}{P(F|C)P(C) + P(F|I)P(I)}$$

$$= \frac{0.9 \times 10^{-5}}{0.9 \times 10^{-5} + 0.01 \times 0.99999} = 0.00009$$

That is, the probability the failed test-taker is the criminal is still less than 1 in 1000. In this example, where the prior probability is so low, some intuition into the result from Bayes' theorem can be gained by noting that a criminal is 90 times more likely to fail the test than an innocent person so the posterior probability that someone is the criminal increases from the prior probability by a factor of approximately 90.

1.5 Introduction to Computing and MATLAB

Most of the features discussed in this section are relevant to any computing language, but specific examples will be presented in MATLAB (and can be run in the freeware alternative version, Octave, or slightly altered for Scilab). Many other tutorials are available for MATLAB online. Here the goal is to lead the newcomer to a sufficient level of understanding that tutorial 2.1 can be completed.

The basic operation of a computer code is to take some numbers (inputs), carry out a series of mathematical operations on those numbers (processing), and produce a new set of numbers (outputs) that contain some desired information. In the original computers, both inputs and outputs would be sequences of 1s and 0s represented as punched holes in cards. Today perhaps the most commonly used codes process inputs that are the coordinates of a touch on a screen, with their outputs being a graphic display. In the simulation codes that we write in this course, we will provide inputs as either the parameters or initial conditions of variables and see the outputs as a graph of the ensuing behavior of those variables, combined with the list of their values. In codes used for analysis, the inputs will be a series of data, with the outputs being a property of the data or a visualization of the data.

1.5.1   Basic Commands

When you run MATLAB, the cursor first appears in the "command window." Anything you type there becomes a command for the computer to do something. When the computer carries out the "command," it is said to "execute" it.

MATLAB's documentation is online and accessed by the software. You can always type doc followed by a word and you might find the answer to your question. If you know the name of a command but need to know how to use it correctly, just type help followed by the command name. The use of most mathematical functions (e.g., exp, log, sin, tanh, sqrt, inv, mean, sum) can be accessed this way.

When writing code, it is important to be aware of how any value is stored in memory for later use and how or when it changes. For example, if we want to store the two numbers 3 and 5, we could write the two commands:

```
x=3;
y=5;
```

If we wanted to use those values later in an operation, we could just refer to the labels. For example,

```
z=x+y;
```

creates a new variable, z, whose value is 8. (If in MATLAB you type the previous three lines of code, then type z without the semicolon

```
z
```

or type

```
disp(z)
```

to display z, you will see its value. However, if you now type

```
x=4;
```

the original value of 3 that you gave to the variable x is no longer present in the computer's memory—it has been overwritten with the new value of 4. This would be a problem if we later needed to use that previous value of 3.

At other times, we may want to keep changing a variable without keeping record of its previous values. For example, any counter has as its basic operation a line of code similar to:

```
x=x+1;
```

While such a line of code makes no sense if we read it as a mathematical equation, in most computer languages, as in MATLAB, it increases the value of x by 1. In general, a single equality sign (=) changes the value of the variable on the left-hand side of the equation to whatever value is on the right-hand side of the equation. If you have typed all the preceding lines of code in MATLAB and now type

```
x
```

you will see its new value of 5.

Sometimes we wish to record all of the previous values of a variable. In particular, if we want to plot how a variable such as membrane potential changes with time, we might want to know its value at every 0.1 ms for 10 s. It would be very inefficient to produce and keep track of a new variable name for the value of the membrane potential at each of these 100,000 time points! Fortunately, we can avoid this difficulty by using an array.

1.5.2 Arrays

> **Index**   An index of an array is a number indicating the position of a particular entry or set of entries in the array, for example by labeling a column number or a row number.

> **Element**   An element of an array is an individual entry or item in the array, which can be accessed by providing its index or indices along with the name of the array.

An array is like a vector or matrix (section 1.4.2), containing a list of numbers that can be acted upon either individually or en masse. The array has a single label or identifier, so it is important when operating with any variable to be aware if its label denotes a single value or an array. Individual entries in an array are accessed via the use of their index, which is denoted by a number within parentheses, after the array's label. Each computing language has its own particular syntax concerning the type of parentheses used and whether the indices of an array can begin with any integer, or always begin with either 0 or 1. The following syntax is applicable to MATLAB.

We can create an array called a as follows:

```
a=[1  3  5  7  9];
```

or

```
a=[1,  3,  5,  7,  9];
```

Notice that the entries are enclosed in square brackets. If we now type

```
b=2*a;
```

we have operated on the whole vector (to see this, type *b*). If we type

```
c=a(4);
```

we have taken the fourth element of the array, a, and created a new variable, c, to store it (to see its value, type c).

Notice that when you type a command followed by a semicolon (;) at the end of the line, any output to the screen is suppressed. A quick and easy way to see what you have changed with a command is to omit the semicolon. However, care should be taken in any code that executes millions of commands repeatedly, or containing huge arrays, since excessive screen output is not only impossible to read but also slows the code by orders of magnitude.

The array, a, that we have just created looks like a row vector. We can also create a column vector by ending a row between each entry with a semicolon as follows:

```
v=[2; 4; 6; 8; 10];
```

notice the difference in output style when you type v.

As we saw in section 1.4.2, vectors are special cases of matrices. In fact, we can think of a as a 1×5 matrix (meaning 1 row and 5 columns) while v is a 5×1 matrix (5 rows and 1 column). You can use the "size" command to see this:

```
size(a)
```

We can just use a single index to access an element of a vector. For example,

```
v(4)
```

returns the fourth element of v. We can also use row number and column number to access the element, so

```
v(4,1)
```

is allowed, as is

```
a(1,4)
```

but not

```
v(1,4)
```

We can produce the transpose of an array—switching all columns into rows and all rows into columns—with the prime symbol ('), as can be seen by typing

```
a'
```

or

```
v'
```

A matrix can be created directly with both rows and columns using comma or blanks space separators within a row, then semicolon separators between rows:

```
M=[1 3 5 7 9; 2 4 6 8 10];
```

produces a 2×5 matrix, M.

Questions   After typing the matrix M preceding:

1.   What is M(2,3)?
2.   Type MP=M'. What will happen if you type MP(2,3)?
3.   Now what will happen if you type MP(4,2)?

1.5.3   Allocation of Memory

In MATLAB, one can "grow" an array or vector one column or row at a time, but it is better—more efficient—whenever possible to create the array at its full size before using it. In many computing languages, the size of an array cannot be changed after it is created. It is common to simply put the value "0" into all entries of the array initially, but other types of initialization are possible. Try the following and see what they produce:

```
A=zeros(5);
B=rand(4,2);
C=eye(3);
D=ones(2,5);
```

Notice that a single integer in the initialization creates a square array of that size, not a vector.

We can also initialize a vector as an empty matrix, then grow it one element at a time. For example, type the following three lines and after each line check the size of the array E.

```
E=[];
E(end+1)=1;
E(end+1)=1;
```

The element end + 1 means one after the last element of the array. Find out what happens if you now type:

```
E(end+2)=1
```

then

```
E(end+3)=1
```

followed by

```
E(end-1)=2
```

1.5.4 Using the Colon (:) Symbol

The colon symbol (:) is useful for generating lists of numbers. For example, type

```
y=31:50;
```

You have created a row vector with 20 entries, the consecutive integers from 31 to 50 inclusive. Now type

```
z=30:5:50;
```

You have now created a row vector that steps in 5s from 30 to 50. The ":" can also be used to denote the indices of an array. For example, we can create a new array with a subset of the values of the array y, created above:

```
x=y(10:15),
```

which, as you see, creates a new array, x, containing the 10th to the 15th values of y.

Within an array, ":" can allow an entire row to be extracted, or any other rectangular subset of the array. For example, if we create an array by typing

```
F = [1, 2, 3, 4, 5; 6, 7, 8, 9, 10; 11, 12, 13, 14, 15];
```

then

```
G = F(2:3,2:4)
```

extracts the second and third rows of the middle three columns of F. If ":" is used on its own without any numbers, then it represents an entire row or an entire column:

```
H = F(2,:)
```

returns the entire second row of F while

```
K = F(:,1:3)
```

returns all of the first three columns of F.

Questions

4. Create a row vector of the numbers 1–100 in reverse order.
5. Create a 10×10 matrix with the number 5 for all diagonal entries and the number -3 for all off-diagonal entries.
6. Create a 5×3 matrix comprising random numbers selected uniformly from the range 10 to 20. (*Hint: think of the effect of multiplying the output of the* rand() *command by a number and adding to it another number*).
7. Create a vector of length 20, whose first 15 numbers are randomly chosen between 2 and 3, and whose last 5 numbers are randomly chosen between -1 and -2.

1.5.5   Saving Your Work

When you have created a lot of arrays that you may want to use later, it is important to save them to a file. The simplest method is to type

```
save mydata
```

where `mydata` can be any name you want to give the file, which contains all of your variables in a file called `mydata.mat`. To see the file in the computer's working directory you can type

```
ls
```

To see what directory you are working in, you can type

```
pwd
```

and to change it you can type

```
cd
```

followed by a directory name.

```
cd ~
```

takes you to your home directory.

```
cd ~/MATLAB/LatestProject
```

takes you to a subfolder of the MATLAB folder, which is in your home directory. To load your variables back into MATLAB, just change to the directory containing the file with your saved data, then type

```
load mydata
```

if `mydata.mat` is the correct filename.

If you wish to save a series of commands, perhaps so they can be carried out with some small changes to the values used, then they should be typed into the "editor" window and then saved (see the "save" icon at the top). Files saved from the MATLAB Editor have by default the extension *.m* and are called "M-files." They can be loaded back into the editor upon restarting MATLAB with the "open" icon.

1.5.6   Plotting Graphs

The documentation for the plot function is the best route to finding more about plotting—as a minimum you will want to use `title`, `xlabel`, and `ylabel` commands as well as plot with different symbols, line styles, and colors (see help `plot`).

By default, the plot command produces a single solid line between consecutive points, which are sent to the plot command as a series of *x*-coordinates in one vector followed by the

y-coordinates in a second vector of identical size. The two vectors can correspond to different rows or columns of the same matrix. For example:

```
g = [0 1 2 3 4];
h = [0 1 4 9 16];
plot(g,h)
```

produces a series of straight lines through the points, whereas

```
plot(g,h,'x')
```

simply produces the points (with a cross at each point). If the command

```
hold on
```

is typed then successive plots can be added to the same graph:

```
plot(g,h)
hold on
plot(h,g)
```

in which case, different colors are used for the successive plot commands. Figures can be saved to a file from the "figure" window.

Questions

8. Plot the line $y = 3x$ for a range of x from -5 to $+5$ with steps of 0.1. On the same graph, in a different color, plot $y = 4x$. Label the axes and indicate with a legend which color corresponds to which line.

9. Write a code that will plot $y = x^2 - 3$ for a range of x from -3 to $+3$ with steps of 0.1. Save the code as an .m file and run that file from the command window.

1.5.7   Vector and Matrix Operations in MATLAB

MATLAB is designed to operate efficiently with arrays. Therefore, mathematical operations on a complete array are permissible and even advisable in MATLAB, whereas in other programming languages the operations would require looping through all elements of the array (see section 1.5.9). In the simplest case, two arrays of the same size can be added together or subtracted:

```
a = [1 2 3 4 5];
b = [2 4 6 8 10];
c = a+b
d = b-a
```

For other simple mathematical functions, care is necessary as their action upon a matrix is ambiguous. If we want to operate on each element of a matrix individually, MATLAB requires us to include a "." before the mathematical operator. For example, following the preceding four lines of code, you can now type:

```
e = a.*b
f = b./a
g = a.^2
```

However, if you type

```
h = a*b
```

which attempts matrix multiplication, you get an error. Why? Compare the results of

```
a*b'
```

and

```
a'*b
```

Why are these two matrix multiplications so different? (See section 1.4.2 if necessary). Many inbuilt mathematical functions can be carried out on arrays, for example:

```
x = 0:0.1:15;
y = sin(x);
plot(x,y);
```

1.5.8 Conditionals

Most computer codes do not just process a series of commands but have a branchlike structure, processing one set of commands in one condition and another set of commands in another condition. Setting up such a branch in the code is via a conditional statement, which tests if an expression is true, in which case it processes the following command, but otherwise does not. The most common of these is the `if` statement; for example, consider the code:

```
b = tan(27);        % b is set as the tan of 27
if (b > 1)          % test if b > 1
  b = 1;                % if above line is true this sets b=1
end                     % after the "end" continue as before
c = 2;              % this line is always carried out
```

In the preceding code, we have used a conditional, because the line $b = 1$ is only performed if the expression in parentheses in the prior line evaluates as true, which it will do if $\tan 27 > 1$. Notice also the use of "%" (which by default appears green in MATLAB) to indicate a comment. Anything that follows the "%" symbol is not evaluated as MATLAB code. It will be important as you write longer codes to include comments to help both you and any other reader understand what the code should be doing.

As an aside, in MATLAB, the effect of the simple conditional used earlier can be written more simply in one line via the `min` function:

```
b = min(b,1);
```

to replace b with the value 1 whenever b > 1. The `if` statement can be extended to include an alternative course of action. For example:

```
b = tan(27);          % b is set as the tan of 27
if (b > 1)            % test if b > 1
  b = 1;              % if above line is true this sets b=1
else                  % if the original (b > 1) test is false
  b = b*b;            % change b to the value of b*b
end                   % after the "end" continue as before
c = 2;                % this line is always carried out
```

It is also possible to have several lines of code between the `if` and the `else` and between the `else` and the `end`. These lines of code can include more conditionals:

```
b = tan(27);          % b is set as the tan of 27
if (b > 1)            % test if b > 1
  b = 1;              % if above line is true this sets b=1
  d = 1;              % then sets d=1
  e = 1; ;            % then sets e=1
else                  % if the original (b > 1) test is false
  if (b < -1)         % test if b < -1
  b = -1;             % if above line is true this sets b=-1
  e = 2;              % and sets e=2
  else                % otherwise carry out the next line
  b = b*b;            % change b to the value of b*b
  e = 3;              % and set e=3
  end                 % perform next line
  d = 2;              % carried out if first "if" was false
end                   % after the "end" continue as before
c = 2;                % this line is always carried out
```

You should try different initial values of b to test the code above until you understand what initial values cause different lines to be processed. Try to predict the final values of c, d, and e before running the code.

Questions

10. Plot a rectified sine wave (negative values are fixed at zero) over the range of values $-\pi \le x \le \pi$.

11. Write a code to indicate whether the cube root of 6 is greater than the square root of 3.

1.5.9 Loops

The power of the computer lies in its ability to perform simple calculations over and over again very quickly without error. When the same calculation is repeated many times, operating on different numbers, rather than writing out the calculation multiple times, a loop can be used. In this section, we consider different uses of `for` loops and `while` loops.

To get an idea of the syntax for using `for` loops, type the following three very brief codes into the editor and save them (with separate filenames), then run them:

```
% for_loop_1.m
for A=[1 2 3 4 5],
 A*A,
end;
% for_loop_2.m
a=[1 2 3 4 5];
for A=a,
 A*A,
end;
% for_loop_3.m
for A=1:5,
 A*A,
end;
```

You should find that each of the `for` loops do the same thing and produce the first five square numbers. In each case the variable A steps through the matrix [1 2 3 4 5], and inside the loop we have asked MATLAB to display the value of A*A.

We have defined the matrix [1 2 3 4 5] three different ways: explicitly; using the variable a; and using the colon operator. The latter method is the most common, but the first two methods also allow you to define a set of values for the variable that do not increase as an arithmetic series.

When the number of cycles through a loop is not known ahead of time it can be better to use a `while` loop instead of a `for` loop. A `while` loop continues to cycle through the loop as long as a conditional expression is true. It is therefore essential that whatever is tested in the conditional is updated within the loop. For example, suppose we wish to find the smallest positive integer whose square is greater than 1000:

```
% test_high_square
i = 1;                        % first positive integer to test
while (i*i < 1000)            % loop only while expression is true
 i = i+1;                     % test the next positive integer
end;                          % finish the loop
display(i);                    % display last value of i reached
```

Questions

Use loops to:

12. Sum all of the cubed positive integers up to 15^3.

13. Find the positive integer n such that $n + n^2 + n^3 + n^4 = 88740$.

1.5.10   Functions

If a sequence of commands is used multiple times by a code or by many different codes, it is good practice to save the sequence of commands as a separate function. The function will be saved as a MATLAB file with an *.m* extension and can be used either by working in the directory (folder) where the file is saved or by adding the path to the directory where it is saved with the addpath command. You have already used built-in functions when carrying out mathematical operations like sin(x). Directories containing many functions are called libraries—these can be added to the path so that MATLAB always finds them when it starts up (see startup).

To write a function, for example that adds the square of a number to the cube of a number—let's call it square_cube_sum—the syntax is:

```
function y = square_cube_sum(x)
 y = x*x + x*x*x;
end
```

To use the function, we "call" it, so if you save the three-line function written above in a file called "square_cube_sum.m" in your current directory, you can then type

```
a = square_cube_sum(14);
```

to set a as the value of $14^2 + 14^3$.

A function can have multiple inputs and multiple outputs. For example:

```
function [ysum, ydiff] = square_cube_sum2(x1, x2)
 ysum = x1*x1 + x2*x2*x2;
 ydiff = x2*x2*x2 - x1*x1;
end
```

should be saved as the file square_cube_sum2.m, in which case the command:

```
[c d] = square_cube_sum2(5, 8);
```

returns $5^2 + 8^3$ as c and $8^3 - 5^2$ as d.

Questions
14. (a) Write a function that takes as input a single vector and returns the sum of the squares of its elements as its single output. (b) Use that function to sum the square of the numbers from 27 to 37 inclusive.
15. Write a function that takes as input a single vector of numbers and returns the mean, the mode, and the median as three separate variables for its output.

Hint: The functions mean, hist, *and* sort *will help.*

1.5.11   Some Operations Useful for Modeling Neurons
Suppose you want to write a code that indicates whenever the tangent function has a value greater than a threshold, here set as 2.

```
% threshold_find.m
clear
thresh = 2;
tmax = 10;
tvector = 0:0.001:tmax;
Nt = length(tvector);
tanval = zeros(size(tvector));          % to store tan of tvector
findhigh = zeros(size(tvector));        % stores when tan > thresh
for i=1:Nt                              % for all values of tvector
  tanval(i) = tan(2*pi*tvector(i));     % set tangent of t
  if (tanval(i) > thresh)                    % if tan is high
  findhigh(i) = 1;                      % store value of t
  end
end
%% Now plot the results
figure(1);
subplot(2,1,1);
plot(tvector,tanval);                   % plot tan(2.pi.t) versus t
axis([0 tmax -5 5]);
subplot(2,1,2);
plot(tvector,findhigh);                 % plot t where tan(2.pi.t)>2
```

```
axis([0 tmax -0.5 1.5]);
%% We can color the portions corresponding to findhigh=1 using
% MATLAB's find command, which extracts the indices of the
% non-zero entries of a matrix:
highindices = find(findhigh);           % indices above threshold
subplot(2,1,1);
hold on
plot(tvector(highindices),tanval(highindices),'r.');
```

We can achieve the same results more efficiently by taking advantage of MATLAB's matrix operations:

```
% threshold_find2.m
clear
thresh = 2;
tmax = 10;
tvector = 0:0.001:tmax;
Nt = length(tvector);
tanval = tan(tvector);              % operates on all values at once
findhigh = tanval>thresh;          % gives 1 or 0 for all entries
%% Now plot the results
figure(1);
clf                                % clears figure for a new plot
subplot(2,1,1);
plot(tvector,tanval);
axis([0 tmax -5 5]);
subplot(2,1,2);
plot(tvector,findhigh);
axis([0 tmax -0.1 1.1]);
```

In most programming languages the first method, which uses a `for` loop with a conditional (`if` statement) inside it, is the only method for solving such problems.

1.5.12   Good Coding Practice
Computer simulations have the advantage that in principle the reason for any result—real or spurious—can be extracted at a later date. Any output of the code depends entirely on the algorithms within the code, any inputs provided, and, in many cases, the seed used to initiate the random number generator. If these attributes are saved then an entire simulation is repeatable with identical results. The route to any interesting or puzzling result can therefore be analyzed at

a later date so long as the data needed to run the code are saved. In computing, there should be no need to go through the frustrating experience of being unable to reproduce prior results.

Tips 1 and 2 following are aimed at ensuring such reproducibility. Tip 3 is aimed at ensuring large amounts of time are not wasted in running "bugged" code. Tip 4 is aimed at ensuring that your code can be used by others and so save having to reinvent the wheel.

1. Whenever you change a code, always assume you will want to run the prior version.

There are good version control systems available (CVS, github etc.) that allow us to go back to prior versions after making changes. If, for some reason—say, because you are beginner—you do not use version control, then at a minimum when you alter a code you should give it a new filename or store it in a new directory so that the old version is not overwritten.

2. Whenever you run a code, assume you will want to obtain that result again one day.

If you run your code with hundreds of parameters or with lots of variations, you may at some point want to retrieve one nice result that you recall or showed to your supervisor. That result should be saved in a file that either has the name with parameter values of the code that produced it, or is in a directory with a unique code and set of parameters, or a combination of the two. This is most easily achieved automatically, by creating a string variable for the filename that includes relevant parameter values. For example, if G_L, V_{th}, and V_{reset} are the parameters that get altered different times a code is run, one could create a filename:

```
filename = strcat(LIF_GL',num2str(G_L),'_Vth',num2str(V_th), … '_
Vreset',num2str(V_reset),'.mat');
```

then at the end of the code have a line

```
save(filename)
```

to save the final output of the code in a clearly labeled filename—and use the same filename in the title of any figure.

3. Always write a code assuming it contains errors (bugs) that need to be found.

As computer codes increase in length, it becomes exceedingly difficult to find any error within the code and even to be sure its output is correct. Wherever possible, use chunks of code that can be easily and thoroughly tested in isolation from the rest of the code. Functions are ideally suited to this purpose. Also, wherever possible, test the output of the code using parameters with known results—that is, ensure the code is correct when you know what the answer is before you trust it to produce a new, unknown answer. It is typical for more time to be spent testing and removing errors, i.e., debugging, than on writing the code in the first place. Careful, patient checking as you write can save many hours of frustration in the debugging process later.

4. Always write a code assuming that others will need to understand how it operates.

Even if nobody else ever looks at or uses your code, when you revisit the code many months or years later—perhaps even days later—you will have forgotten what you were thinking when you wrote the code. Therefore, you should ensure it is clearly written, adding comments whenever you include something unobvious—and even if it is obvious at the time it may be unobvious later—and using variable names that indicate the meaning of the variable. If we were to use x for membrane potential, y for conductance, and z for time, our codes would quickly become impenetrable!

1.6 Solving Ordinary Differential Equations (ODEs)

1.6.1 Forward Euler Method

In general, we wish to solve numerically (which means via a computer simulation—rather than analytically, which means via a mathematical formula) an equation of the form:

$$\frac{dx}{dt} = f(x,t),$$

where $f(x,t)$ is a function that can be calculated for any value of x and time, t. An initial value of x must be provided and that value is usually taken to correspond to a time of zero, i.e.,

$$x_0 = x(t=0).$$

To simulate such an equation, we must evaluate x at discrete time-points. The interval between time-points is fixed in the Euler method—here we denote the interval as Δt. To simplify the notation, we write $x_n = x(t_n = n\Delta t)$.

Using the Forward Euler method, we calculate the rate of change of x at one point in time and use that rate of change to extrapolate to the value of x at a later point in time (figure 1.16). That is, by setting the change in x over the change in time as the instantaneous gradient:

$$\frac{x_{n+1} - x_n}{\Delta t} = f(x_n, t_n)$$

we obtain the new value of x as x_{n+1} from the old value, x_n:

$$x_{n+1} = x_n + f(x_n, t_n)\Delta t.$$

The method is an approximation that is accurate to the extent that $f(x,t)$ remains constant during the time interval, Δt. If $f(x,t)$ changes—either because the function explicitly depends on t, or because x changes—in a timescale of T, then the forward Euler method is accurate to the order of $\Delta t / T$. That means to reduce the accumulated error by a factor of 10, the timestep should be ten times smaller.

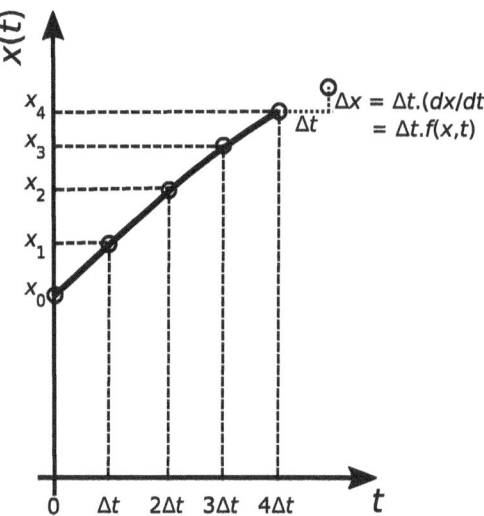

Figure 1.16
Euler method. At each time point, the gradient of the curve of x versus t is evaluated as $dx/dt = f(x,t)$ from the relevant ordinary differential equation. Then a small step in time, of size Δt is taken and the value of x is increased by an amount $\Delta x = \Delta t\left(\dfrac{dx}{dt}\right) = \Delta t \times f(x,t)$.

The Forward Euler method has the advantage of being straightforward and that it allows noise to be added in a relatively simple and scalable manner (see the Euler-Mayamara method that follows). However, it is not always robust and is less efficient than other methods. The Backward Euler method is more robust and both second-order and fourth-order Runge-Kutta methods are more accurate, perfectly accounting for quadratic and quartic variation in the variable respectively, and being accurate to the order of $(\Delta t / T)^2$ and $(\Delta t / T)^4$ respectively. An example of fourth-order Runge-Kutta is presented in the online code `integration_test_sho.m`, since it is often the most efficient method for solving ODEs. For further reading, see *Numerical Recipes* by W. H. Press et al. and *Modeling in Biology: Differential Equations* by Clifford Taubes.

1.6.2 Simulating ODEs with MATLAB

Here we consider two methods for simulating the differential equation: $dx/dt = 2sin(t^2 /10)$ with initial condition $x(0) = 2$.

To simulate any differential equation, we must specify the range of time across which we want to record all the values of the variables. Therefore, we first create a vector of time points. We then set the initial conditions. In the first method we loop through time to integrate up the differential equation (as we will do in tutorial 2.1 on the leaky integrate-and-fire neuron):

```
% test_euler.m
clear
x0 = 2;                            % the first value of x, the variable
t0 = 0;                            % the first value of time to be used
dt = 0.001;                        % the time step for simulations
tmax = 20;                         % the maximum time used
tvec = t0:dt:tmax;                 % creates a vector of time points
x = zeros(size(tvec));             % creates vector of same size
x(1) = x0;                         % value of x at t=0, as tvec(1)=0
for i = 2:length(tvec)             % loop through time points
  tval = tvec(i);                  % value of time
  dxdt = 2*sin(0.1*tval*tval);     % rate of change of x, dx/dt
  x(i) = x(i-1) + dt*dxdt;         % update of x (Forward Euler)
end
plot(tvec,x);                      % plot x against t
```

You can stick with the above method for solving ordinary differential equations throughout the class. Alternatively, you can use the inbuilt MATLAB solvers. At least try this one (you will create 2 files, odepractice.m and test_ode.m in the same directory):

```
% odepractice.m
clear
x0 = 2;                      % the first value of x, the variable
t0 = 0;                      % the first value of time to be used
dt = 0.001;                  % the time step for simulations
tmax = 20;                   % the maximum time used
tvec = t0:dt:tmax;  % creates a vector of time points
[t, x] = ode45(@test_ode, tvec, x0)          % use built-in function
plot(t,x)                    % plot x against t
```

which requires that we write the function test_ode saved as test_ode.m:

```
function [ dydt ] = test_ode(t, y)
%integrate_sinx Used to demonstrate ODE solvers
dydt = 2*sin(0.1*t*t);end
```

Have no fear in using doc or help commands to find out how to use MATLAB functions (e.g., help ode45 will tell you about the built-in MATLAB differential equation solver used earlier).

1.6.3   Solving Coupled ODEs with Multiple Variables

In most systems of interest, variables interact with each other, which means they are coupled. When the dependence of one variable on another is direct and instantaneous, the interaction can be included via a formula directly, as, for example, the current can be calculated through a channel if the potential difference, reversal potential, and conductance are known. However, in many cases the value of one variable impacts the rate of change of another so that the respective ODEs are coupled. In that case, the rate of change of each variable is updated on each timestep using the values of all other variables at that point in time, then the set of variables is updated for the next time point.

For example, if we want to solve the system of equations:

$$\frac{dx(t)}{dt} = v(t)$$

$$\frac{dv(t)}{dt} = -\omega^2 x(t) \tag{1.33}$$

with initial conditions $x(0) = 1$, $v(0) = 0$, we would write a code using the Forward Euler method as follows:

```
tmax = 100;                          % time to integrate over
x0 = 1.0;                            % initial position, x at t=0
v0 = 0.0;                            % initial velocity, v at t=0
omega = 3;                           % angular frequency
omega_sqr = omega*omega;             % square of angular frequency
dt = 0.00001;                        % small dt for Forward Euler
tvector = 0:dt:tmax;                 % vector of time points
x1 = zeros(size(tvector));           % x1: values of position
x2 = zeros(size(tvector));           % x2: values of velocity
x1(1) = x0;                          % set initial value of x1
x2(1) = v0;                          % set initial value of x2
for i = 2:length(tvector);           % integrate over time
  x1(i) = x1(i-1) + x2(i-1)*dt;      % update x1
  x2(i) = x2(i-1) - omega_sqr*x1(i-1)*dt;      % update x2
end
```

Notice that all of the computation required to solve the equations is contained in the two lines of code that are followed by the comments `update x1` and `update x2` within the `for` loop. All of the other lines of code are setting up the parameters used, the array to store the results, and the initial conditions.

1.6.4 Solving ODEs with Nested `for` Loops

We will often want to find out how changing a parameter affects the behavior of an ODE. Indeed, most of the tutorials in this book will require a code that will reveal such parameter-dependence. Such codes typically require nested `for` loops: an outer loop for variation of the parameter and an inner loop for the simulation with a particular value of the parameter. Care must be taken with saving a result produced with one value of the parameter when it could be overwritten by a simulation with the next value of the parameter.

For example, following section 1.6.2, suppose we wanted to simulate $dx/dt = 2sin(t^2 / K)$, where the parameter K can vary from 1 to 20, and our goal is to find how the value of x at the time $t = 3$, i.e., $x(3)$, depends on the parameter K, when $x(0) = 0$. A code to achieve this follows:

```
clear;                                  % clear prior variables from memory
dt = 0.001;                             % time step for the simulation
tmax = 3;                               % maximum time for the simulation
t = 0:dt:tmax;                          % vector of time points
Kvals = 0.1:0.01:20;                    % Set of values used for K
xfinal = zeros(size(Kvals));           % To record x(3) for each K
for simulation = 1:length(Kvals)       % Outer loop alters K
K = Kvals(simulation);                 % Select K for a simulation
x = zeros(size(t));                    % Initialize x each new K
for i = 2:length(t)                    % Inner loop through time
x(i) = x(i-1) + dt*2*sin(t(i)*t(i)/K); % Update x
end;                                    % End the inner for loop
xfinal(simulation) = x(end);           % Record x(3) for this K
end;                                    % End the outer for loop
figure(1)
plot(Kvals,xfinal)                     % Plot x(3) versus K
```

1.6.5 Comparing Simulation Methods

It can be useful to use the commands `tic` and `toc` to test precisely how long a piece of code takes to run. The online code (not copied here) `integration_test_sho.m` times different methods for integrating the coupled ODEs, equations 1.33 of section 1.6.3. In this case the exact solution is known, so the mean-square error is evaluated for each method as well as the total time. In this case, the fourth-order Runge-Kutta method does better than all others, though Forward Euler is comparable with MATLAB's inbuilt ODE solver `ode45`. Indeed, in this example the Forward Euler method is more accurate if the line of code updating `x2` is altered to make use of the previously updated version of `x1`:

```
x2(i) = x2(i-1) - omega_sqr*x1(i)*dt; % update x2
```

However, this is not always the case. In general, when exact solutions are not known, always test results with a simulation using a smaller value of dt (e.g., by a factor of 10) to ensure you do not have significant numerical error.

You should find that the codes in this introductory course run in short enough time that you do not need to worry about optimizing efficiency. However, for most computational research, or for analysis of large datasets, it is worth devoting a fraction of your time to improve the efficiency of your code (reducing time taken for the same accuracy).

1.6.6   Euler-Mayamara Method: Forward Euler with White Noise

Random fluctuations due to processes that are not accounted for explicitly in a simulation can be included as "noise" using the random number generator. Such noise is often incorporated by adding to the input current of a cell a white noise process, $w(t)$, which is defined as having zero mean. It is important to recognize that even though the mean of the added current is zero, the effect of such a current on the mean firing rate of a cell—or the mean of any nonlinear function of the current—is not zero. The white-noise process is defined such that its variance is a delta function in time, so that

$$\langle w(t) \rangle = 0 \text{ and } \langle w(t)w(t') \rangle = \delta(t - t'),$$

where $<\cdots>$ means the expected value when averaging over time. The delta-function is defined to be zero unless $t = t'$, which means that the value of $w(t)$ is uncorrelated with the value of $w(t')$ unless $t = t'$ (in which case, of course $w(t) = w(t')$).

A second property of the delta-function is that its integral over all time is equal to one:

$$\int_{-\infty}^{\infty} \delta(t - t')dt = 1$$

so that the delta-function has units of inverse-time (s^{-1}), which means the white noise function has units of the square-root of inverse-time ($s^{-0.5}$). This is important in simulations as it gives us an indication of how to scale the term when the timestep changes.

The form of the mathematical equation to simulate is:

$$\frac{dx}{dt} = f(x,t) + \sigma w(t)$$

where σ scales the level of noise. Notice that if x were to represent membrane potential then the units of σ would be $Vs^{-0.5}$, since the whole equation would have units of Vs^{-1}.

When we simulate the equation, we update the variable using the methods already described in this chapter, but must include an additional noise term:

$$x_{n+1} = x_n + f(x_n)\Delta t + \sigma \tilde{w}_n \sqrt{\Delta t}$$

where $\tilde{w}_n$ is a random number selected from a distribution with zero mean and unit variance. The Gaussian is the standard choice (the exact distribution does not matter if Δt is small enough, because of the central limit theorem), in which case the probability of any value is given by

$$P(\tilde{w}_n) = \frac{1}{\sqrt{2\pi}} e^{-\tilde{w}_n^2/2}.$$

Such a unit-variance, Gaussian-distributed random number is obtained in MATLAB via the function randn().

2

The Neuron and Minimal Spiking Models

Neurons stand out from other cells because of their extensive structures, which allow them to target and interact with specific cells located far away, and because of their active electrical properties, which allow them to send rapid signals to those distal cells. In chapters 2 and 4 we focus on the electrical responses of neurons, while connections between neurons are the subject of chapter 5.

At the end of this chapter, table 2.2 contains the meaning and a description of each of the many variables used throughout chapter 2.

2.1   The Nernst Equilibrium Potential

Our brains consume energy—one-fourth of the total energy of the body, despite being one-fiftieth of the body's mass in the average human. Much of this energy consumption is in the maintenance of ongoing electrical activity in neurons. The electrical activity arises from the rapid flow of ions back and forth across the cell membrane. The flows of different types of ion conspire to produce spikes in the membrane potential—an electrical potential difference between the inside and outside of the cell—that can travel down axons of neurons. Since ions are electrically charged, their movement across the membrane must be propelled by a driving force that acts like the electromotive force of a battery. It is the continual recharging of tens of billions of such tiny effective batteries that requires so much energy.

Membrane potential   The potential difference across the cell membrane, which can vary across a range of tens of millivolts in neurons.

Voltage spike   Also called an action potential, a rapid upswing in the membrane potential that can propagate down the axon of a neuron and enable fast transmission of information throughout the brain and the body.

Ion channel   A channel allowing entry or exit of a particular type of ion into or out of the cell through the cell membrane. Ion channels can be passive, with fixed conductance, or active, with a conductance that depends on conditions such as the membrane potential.

An effective battery is generated by the differences in the concentration of ions across the membranes of neurons. When an ion channel opens in the membrane, it permits a specific type of ion to flow through it. The ions flow preferentially from high concentration to low concentration, tending to reduce the concentration difference. However, the concentration difference is maintained by the ATP-dependent action of ion pumps and exchangers in the neural cell membrane. The pumps are constantly expending energy to move ions "upstream" against the concentration gradient they have produced. By doing this, the pumps are recharging the effective batteries that drive the flow of electrical current into and out of active neurons (see section 1.3.3).

The flow of ions through a channel can be combatted by an electrical potential difference that drives the ions in the opposite direction of their concentration gradient. The value of such a potential difference at which the electrical force exactly matches the force due to the concentration gradient across the cell membrane of a specific type of ion is the Nernst potential for that ion. If the potential difference between the inside and the outside of a cell is equal to the Nernst potential of a particular ion, then there is no net flow of that ion across the membrane through an open channel. Any inward flow is balanced by outward flow. Such balance leads to the Nernst potential, also called the Nernst equilibrium.

Nernst potential   Also called reversal potential, the membrane potential at which the flow of a particular ion is in a dynamic equilibrium, meaning the outflow is precisely matched by the inflow of that ion.

If the concentration of ions within the cell is identical to that outside the cell, then the Nernst potential is zero, since any drift of ions by chemical diffusion into the cell is balanced by chemical diffusion out of the cell and any potential difference would break that balance.

The derivation of the formula for the Nernst potential depends on some physics and is left to the appendix (equations 2.23–2.27), but its properties can be understood without the details of the derivation. For an ion, A, of charge z_A, with intracellular concentration $[A_{in}]$ and extracellular concentration $[A_{out}]$, the Nernst potential for that ion, E_A, is given by:

$$E_A = \frac{k_B T}{z_A q_e} \ln\left(\frac{[A_{out}]}{[A_{in}]}\right), \tag{2.1}$$

Table 2.1
Ions and their properties

Ion	Charge	Internal Concentration	External Concentration	Nernst Potential
Sodium	+1	15 mM	120 mM	55.5 mV
Potassium	+1	150 mM	6 mM	−86.0 mV
Calcium	+2	50 nM	2 mM	141.5 mV
Chloride	−1	10 mM	120 mM	−66.4 mV

Note: The most common species of ions that flow through ion channels in neurons with some examples of intracellular and extracellular concentrations within the ranges found in mammals alongside the corresponding Nernst potentials produced at 37°C (310K).

where T is the temperature in absolute units of Kelvin, k_B is the Boltzmann constant $(1.38 \times 10^{-23} \, JK^{-1})$, which converts units of temperature to units of thermal energy, and q_e is the fundamental electronic charge (i.e., the charge of an electron, 1.60×10^{-19} C).

For an ion such as sodium with unit positive charge, a quick calculation of the factor in front of the natural logarithm yields a value of 25.1 mV at a typical lab temperature of 18°C (291K) and a slightly higher value of 26.7 mV at human body temperature (37°C = 310K). This sets the scale of typical changes in neural membrane potentials to be in the tens of millivolts, since even a hundredfold concentration ratio would lead to an additional factor of only 4.6 after taking the natural log in equation 2.1.

The temperature-dependence of the Nernst potential is a consequence of the kinetic energy of ions being proportional to absolute temperature: A greater electrical potential difference is needed to counteract diffusion when the ions move more rapidly. It is important to bear this in mind when measuring the membrane potentials of neurons in the laboratory and extrapolating to their operation at body temperatures in vivo.

The Nernst potential is inversely proportional to the charge on an ion. Intuitively, this is because the electrical force on an ion is proportional to its electrical charge and to the potential difference, so the electrical force needed to counteract diffusion is achieved at a smaller potential difference if the ion has greater charge.

Since the membrane potential is measured as the internal potential minus the external potential, a positive membrane potential acts to drive positive ions out of the cell. Therefore, a positively charged ion has a positive Nernst potential when the outside of the cell has greater ionic concentration than the inside of the cell. In such a case, at the Nernst potential, the inward diffusive flow is balanced by the outward electrically driven flow of the positive ions.

The logarithmic dependence on the concentration ratio (equation 2.1) produces a potential difference of zero when concentrations are equal ($\log(1) = 0$). The logarithmic dependence arises from the exponential reduction in the probability of an ion moving against an electrical potential gradient as that gradient increases. The dependence on the concentration ratio is important in experiments in vitro, where the extracellular medium (artificial cerebro-spinal fluid, aCSF) is controlled by the scientist. When different laboratories measure the properties of the same neuron, but with different concoctions for their aCSF, they will arrive at different results. In vivo, ionic concentrations are well regulated. In cases such as epileptic seizures, during which neural activity is so strong that the ion pumps cannot maintain standard operating conditions, the resulting imbalance in ionic concentration and change in Nernst potentials may play an important role in bringing each seizure to an end.

Equilibrium potential   Also called resting potential, the membrane potential at which the flow of electrical current from all types of ions into and out of the cell is balanced, so there is no net current and the membrane potential is not caused to change.

The potential at which total currents into and out of the cell are balanced is the "equilibrium potential" (also called the resting membrane potential or leak potential). The membrane potential moves toward this balance point when not driven away from it by sources of electrical current. The equilibrium potential is negative for most neurons, from which we can deduce that the dominant channels transmit either positive ions with a high internal concentration or negative ions with a high external concentration. These conditions are satisfied by potassium ions and chloride ions (respectively), which both have negative Nernst potentials (table 2.1). Sodium ions and calcium ions have positive Nernst potentials, but their corresponding channels, while able to produce strong current influx to the neuron, are almost entirely closed at equilibrium.

2.2   An Equivalent Circuit Model of the Neural Membrane

Membranes of neurons, like other cells, are formed by a lipid bilayer, which is a hydrophobic electrical insulator that does not permit ions to flow across it. Rather, charge can be stored on its surface, so the membrane acts as a capacitor with a capacitance (capacity to store charge) that is proportional to its area.

Within the lipid bilayer are implanted ion pumps and ion channels. The action of the pumps is to create concentration differences across the membrane for specific ions. The action of pumps produces the separate Nernst potentials for each ion (table 2.1) and does not have to be explicitly modeled.

Ion channels are selective, often permitting the flow of only one species of ion. When open, they pass electrical current in the direction needed to bring the membrane potential toward the Nernst potential of the specific ions that flow through the channel. That is, if the Nernst potential is more positive than the membrane potential, then positive ions flow into the cell or negative ions flow out of the cell, either of which would make the membrane potential more positive. Conversely, if the Nernst potential is more negative than the membrane potential, then positive ions flow out of the cell or negative ions flow into the cell, either of which would make the membrane potential more negative. Because of this switch in the direction of current flow, the Nernst potential for a channel is also called its *reversal potential.*

Each type of ion channel can be modeled as a battery in series with a resistance (figure 2.1). The inverse of the resistance—the conductance—is proportional to the number of open channels. In many cases this conductance is both temperature- and voltage-dependent. The capacitance of the membrane—its ability to store charge—is not affected by channels opening and closing because these channels make up only a tiny fraction of the membrane's total surface area. The capacitance is therefore treated as a constant in all standard models.

We can use the equivalent circuit to determine the resting membrane potential, which is the potential at which the electrical charge on the capacitor remains constant, i.e., at which there is no net flow of current into or out of the cell. When the potential difference across the circuit is equal to the resting potential, the currents through each of the resistors must cancel, so that inward currents through sodium and calcium channels are matched by outward currents through the potassium and leak channels. The current through a channel is given by:

$$I_i = G_i \left(V_m - E_i \right)$$

where the index, i, represents the type of channel. In this terminology, which is standard, a positive current is one that flows out of the cell. When the cell is at equilibrium the different currents balance each other out and sum to zero:

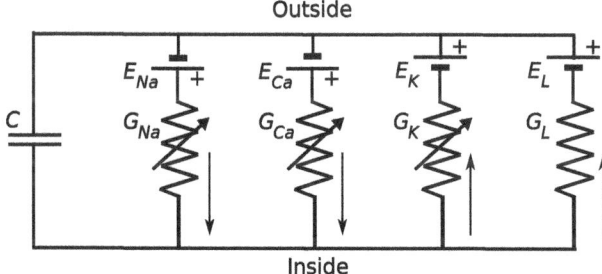

Figure 2.1
The equivalent circuit representation of the neuron's membrane. Various ion channels, each type represented by a resistor in series with a battery, span the cell membrane. All operate in parallel with each other—so their conductances sum—and with a capacitor, which represents the insulating cell membrane. The battery for a type of ion channel is directed to bring the membrane potential (electrical potential on the inside of the membrane minus the potential on the outside) to the Nernst equilibrium potential for that ion. Types of ion channels with a variable total conductance—sodium (G_{Na}), calcium (G_{Ca}), and potassium (G_K)—are represented by the arrow through the resistor, whereas the leak conductance, G_L, is defined as a constant. The vertically directed arrows indicate the direction of current flow when the membrane potential is at zero or another intermediate value. If all channels with variable conductance are closed so that no current flows through the central parallel paths in the above circuit, then current will flow through the leak channels (charging the capacitor) until the inside of the cell membrane is at the leak potential E_L.

$$I_m = \sum_i I_i = \sum_i G_i (V_m - E_i) = 0,$$

where I_m is the total membrane current.

In the specific circuit of figure 2.1, the equilibrium condition in equation 1.2 becomes:

$$G_{Na}(V_m - E_{Na}) + G_{Ca}(V_m - E_{Ca}) + G_K(V_m - E_K) + G_L(V_m - E_L) = 0. \tag{2.3}$$

Solving equation 2.3 for V_m, we find that the resting membrane potential, at which no net current flows, is a weighted average of the individual Nernst potentials:

$$V_m = \frac{G_{Na}E_{Na} + G_{Ca}E_{Ca} + G_K E_K + G_L E_L}{G_{Na} + G_{Ca} + G_K + G_L}. \tag{2.4}$$

The derivation of the resting membrane potential is typically more complicated, because the conductance values themselves depend on the membrane potential, However, typically the membrane potential of the cell moves toward the Nernst potential of the ion channel whose conductance is greatest.

We can begin to understand the dynamics of the changing membrane potential by first focusing on the passive properties of the cell, i.e., its properties when all ion channels have fixed conductance. These constant values of conductance can be gathered together and combined as a single leak conductance. The resting potential then acts as the leak potential and the entire circuit reduces to the outer loop of figure 2.1.

The membrane potential is generated by the charge stored on the membrane. It depends on both the stored charge and the membrane's capacitance, C_m, via the standard equation for a capacitor (section 1.3.1):

$$Q = C_m V_m. \tag{2.5}$$

We use the convention that positive charge on the inner surface produces a positive membrane potential. Recalling that current is defined as positive when it flows out of the cell, the rate of change of charge on the inside of the cell's membrane is the negative of the total membrane current, which can be written as:

$$\frac{dQ}{dt} = -I_m = -G_L(V_m - E_L). \tag{2.6}$$

Since the capacitance is fixed, equations 2.5 and 2.6 can be combined to yield the dynamics of the membrane potential as:

$$C_m \frac{dV_m}{dt} = G_L(E_L - V_m). \tag{2.7}$$

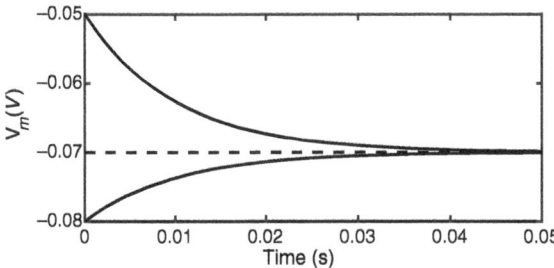

Figure 2.2
Exponential decay of the membrane potential. A leak potential of –70 mV (dashed line) is reached by exponential decay with a time constant of 10 ms. The solid curves follow equation 2.8, but differ in their initial conditions, $V_m(0)$, which is –50 mV (upper curve) or –80 mv (lower curve). This figure is created by the online code `exponential.m`.

This is a linear, first-order ordinary differential equation—the one differential equation whose solution all scientists should know (section 1.4.1)—which tells us that the membrane potential follows an exponential decay from any initial value to its steady state value, E_L. The time constant of the decay is $\tau_m = C_m / G_L$ or $\tau_m = R_m C_m$ where $R_m = 1/G_L$ is the total membrane resistance. The equation for the time-dependence of the membrane potential, $V_m(t)$, when it is initially at a value of $V_m(0)$ is:

$$V_m(t) = E_L + [V_m(0) - E_L]\exp(-t/\tau_m). \tag{2.8}$$

We will use the term exponential decay for any such exponential function whose rate of change approaches zero at large times, even if the function increases with time rather than decreases (figure 2.2).

2.2.1 Depolarization versus Hyperpolarization
In simple models, the resting potential—the neuron's equilibrium in the absence of inputs—is at the leak potential, E_L, in the range of –60 mV to –75 mV. The negative resting membrane potential means the interior of the neuron is at a negative potential compared to the exterior, because of a net removal of positive ions from the cell. Since like charges repel each other, the excess internal negative charge builds up on the cell's membrane. The membrane is therefore polarized, with negative charge on the inner surface and positive charge on the outer surface. An inward current (positive charge flowing into the cell) reduces the charge imbalance across the membrane. Since the reduction in charge imbalance reduces the polarization of the membrane, such an inward current is termed a depolarizing current. Conversely, an outward current that removes positive charge from the cell (or injects negative charge) and increases the membrane's polarization is termed a hyperpolarizing current.

Depolarization   Increase of the membrane potential toward a more positive value through inward electrical current.

Hyperpolarization   Decrease of the membrane potential toward a more negative value through outward electrical current.

Table 2.1 reveals that when sodium and calcium channels are open they produce a depolarizing, inward current, whereas when potassium channels open they produce a hyperpolarizing outward current. Chloride channels are interesting as the current they produce can be depolarizing or hyperpolarizing depending on the exact reversal potential of chloride ions and the value of the membrane potential. Late in development the current through chloride channels tends to be hyperpolarizing, or at least sufficient to inhibit depolarization. However, early in development, the difference in chloride ion concentrations across the membrane is not quite so stark, so the reversal potential for chloride ions is less negative, making the chloride current depolarizing.

2.3   The Leaky Integrate-and-Fire Model

The leaky integrate-and-fire (LIF) model[1] is a good starting point for simulating neurons, because it reproduces some of the qualitative features of the membrane potential's dynamics, it emits spikes at a rate that increases with injected current, and it introduces a framework upon which we can build more realistic models of neurons.

The LIF model is a differential equation for the membrane potential of a neuron with a capacitance and a leak term (the outer loop of the equivalent circuit in figure 2.1) combined with an additional caveat: When the membrane potential reaches a particular value—the threshold—a spike is emitted and the membrane potential is returned to a low, reset value. In this manner, spikes are added to the model artificially to make up for the neglect of the voltage-dependent sodium and potassium channels that would otherwise produce a biophysical spike. We hold off our discussion of the mechanisms of biophysical spikes until chapter 4.

One of the most common electrophysiological measurements of a neuron is its response to injected current. To simulate such externally applied current, we include an additional term, I_{app}, in the dynamical equation for the membrane potential, which follows:

$$C_m \frac{dV_m}{dt} = G_L(E_L - V_m) + I_{app}; \text{ if } V_m > V_{th} \text{ then } V_m \mapsto V_{reset}. \tag{2.9}$$

The first part of equation 2.9 produces leaky integration. In the absence of the leak conductance, the membrane potential would perfectly integrate the applied current as charge is added to the cell's membrane. However, the conductance term causes charge to leak out and the voltage to decay with the time constant of C_m / G_L. The second part is the "fire," which includes a reset. Notice that the spike itself does not appear in the dynamics of the membrane potential—to see a spike of the membrane potential in any simulation of the LIF model it must be put in by hand before the membrane potential is reset. Spike times are recorded at the time when the membrane potential crosses the threshold.

Threshold current   The amount of inward current needed before a neuron or a model neuron produces spikes.

The behavior of the leaky integrate-and-fire neuron in response to three different levels of applied current is shown in figure 2.3. It can be seen that if the applied current is insufficient then the membrane potential does not reach threshold and the model neuron does not produce any

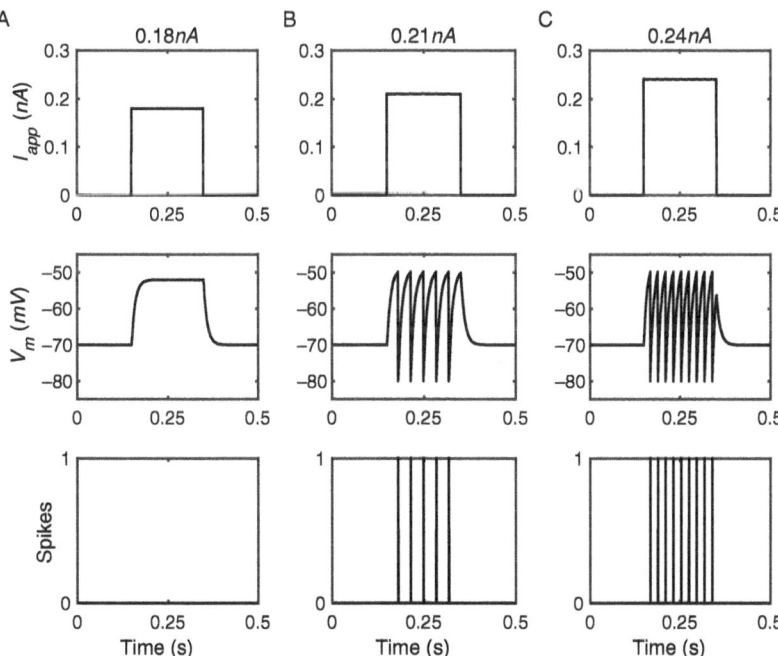

Figure 2.3
Behavior of the leaky integrate-and-fire neuron. Membrane potential of the model neuron (equation 2.9) in response to three different 200 ms current pulses (top) is depicted in the middle row, with spike times indicated by vertical lines in the bottom row. Parameters of the model are: $C_m = 100$ pF, $G_L = 10$ nS, $E_L = -70$ mV, $V_{th} = -50$ mV, and $V_{reset} = -80$ mV. Combining these parameters yields a time constant of $\tau_m = 10$ ms and $I_{th} = 200$ pA. This figure was created by the online code `LIF_model.m`.

spikes (figure 2.3A). The threshold current needed to produce spikes can be determined by calculating the membrane potential reached by the neuron in the presence of a fixed applied current, while ignoring the "fire" and reset mechanism. Setting $dV_m/dt = 0$ in equation 2.9 reveals the steady state membrane potential, V_m^{ss}, to be

$$V_m^{ss} = E_L + I_{app}/G_L. \tag{2.10}$$

If the steady state is below threshold then the model neuron does not fire. The threshold current, I_{th}, is the applied current needed to ensure steady-state membrane potential reaches the threshold (i.e., $V_m^{ss} = V_{th}$) so that spikes can be produced. This requirement, when combined with equation 2.10, leads to

$$I_{th} = G_L (V_{th} - E_L). \tag{2.11}$$

Steady state   The state of a system, as described by the values of all of its variables, which the system approaches and at which it remains without further change.

2.3.1   Specific versus Absolute Properties of the Cell

In the equivalent circuit and the formulae derived from it so far, we have included the cell's properties as parameters that depend on the ability of the total membrane of the cell to either store charge or allow its passage. These properties depend on the cell's surface area. An equivalent formalism is possible using the specific properties of the cell membrane. For example, the resistivity or specific resistance of the cell's membrane is the resistance of a unit area of membrane. Similarly, the specific capacitance is the capacitance per unit area of membrane. We use lowercase letters to denote these specific or intrinsic properties of the cell's membrane and uppercase letters to denote the total value for the cell. The relationships between specific and absolute properties are:

$$C_m = Ac_m; R_m = r_m / A; G_L = Ag_L.$$

The equation for the leaky integrate-and-fire neuron is unchanged if specific units are used, so long as in place of total injected current, I_{app}, the current per unit area, $i_{app} = I_{app} / A$, is used. In neural circuits, current enters each cell through ion channels, which are likely to become more numerous as the cell's size increases. Therefore, the total current is likely to increase with the size of the cell, making current per unit area a quantity that could be less variable across cells. The dynamics of the membrane potential then follows:

$$c_m \frac{dV_m}{dt} = g_L (E_L - V_m) + i_{app}, \tag{2.12}$$

which is obtained from the previous equation for the LIF model (equation 2.9), once the cell's surface area is divided out on both sides.

In the simplest neural models, such as the LIF model, the cell's surface area does not impact the membrane potential's dynamics as long as the applied currents are scaled accordingly. In more sophisticated models in which the spatial structure of the cell is considered, the relative areas of different sections of the cell are important. When charge flows from one part of the cell to another, its effect on the local membrane potential depends on the surface area over which it is distributed—the smaller the region, the bigger the effect for a given amount of charge transfer. In models that incorporate any effect of the concentration of an ion—in particular the concentration of calcium—then it is important to consider that the ions of a particular species spread out over a volume (concentration depends on inverse volume) while the excess charge in a region is stored on the surface, so a factor converting surface area to volume is needed. Such a factor would be size- (and shape-) dependent.

2.3.2 Firing Rate as a Function of Current (f-I Curve) of the Leaky Integrate-and-Fire Model

When the applied current is held fixed, the time for the neuron's membrane potential to increase from its reset value to the threshold can be calculated. We use the formula for the passive properties of the cell to obtain the equation for the membrane potential as a function of time while it is below threshold. We set the initial value of the membrane potential to the reset level and use equation 2.10 for the steady state of the membrane potential, V_m^{ss}, to obtain the solution of the leaky-integration part of equation 2.9 as

$$V_m(t) = V_m^{ss} + \left[V_{reset} - V_m^{ss}\right]\exp(-t/\tau_m). \tag{2.13}$$

In equation 2.13, the time, t, corresponds to the time since the previous spike when the membrane potential was last reset to V_{reset}.

We want to find the time, T, when the next spike is produced. We know that by time, T, the membrane potential has increased to the threshold, so $V_m(T) = V_{th}$. Therefore, the equation to solve for T is:

$$V_{th} = V_m(T) = V_m^{ss} + \left[V_{reset} - V_m^{ss}\right]\exp(-T/\tau_m). \tag{2.14}$$

Equation 2.14 can be rearranged as:

$$\exp(-T/\tau_m) = \frac{V_m^{ss} - V_{th}}{V_m^{ss} - V_{reset}}. \tag{2.15}$$

The term on the right of equation 2.15 must be positive and less than 1 for a solution to exist with $T > 0$ (since $0 < e^{-x} < 1$ for all real $x > 0$), which corresponds to a requirement that $V_m^{ss} > V_{th}$. Such a requirement is just the mathematical way to indicate that we can only calculate the time between spikes if the membrane potential increases above threshold so as to produce spikes.

The solution of equation 2.15 for the time from one spike until the next—that is, the interspike interval, ISI—is found as:

$$ISI = T = -\tau_m \ln\left(\frac{V_m^{ss} - V_{th}}{V_m^{ss} - V_{reset}}\right) = \tau_m \ln\left(\frac{V_m^{ss} - V_{reset}}{V_m^{ss} - V_{th}}\right). \tag{2.16}$$

Interspike interval (ISI)   The time between two successive voltage spikes.

Therefore, the firing rate of the neuron, the inverse of the interspike interval, is:

$$f\left(I_{app}\right) = \frac{1}{ISI} = \frac{1}{\tau_m \ln\left(\dfrac{V_m^{ss} - V_{reset}}{V_m^{ss} - V_{th}}\right)} = \frac{1}{\tau_m \ln\left(\dfrac{E_L + I_{app}/G_L - V_{reset}}{E_L + I_{app}/G_L - V_{th}}\right)}. \tag{2.17}$$

While the formula looks complicated, this is a rare case where the firing rate curve can be written in terms of standard mathematical functions, and so can be plotted easily following a calculation rather than requiring a simulation. In tutorial 2.1, the calculated curve will be compared with the simulated curve.

2.4   Tutorial 2.1: The f-I Curve of the Leaky Integrate-and-Fire Neuron

You should be familiar with the material in sections 1.3, 1.4.1, 1.5, and 1.6 before completing this tutorial.

Neuroscience goals: Understand why firing rate increases with current, why the increase is sharp at threshold in the LIF model, and how noise causes a smoothing of the f-I curve.

Computational goals: Gain experience using the forward Euler method, recording data from nested `for` loops, and using the random number generator to add noise.

1a.   Write a code to simulate a model leaky integrate-and-fire neuron from the equation

$$C_m \frac{dV_m}{dt} = (E_L - V_m)/R_m + I_{app}$$

with the condition if $V_m > V_{th}$ then $V_m \mapsto V_{reset}$, by following the steps below:

> (i)   Define parameters with your choice of variable names that are understandable, setting the values to be: $E_L = -70$ mV, $R_m = 5$ MΩ and $C_m = 2$n F. Assume the spike threshold for the neuron is at $V_{th} = -50$ mV and the reset potential is $V_{reset} = -65$ mV.

(ii)   Create a time vector with steps of $\Delta t = 0.1$ ms from $t = 0$ to a maximum time of $t_{max} = 2$ s.

(iii)   Create a vector for the membrane potential, V, of identical size to the time vector.

(iv)   Set the initial value of the membrane potential (the first value in its array) to the leak potential, E_L.

(v)   Create a vector for the applied current, I_{app}, of identical size to the time vector with each entry set to a constant value, I_0, to be determined later.

(vi)   Set up a `for` loop with an index running from 2 to the number of points in the time vector. This loop will be used to integrate through time.

(vii)   Within the `for` loop, update the membrane potential using the Forward Euler method (section 1.6.1) or your preferred method of integration.

(viii) Within the `for` loop at each time point, test if the membrane potential is above threshold (use an `if` statement), and if it is so, reset the membrane potential to V_{reset}.

1b. What is the minimum applied current needed for the neuron to produce spikes? Calculate this current from equation 2.11, then simulate the model with applied currents (i) slightly lower than and (ii) slightly higher than this value to check you are correct. In each case, plot the membrane potential, $V(t)$, over a time interval of 200 ms (or a complete interspike interval if that is longer).

1c. Make another `for ... end` loop to use at least 10 different values of I_{app}, one value for each 2 s simulation (a "trial") such that the average firing rate (f) varies in the range from 0 to 100 Hz. Plot the resulting firing rate as a function of injected current (called the firing-rate curve or f-I curve).

Hint: You will need to create one vector that stores each value of I_{app} and a vector of the same size to store the corresponding firing rates. In each trial, you need to count spikes and convert to a firing rate. You will plot these two vectors against each other after simulating all trials.

1d. Compare, by plotting with different symbols on the same graph produced in 1c, the curve you obtain from the equation below for the firing rate of the neuron as you vary injected current:

$$\frac{1}{f} = \tau_m \ln\left(I_{app} R_m + E_L - V_{reset}\right) - \tau_m \ln\left(I_{app} R_m + E_L - V_{th}\right)$$

Note: You will need to use an `if` statement to ensure natural log of negative numbers are not taken when you plot over a wide range of I_{app}.

2a. Add a noise term to the simulation of Q.1 by adding to the total membrane potential change at each time step (see section 1.6.6 for an explanation of the square-root term):

```
randn(1)*sigma_I*sqrt(dt),
```

or you can use a value taken from an entire vector of noise for each time point, generated as:

```
noise_vec = randn(size(t))*sigma_I*sqrt(dt).
```

Make sure you use a different set of random numbers for each simulation with a different current.

2b. Plot the firing-rate curve (firing-rate as a function of I_{app}) for at least 2 different values of sigma_I (increase sigma_I until you notice a change from the result with a value of zero, but keep sigma_I fixed in each firing-rate curve). Explain the effect of increasing sigma_I, which is proportional to the standard deviation of the voltage noise.

2c. Test that your code is correct by repeating a simulation with your timestep, dt, a factor of ten smaller. Are the results significantly different?

2.5 Extensions of the Leaky Integrate-and-Fire Model

2.5.1 Refractory Period

The LIF model replaces the dynamics of opening and closing of ion channels that produce a spike in the membrane potential with the simple rules, "record the spike time" and "reset the membrane potential." Additional rules can be added to account for other observed features of real spikes, also called action potentials. One of the observed features is a refractory period—immediately after a spike, the neuron cannot produce another spike for a short period of time called the refractory period. The refractory period, τ_{ref}, is related to the time it takes for ion channels to return to their steady state configurations at the resting membrane potential. While this time-course is cell-type specific, a typical refractory period of $\tau_{ref} = 2$ ms, when added to models, limits the maximum firing rate to 500 Hz (a rate still much higher than that ever observed in most real neurons). The refractory period can be included in models of neurons in a number of ways.

Refractory period   The minimum time interval between one spike and another spike from a neuron, during which depolarizing ion channels can no longer open while they return back to their baseline state.

Method (1): Forced Voltage Clamp   Perhaps the simplest method clamps or fixes the voltage at its reset value following a spike for the duration of the refractory period (figure 2.4A). One disadvantage of this method is that as the firing rate of the neuron increases, the neuron spends a greater proportion of its time in the refractory period with the membrane potential at its low reset value. Therefore, the mean membrane potential can decrease with increased input in such a model, in contrast to the behavior of real neurons.

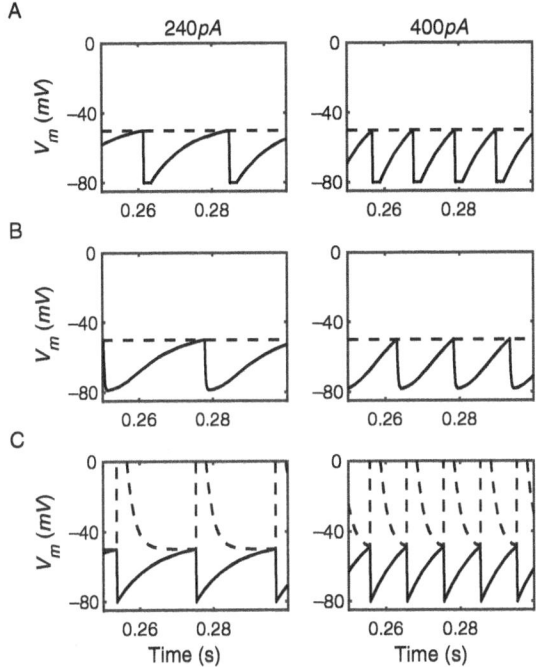

Figure 2.4
Three methods for incorporating a refractory period in an LIF model. (A) Method 1, a hard reset with the membrane potential fixed at the reset value for the refractory period. (B) Method 2, a refractory conductance brings the membrane potential to reset. Its effect lingers as the conductance decays while the membrane potential rises, slowing the rate of spikes. (C) Method 3, following a hard reset the membrane potential can rise immediately, with the refractory period ensured by a raised threshold (dashed line). Solid line = membrane potential; dashed line = threshold voltage. Left column, input current is 240 pA. Right column, input current is 400 pA. This figure was created by the online code LIF_model_3trefs.m.

Method (2): Refractory Conductance The refractory period can be mimicked by addition of a large conductance that produces an outward (hyperpolarizing) potassium current.[2] Following this method, the refractory conductance increases at the time of each spike and decays between spike times with a short time constant:

$$\frac{dG_{ref}(t)}{dt} = -\frac{G_{ref}(t)}{\tau_{ref}} \text{ and after a spike } G_{ref} \mapsto G_{ref} + \Delta G. \tag{2.18}$$

The potassium current produced by this refractory conductance yields an additional term of

$$G_{ref}(t)[E_K - V_m(t)]$$

on the right-hand side of the dynamical equation for the LIF neuron (2.9). When $G_{ref}(t)$ is much greater than the leak conductance, this extra term clamps the membrane potential at the Nernst potential for potassium ions, E_K. Therefore, the reset step of the LIF model can be omitted in simulations using method (2), because the step increase in $G_{ref}(t)$ at the time of a spike causes the desired drop in membrane potential.

Since the refractory conductance persists beyond its time constant while it decays exponentially, it can delay later spikes (figure 2.4B). Unlike method 1, method 2 has the benefit that, like real neurons, the time spent at the reset value depends on the strength of other currents entering the neuron—the stronger the other inputs, the more quickly they overcome the decaying refractory current.

Method (3): Raised Threshold   The voltage threshold for producing a spike can be raised immediately following a spike and allowed to decay back to its baseline level with a short time constant:

$$\frac{dV_{th}(t)}{dt} = \frac{V_{th}^{(0)} - V_{th}(t)}{\tau_{ref}} \text{ and after a spike } V_{th} \mapsto V_{th} + \Delta V. \tag{2.19}$$

Method 3 allows the mean membrane potential to increase with firing rate—as is typically observed—while preventing a spike during a refractory period. As with method 2, this method has the advantage that the refractory period is not absolute—the greater the input, the sooner the membrane potential will reach the decaying threshold. The raised threshold can be combined either with a hard reset of the membrane potential following a spike (figure 2.4C), as in the standard LIF model (equation 2.9), or with a refractory conductance that produces a temporary drop in the membrane potential, as in method 2.

2.5.2 Spike-Rate Adaptation (SRA)

Many neurons respond to a pulse of constant current with spikes that appear at a decreasing rate over the course of the current pulse, an effect known as spike-rate (or spike-frequency) adaptation. The time interval between the first and the second spike—the first interspike interval—can be substantially shorter than the time interval between later spikes when the response eventually becomes periodic (figure 2.5). Such adaptation is particularly valuable and noticeable in sensory systems—for example, we have all noticed how a strong smell when we enter a room becomes less noticeable as we adapt to it. The value of adaptation is that the neuron uses more spikes (and hence more energy) to transmit new information compared with older information that has already been processed by the brain.

Spike-rate adaptation   A slowing down of the spike-rate (increase in ISIs) following the first spike when a neuron responds to depolarizing input.

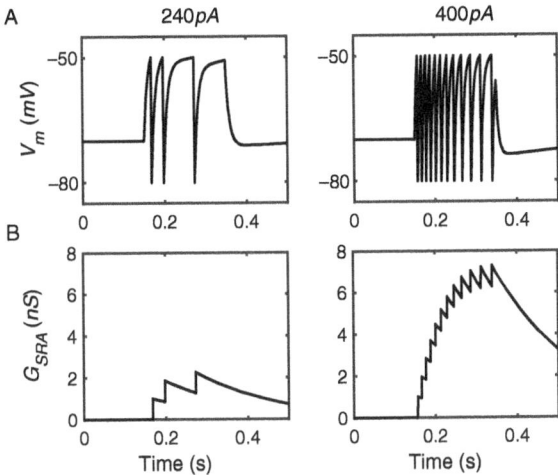

Figure 2.5
Spike-rate adaptation in a model LIF neuron. (A) The membrane potential as a function of time in response to applied current steps of duration 200 ms and of amplitude 240 pA (left) or 400 pA (right). Note the increasing separation between spike times. (B) The potassium conductance accumulates over successive spikes until its decay between spikes matches its increment following a spike (right), in which case a regular interspike interval is reached. This figure is created by the online code, LIF_model_sra.m.

Spike-rate adaptation can be simulated in the LIF model by incrementing a potassium conductance following each spike in the same manner as method 2 for adding a refractory conductance.[2, 3] Also, as with a refractory conductance, the conductance producing spike-rate adaptation should decay between spikes. Therefore, the changes in the LIF equations necessary for adding spike-rate adaptation are *qualitatively* identical to those necessary for adding a refractory conductance.

Yet, since spike-rate adaptation differs from a refractory period, its simulation must also differ. The difference lies in the values of two parameters. First, the increment of the conductance following each spike is much smaller for spike-rate adaptation, which does not prevent spikes, but just slows down their rate. Second, the timescale for the decay of the conductance is much longer—longer than typical intervals between spikes—so that the conductance can accumulate over a spike train (figure 2.5B).

When we include a refractory conductance or a spike-rate-adaptation conductance in the standard LIF model, we produce a two-variable model. In these two-variable models the time-varying conductance impacts the dynamics of the membrane potential, while in return the value of the membrane potential impacts the dynamics of the conductance. In these initial models, the conductance is only affected by the membrane potential through discrete changes at the discrete times of spikes. In more general models—such as the conductance-based models of chapter 4—the rate of change of conductance is a continuous function of the membrane potential.

2.6 Tutorial 2.2: Modeling the Refractory Period

Neuroscience goals: Understand how a biological feature like the refractory period can be implemented in multiple simplified ways in a model, and be aware of the consequences of different simplifications relative to the behavior of real neurons.

Computational goals: Further practice with nested `for` loops and use of conditionals to control the path of execution of the code within each `for` loop. Understand how relevant behaviors can be added into the LIF model using such computational tools. Learn how to affect the output graph to view spikes, without impacting the simulation.

In this tutorial, we will simulate the refractory period via the three different methods shown in figure 2.4. The refractory period ensures that immediately after a spike the neuron cannot produce a second spike, so the refractory period limits the maximum firing rate of a cell.

Each of the following questions, 1–3, requires you to use just one of the different methods for producing a refractory period. In each question, you will simulate a leaky integrate-and-fire neuron for 2 s using a range of input currents, I_{app}, from 100 to 600 pA. Use the following parameters: $E_L = -70$ mV, $R_m = 100$ MΩ, $C_m = 0.1$ nF, such that the membrane potential follows

$$C_m \frac{dV_m}{dt} = (E_L - V_m) / R_m + I_{app}.$$

For each of the three questions, calculate and plot on the same graph the mean firing rate as a function of the input current (results for all three questions plotted on the same graph 1). On a separate graph plot the mean membrane potential as a function of the input current for each of the three questions (results for all three questions plotted on the same graph 2). On a third graph plot the mean membrane potential as a function of firing rate for each of the three questions (results for all three questions plotted on the same graph 3). That is, you will produce three graphs in total, each of which will contain three curves, with differences between the curves arising from the different models used in each question. Explain any trends and any nonmonotonic behavior, as well as the differences between the curves in each figure.

Also, for each question, separately plot the membrane potential as a function of time for 100 ms when $I_{app} = 220$ pA and on the same trace in a different color, when $I_{app} = 600$ pA. For the membrane potential trace, use a different graph for each question. For each membrane potential trace, make each spike uptick visible by plotting a rise in the membrane potential up to a peak value, $V_{peak} = 50$ mV, for a single time-point. You should only add the peak to each spike, by altering every voltage above threshold to be V_{peak}, after the simulation through time is complete, as otherwise you will disrupt the voltage-update rule.

Q.1. Forced Voltage Clamp Assume the spike threshold for the neuron is fixed at $V_{th} = -50$ mV and the reset potential is $V_{reset} = -65$ mV. If $V_m > V_{th}$ then set $V_m = V_{reset}$ and record the spike time. Following a spike, until the time is greater than the time of the last spike plus the refractory period, $\tau_{ref} = 2.5$ ms, fix $V_m = V_{reset}$ instead of following the previous differential equation (you will need to incorporate an `if` statement to check whether sufficient time since the prior spike has elapsed).

Notes: (i) You will need to initialize the time of last spike to a value less than $-\tau_{ref}$ to ensure the simulation does not commence in a refractory period. (ii) To plot the spikes in the membrane potential you can either plot a single line that increases along the y-axis from V_{th} to V_{peak} at every spike time (using a `for` loop), or, after the simulation change each above-threshold value of the membrane potential to V_{peak}. Finally, via a "dirtier" method, when you reset the membrane potential at the time of a spike to V_{reset}, you can set the value on the previous time-point to V_{peak}. The previous value is never used again in the simulation, so does not affect results. The shift of the time of the spike by the timestep, Δt, will be unnoticeable in any figure.

Q.2. Threshold Increase Make the voltage threshold dynamic so that it follows

$$\frac{dV_{th}(t)}{dt} = \frac{V_{th}^{(0)} - V_{th}(t)}{\tau_{Vth}},$$

with the baseline, $V_{th}^{(0)} = -50$ mV, an initial condition $V_{th}(0) = V_{th}^{(0)}$ and a refractory time constant, $\tau_{Vth} = 1$ ms.

If $V_m(t) > V_{th}(t)$, a spike occurs. At the time of the spike set $V_m = V_{reset}$ and increase the threshold to the value, $V_{th}^{(max)} = 200$ mV. Immediately after the spike the dynamical equations are followed for both $V_m(t)$ and $V_{th}(t)$.

Notes: Create a vector for $V_{th}(t)$ of the same size as $V_m(t)$ and update $V_{th}(t)$ in the same `for` loop as you update $V_m(t)$.

Q.3. Refractory Conductance with Threshold Increase Simulate the leaky integrate-and-fire dynamics with the additional potassium conductance term as follows:

$$C_m \frac{dV_m(t)}{dt} = [E_L - V_m(t)]/R_m + G_{ref}(t)[E_K - V_m(t)] + I_{app},$$

where $E_K = -80$ mV and the refractory conductance follows:

$$\frac{dG_{ref}(t)}{dt} = -\frac{G_{ref}(t)}{\tau_{Gref}} \text{ and at the time of a spike } G_{ref} \mapsto G_{ref} + \Delta G.$$

Initialize the refractory conductance to zero, $G_{ref}(0) = 0$. Use the parameters $\tau_{Gref} = 0.2$ ms and $\Delta G = 2$ μS. Let the voltage threshold vary in the same manner as Q.2, increasing $V_{th}(t)$ to $V_{th}^{(max)} = 200$ mV at the time of a spike and decaying back to $V_{th}^{(0)} = -50$ mV between spikes. Do not change $V_m(t)$ at the spike-time—rather you should see a reduction in $V_m(t)$ arise (quickly but not instantaneously) because of the sudden increase in refractory conductance.

Notes: You will need to produce another vector, for $G_{ref}(t)$, of the same size as $V_m(t)$ and update it within the same for loop as you update $V_m(t)$ and $V_{th}(t)$.

2.7 Further Extensions of the Leaky Integrate-and-Fire Model

2.7.1 Exponential Leaky Integrate-and-Fire (ELIF) Model

Perhaps the most obvious failing of the leaky integrate-and-fire model neuron is its lack of a spiking mechanism. Spikes in the membrane potential can be added to the model for aesthetic purposes, simply by setting the membrane potential to a high value such as the Nernst potential for sodium whenever it has crossed threshold, before returning it to reset (as in tutorial 2.2). However, any electrophysiologist would be disturbed by the manner in which the membrane potential changes more and more slowly as it approaches threshold. In reality, an inflexion point—at which the decreasing gradient begins increasing—is visible in the plot of membrane potential against time for a real cell, as the voltage accelerates up past a threshold.

Moreover, careful experiments indicate there is no fixed threshold of the membrane potential. That is, neurons do not have a point of no return for production of a spike. Rather, there is a threshold range, within which a spike may or may not be produced, depending on prior inputs.

The exponential leaky integrate-and-fire (ELIF) model[4] incorporates an additional spike-generating term in the dynamics of the membrane potential to address these failings. The differential equation becomes:

$$C_m \frac{dV_m}{dt} = G_L \left[E_L - V_m + \Delta_{th} \exp\left(\frac{V_m - V_{th}}{\Delta_{th}} \right) \right] + I_{app}, \tag{2.20}$$

where V_{th} is the threshold as before, and Δ_{th} denotes the voltage-range over which the spike-generating term becomes important and harder to overcome with inhibition.

The additional exponential term produces an inward, depolarizing current that arises when the membrane potential reaches the vicinity of the standard threshold. Thereafter the inward current increases as the membrane potential increases, creating positive feedback: The current causes the membrane potential to rise, which increases the current further. Such positive feedback generates a rapidly accelerating increase in membrane potential (figure 2.6), producing a spike with an interesting property: The membrane potential becomes infinite in a short, finite time. That may seem bizarre, but mathematically it is fine—for example, the function $\tan x$ reaches infinity over a finite range of x.

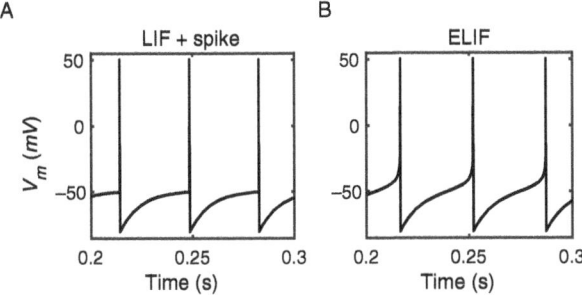

A   LIF + spike   B   ELIF

Figure 2.6
Spike generation in the ELIF model. The ELIF model (right) produces the typical inflexion of the voltage trace that is absent in the LIF model (left). Parameters are as in figure 2.3 with $I_{app} = 210$ pA. For the ELIF model $\Delta_{th} = 5$ mV and $V_{max} = 50$ mV. This figure was created by the online code `LIFspike_ELIF.m`.

> **Positive feedback** A situation whereby a change in a system causes processes to come into play that produce an enhancement of the change.

Since computers have a problem with infinity and any timestep would need to become infinitesimal to capture the rapid change, in simulations one sets a maximum level of the membrane potential, V_{max} (such as $V_{max} = 50$ mV), beyond which it is reset in the standard manner. Therefore, the model includes an additional condition:

if $V_m > V_{max}$ then $V_m \mapsto V_{reset}$.

It turns out that the rate of change is so rapid that it makes little difference in simulations whether spikes are clipped to +10 mV, +50 mV, +200 mV, etc., so long as a small enough timestep is used to keep the simulation numerically stable.

2.7.2 Two-Variable Models: The Adaptive Exponential Leaky Integrate-and-Fire (AELIF) Neuron

The adaptive exponential leaky integrate-and-fire (AELIF) model[5] adds to the ELIF model an outward (hyperpolarizing) spike-rate adaptation current (figure 2.7). The adaptation current is similar to that produced in the previous section with an adaptation conductance, but with a twist—the rate of decay of the adaptation current depends on the membrane potential. Therefore, the membrane potential impacts the adaptation current in two ways. First, it produces a step increase in the current at the time of a spike. Second, while the membrane potential is high, even below threshold, the rate of decay of the current is slowed and its steady state is not zero—the current persists in the absence of spikes.

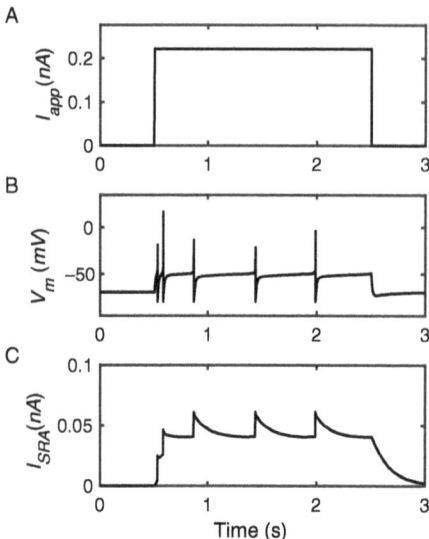

Figure 2.7
Response of the AELIF model to a current step. (A) The applied current. (B) The membrane potential. (C) The adaptation current. Note the very strong lengthening of the interspike interval following the first two spikes. Parameters used are: $C_m = 100$ pF, $G_L = 10$ nS, $E_L = -70$ mV, $V_{th} = -50$ mV, $V_{reset} = -80$ mV, $\Delta_{th} = 2$ mV, $V_{max} = 50$ mV, $a = 2$ nS, $b = 20$ pA, $\tau_{SRA} = 200$ ms, and $I_{app} = 0.221$ nA. This figure was created by the online code AELIF_code.m.

The full model is the pair of coupled differential equations:

$$C_m \frac{dV_m}{dt} = G_L \left[E_L - V_m + \Delta_{th} \exp\left(\frac{V_m - V_{th}}{\Delta_{th}} \right) \right] - I_{SRA} + I_{app} \tag{2.21}$$

$$\tau_{SRA} \frac{dI_{SRA}}{dt} = a(V_m - E_L) - I_{SRA}, \tag{2.22}$$

combined with the two rules:

if $V_m > V_{max}$ then $V_m \mapsto V_{reset}$ and $I_{SRA} \mapsto I_{SRA} + b$.

The full model presented here is surprisingly versatile. The equation includes two additional parameters: a, possessing units of conductance, and b, possessing units of current. a and b are both positive if the I_{SRA} term is to represent the magnitude of an adaptation current. In the standard nomenclature, although I_{SRA} is defined to be positive, it is an outward current acting to hyperpolarize (make more negative) the membrane potential (it enters the membrane potential's dynamics with a minus sign). The AELIF model can be simulated with either or both of the terms a and b being negative, in which case the model can be used to simulate other classes of neurons, although the biophysical meaning of the additional terms is lost.

2.7.3 Limitations of the LIF Formalism

We have seen that the leaky integrate-and-fire model can be extended to incorporate a realistic upswing of the spike, a refractory period without any clamping of the membrane potential, and spike-rate adaptation. In principle, any observed behavior of the membrane potential in response to inputs could be reproduced in a model based on the LIF neuron if appropriate extra terms are added to the set of ordinary differential equations. One might ask then, why simulate more complex models based on active conductances (as we will in chapter 4)?

First, if we want to understand the effect of any change in the ion channels in a cell—either addition of a new type of channel, or modulation of the properties of any existing channel—models based on the LIF are of little value. For example, it is unclear which parameters in the model AELIF neuron would change and by how much, if the total potassium conductance were to increase.

Second, if we wanted to consider any properties of a neuron that do not just depend on the membrane potential—for example, any changes wrought by an influx of calcium ions—then the corresponding ion channels should be modeled explicitly.

Third, if we need to consider any spatial variation in the membrane potential of a cell, then a more realistic depiction of a spike becomes important. For example, suppose we want to know how the membrane potential in the dendrites depends on the membrane potential in the soma. The actual shape of the spike determines how much current flows between soma and dendrite, so just knowing the spike time is insufficient.

In chapter 4 we consider how to implement these behaviors in more realistic conductance-based models of the neuron. The basis of such models is the Hodgkin-Huxley formalism, with which we introduce that chapter.

2.8 Tutorial 2.3: Models Based on Extensions of the LIF Neuron

Neuroscience goal: Understand the reciprocal impacts of adaptation currents on spike trains and spike trains on adaptation currents; understand how limits on the firing rate and mean membrane potential depend on the spiking mechanism.

Computational goal: Simulate coupled differential equations with two variables; analyze simulation results and plot appropriate features of the simulation, for example, depending on the number of spikes produced.

In this tutorial, we will analyze the impact of adaptation currents on the f-I curve of a neuron.

1. Write a code to simulate the LIF model with an adaptation current so that the full model becomes:

$$C_m \frac{dV}{dt} = (E_L - V)/R_m + G_{SRA}(E_K - V) + I_{app}$$

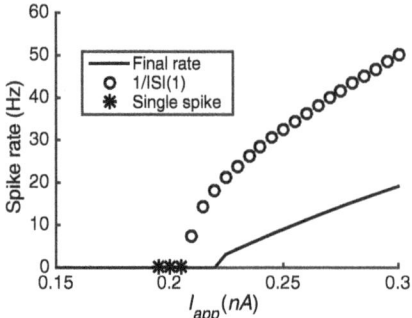

Figure 2.8
Firing rate curve for AELIF model. The solid line is the firing rate after any initial transient when the cell fires periodically. Circles denote the reciprocal of the first interspike interval, effectively the initial firing rate, evaluated when a current step produces two or more spikes. Asterisks correspond to current steps that produce a single spike. Parameters are as in figure 2.7. This figure was produced by the online code AELIF_code_loop.m.

and

$$\frac{dG_{SRA}}{dt} = -G_{SRA} / \tau_{SRA},$$

with the rule that if $V > V_{th}$ then $V \mapsto V_{reset}$ and $G_{SRA} \mapsto G_{SRA} + \Delta G_{SRA}$.

Use the parameters: $E_L = -75$ mV, $V_{th} = -50$ mV, $V_{reset} = -80$ mV, $R_m = 100$ MΩ, $C_m = 100$ pF, $E_K = -80$ mV, $\Delta G_{SRA} = 1$ nS, and $\tau_{SRA} = 200$ ms. Initially set $V = E_L$ and $G_{SRA} = 0$.

a. Simulate the model neuron for 1.5 s, with a current pulse of $I_{app} = 500$ pA applied from 0.5 s until 1.0 s. Plot your results in a graph, using three subplots with the current as a function of time, the membrane potential as a function of time and the adaptation conductance as a function of time stacked on top of each other, as in figure 2.7.

b. Now simulate the model for 5 s with a range of 20 different levels of constant applied current (i.e., without a step pulse) such that the steady state firing rate of the cell varies from zero to 50 Hz. For each applied current calculate the first interspike interval and the steady-state interspike interval. On a graph, plot the inverse of the steady-state interspike interval against applied current to produce an f-I curve. On the same graph plot as individual points (e.g., as crosses or squares) the inverse of the initial interspike interval as in figure 2.8. Comment on your results.

2. Write a code to simulate the AELIF model:

$$C_m \frac{dV_m}{dt} = G_L \left[E_L - V_m + \Delta_{th} \exp\left(\frac{V_m - V_{th}}{\Delta_{th}} \right) \right] - I_{SRA} + I_{app}$$

$$\tau_{SRA} \frac{dI_{SRA}}{dt} = a(V_m - E_L) - I_{SRA},$$

while

if $V_m > V_{max}$ then $V_m \mapsto V_{reset}$ and $I_{SRA} \mapsto I_{SRA} + b.$

Use the parameters: $E_L = -75$ mV, $V_{th} = -50$ mV, $V_{max} = 100$ mV, $V_{reset} = -80$ mV, $\Delta_{th} = 2$ mV, $G_L = 10$ nS, $C_m = 100$ pF, $a = 2$ nS, $b = 0.02$ nA and $\tau_{SRA} = 200$ ms. Initially set $V = E_L$ and $I_{SRA} = 0$.

  a. Simulate the model neuron for 1.5 s, with a current pulse of $I_{app} = 500$ pA applied from 0.5 s until 1.0 s. Plot your results in a graph, using two subplots with the current as a function of time plotted above the membrane potential as a function of time.

  b. Now simulate the model for 5 s with a range of 20 different levels of constant applied current such that the steady state firing rate of the cell varies from zero to 50 Hz. For each applied current calculate the first interspike interval and the steady-state interspike interval. On a graph plot the inverse of the steady-state interspike interval against applied current to produce an f-I curve. On the same graph plot as individual points (e.g., as crosses or squares) the inverse of the initial interspike interval. Comment on your results.

3. **Optional Challenge Question.** Alter the AELIF model to produce a refractory period using a combination of the methods of tutorial 2.2—in particular, change the spike-rate-adaptation *current* into a refractory *conductance* and make the neuron's spiking threshold dynamic. The goal is to produce a cell whose average membrane potential increases monotonically with input and firing rate, while producing reasonable spikes at a firing rate that never exceeds 500 Hz, even with excessively large input currents. Do not change the basic cell properties (i.e., R_m, C_m, E_L, E_K are the same as in Q.2). However, parameters producing the spiking mechanism can be altered and V_m can be reset to any desired value following each spike. Also, the steady-state of the refractory conductance can be made into any nonlinear function of the membrane potential (i.e., the term $a(V_m - E_L)$ can be altered in the dynamics of I_{SRA}). Produce a figure with four different subplots: (a) an example of the membrane potential dynamics; (b) firing rate as a function of input current; (c) mean membrane potential as a function of input current; and (d) mean membrane potential as a function of firing rate. Figure 2.9 provides an example of a solution to this question.

Note: In chapter 4 we will see that in more realistic models of neurons, as the injected current to a cell is increased, a point is reached where the spiking mechanism breaks down, so the neuron's firing rate is fundamentally limited.

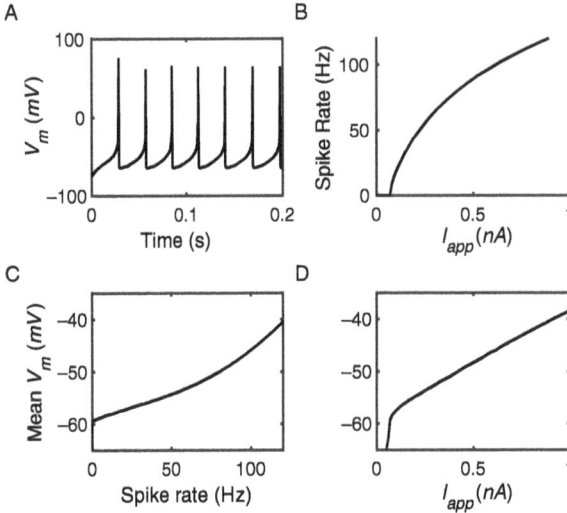

Figure 2.9
Response of an altered AELIF model to produce a refractory period. Possible solution of tutorial 2.3, question 3.
(A) Response of membrane potential to applied current of 160 pA. (B) Firing rate as a function of applied current
(currents in the tens of nA produce failure rather than high rates in this model). (C, D) Mean membrane potential
increases monotonically with firing rate (C) and with applied current (D). This figure is produced by the online code
`Tutorial_2_3_Q3.m`.

Table 2.2
List of symbols used in chapter 2

Symbol	Meaning	Description
E_A	Reversal potential of ion-channel "A"	Membrane potential at which the ionic current through a particular channel is zero
k_B	Boltzmann constant	Converts temperature to thermal energy
T	Temperature	In units of Kelvin (degrees Celsius + 273)
q_e	Charge of an electron.	
z_A	Valence of an ion "A"	Total charge of ion in units of q_e
C_m	Total membrane capacitance	Cell's capacity to store charge on its surface
c_m	Specific membrane capacitance	Capacitance per unit area of membrane
G_A	Total conductance of ion-channel "A"	Ability of ions to flow in or out of the cell through channels of type "A"
g_A	Specific conductance of "A"	Conductance per unit area of membrane
I_m	Total membrane current	Rate of flow of charge through the membrane
i_m	Specific membrane current	Current per unit area of membrane
V_m	Membrane potential	Interior potential minus exterior potential
τ_m	Membrane time constant	Timescale for changes in V_m
t	Time	Time is essential for change and dynamics!
Q	Electrical charge	Used for amount stored on total membrane

Table 2.2 (continued)

Symbol	Meaning	Description
G_L	Leak conductance	Fixed value not ascribed to a particular ion
R_m	Total membrane resistance	The fixed component at rest with $R_m = 1/G_L$
E_L	Leak reversal potential	Value of V_m where leak current is zero
V_{th}	Threshold potential	Value of V_m at which a spike is produced
V_{reset}	Reset potential	Value to which V_m is set after a spike
I_{th}	Threshold current	Value of applied current required for spiking
I_{app}	Applied current	Current supplied by experimental apparatus
V_m^{ss}	Steady state of V_m	Asymptotic value of V_m with fixed conductance or applied current
$f(I)$	Firing rate at a given current	Defines the response of a neuron to input
ISI	Interspike interval	Time between spikes, $ISI = 1/f$
τ_{ref}	Refractory period	Time following a spike during which generation of a new spike is impossible
V_{peak}	Peak membrane potential	Used to plot a spike of membrane potential up to a particular value, V_{peak}, if the spike itself is not generated within the code
$V_{th}^{(0)}$	Baseline of threshold potential	If V_{th} increases following a spike, $V_{th}^{(0)}$ is the value it returns toward between spikes
τ_{Vth}	Time constant of threshold potential	Less than a few ms to describe the rate of change of V_{th} back to baseline over the refractory period
G_{ref}	Refractory conductance	Temporary conductance of a hyperpolarizing channel that opens after a spike
ΔG_{ref}	Change in refractory conductance	Sufficiently large increment in G_{ref} following a spike to temporarily prevent another spike
τ_{Gref}	Time constant of refractory conductance	Fast time constant (up to a few ms) for decay of G_{ref} soon after a spike before the next spike is possible
Δ_{th}	Voltage-range for spike uptick	Used in the ELIF model to simulate a spike that accelerates over this range of membrane potential as it approaches threshold
V_{max}	Maximum membrane potential	Used to limit the membrane potential's rapid rise to infinity in the ELIF model by simulating only until V_m reaches V_{max}
I_{SRA}	Spike-rate adaptation current in the AELIF model	Typically defined as a positive quantity representing a hyperpolarizing outward current that grows with increasing spike rate
a	I_{SRA} control term	Determines the voltage-dependent decay rate and steady-state of in the AELIF model
b	I_{SRA} current step	Step change in following each spike in the AELIF model
G_{SRA}	Spike-rate adaptation conductance G_{SRA}	Time-varying conductance of hyperpolarizing potassium channels
ΔG_{SRA}	Step change in G_{SRA}	Increment in G_{SRA} following each spike
τ_{SRA}	Time-constant for G_{SRA}	Slow (a few hundred ms) time constant for decay of G_{SRA} between spikes

Questions for Chapter 2

1. If the extracellular concentration of potassium gets reduced, what do you expect to happen to the resting membrane potential of a neuron, and why?

2. Consider an LIF neuron (equation 2.9) with parameters: $E_L = -75\,\text{mV}$, $G_L = 400\,\text{nS}$, $C_m = 2$ nF, $V_{th} = -50$ mV and $V_{reset} = -70$ mV.
   a. If it receives an applied current of 6 nA, what is its steady state membrane potential?
   b. If it receives an applied current of 15 nA, what is its firing rate?

3. Explain two similarities and two differences between how you incorporate a refractory period and how you incorporate an adaptation conductance into a simulation of an LIF neuron.

4. In the models of spike-rate adaptation which we have produced, is there any possibility that the model neuron can respond to a prolonged current step with a small number of spikes followed by complete quiescence (an absence of spiking) while the applied current remains, because of the adaptation current? Explain your reasoning.

2.9   Appendix: Calculation of the Nernst Potential

In this appendix, we will calculate the value of the membrane potential, i.e., the potential difference across the cell membrane, such that any flux of ions due to a concentration difference across the membrane is exactly matched by a flux due to the potential difference.

To make progress, we consider potassium ions, which have greater concentration inside than outside the cell, a positive charge, and a negative reversal potential (figure 2.10).

For the net flow through a channel to be zero for a particular ion, the rate of ions reaching the channel opening multiplied by their probability of passing through the channel once they reach the opening, must be the same in both directions, i.e.,

$$(Outside\ arrival\ rate) \bullet P(entry) = (Inside\ arrival\ rate) \bullet P(exit). \tag{2.23}$$

The rate of ions reaching the channel opening is simply proportional to the concentration of ions—one can consider a small ion-size volume adjacent to the channel opening and the probability of an ion occupying that volume in any small time-interval is simply the concentration of ions multiplied by the volume. Therefore, the ratio of arrival rates is the ratio of concentrations:

$$\frac{(Outside\ arrival\ rate)}{(Inside\ arrival\ rate)} = \frac{[K_{out}]}{[K_{in}]}, \tag{2.24}$$

where $[K_{out}]$ and $[K_{in}]$ are respectively the external and internal concentrations of the ion under consideration, in this case potassium.

For the equilibrium of equation 2.23 to hold, then equation 2.24 can be combined with equation 2.23 to give:

A

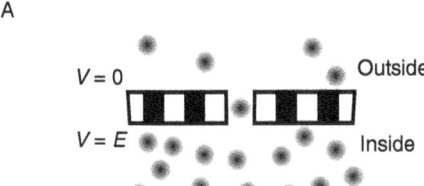

B

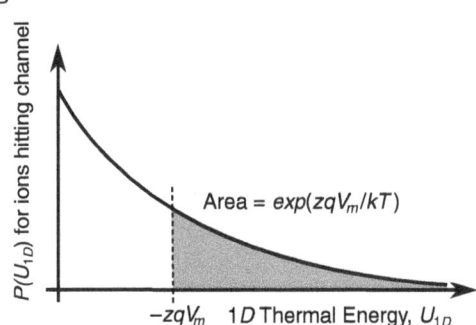

Figure 2.10
Calculation of Nernst equilibrium potential. (A) When a species of ion, such as potassium, has a greater concentration inside than outside the neuron, there would be a net outward flow of ions through any open channel (i.e., upward in the diagram) unless countered by a flow in the opposite direction due to a potential difference (in this case the outside of the cell should be at a higher electrical potential than the inside, in order to drive positive ions back into the cell). (B) In a simplified model, when the outside of the cell is at a higher potential than the inside, external ions that reach the channel opening will pass through it, but internal ions that reach the channel opening will pass through only if they have enough velocity, v_{1D}, in the direction perpendicular to the channel opening. The speed criterion for exiting the cell is that the component of their kinetic energy perpendicular to the channel opening, $mv_{1D}^2 / 2 = U_{1D}$, is greater than the potential energy the ion would acquire when passing through the channel, $-zq_eV_m$, where zq is the total charge on the ion and V_m is the membrane potential. The distribution of U_{1D} is exponential, such that the probability any ion has $U_{1D} > -zq_eV_m$ is $\exp(zq_eV_m / kT)$, for $V_m < 0$.

$$\frac{P(exit)}{P(entry)} = \frac{(Outside\ arrival\ rate)}{(Inside\ arrival\ rate)} = \frac{[K_{out}]}{[K_{in}]}. \tag{2.25}$$

We can calculate the ratio of probabilities of exit to entry from the total change in energy of an ion when it crosses the channel. If biological details are included, this should be evaluated as a multistep process with forward and backward transition rates. However, the resulting probability ratio only depends on the total change of energy, so can be evaluated by ignoring any complexities of the channel and simply thinking of it as a "tunnel" through the membrane, as we do in the next paragraph.

Equation 2.25 tells us that at equilibrium, the electrical potential difference between the ends of the channel must reduce the probability of ions exiting the cell when the internal ion concentration exceeds the external concentration. Therefore, for potassium ions, the reversal potential must be negative, so that when ions from inside the cell reach the channel entrance they have a lower propensity to exit the channel. By way of analogy, the potassium channel is then equivalent to a vertical chimney through the roof of a house. A high density of rapidly moving balls (cf. ions) inside the house generates a net upward flow through the chimney that is considerably reduced by the requirement that a ball entering the chimney has sufficient vertical velocity to exit. Such upward movement could be matched by a much lower density of balls outside the chimney, where they more rarely reach the chimney opening, but then always flow through it.

A result from statistical mechanics and thermodynamics shows that the component of kinetic energy in a particular direction—which, for ions is dependent on their thermal energy—is an exponential function that depends on temperature. The proportion of ions with sufficient energy to escape an energy barrier in the manner suggested is $\exp(-\Delta U / k_B T)$ where ΔU is the energy barrier, k_B is Boltzmann's constant and T is temperature. For ions of charge, zq_e, moving across an electrical potential difference of $-V_m$, the energy barrier is $-zq_e V_m$. Combining these results and setting $V_m = E_K$ at the Nernst equilibrium (the reversal potential) for potassium ions, where equation 2.23 and equation 2.25 are satisfied, we find:

$$\frac{P(exit)}{P(entry)} = \exp(zq_e E_K / kT) = \frac{[K_{out}]}{[K_{in}]}. \tag{2.26}$$

Rearrangement of equation 2.26 to solve for E_K leads to the Nernst equation (see also equation 2.1):

$$E_K = \frac{kT}{zq_e} \ln\left(\frac{[K_{out}]}{[K_{in}]}\right). \tag{2.27}$$

3

Analysis of Individual Spike Trains

Action potentials, or more simply put, spikes, are the main units of information transfer in mammalian brains, because spikes can propagate down the axon away from one neuron's cell body to impact other neurons that may be far away. The rapid transient deflection of the membrane potential that comprises a spike can be detectable by an electrode inserted into the surrounding neural tissue. Such measurements of spikes, carried out in vivo, provided our first insights into how neurons in the brains of living mammals respond to sensory stimuli. In this chapter, we will ignore the particular dynamics of the membrane potential that generate a spike and the manner in which spikes are detected, but will simply treat the spikes of each cell according to the discrete points in time at which they occur.

3.1 Responses of Single Neurons

3.1.1 Receptive Fields

A receptive field is a description of the stimulus or range of stimuli that generates a vigorous response in a neuron. The clearest response of a neuron to a stimulus—and often the only one considered—is a change in its firing rate. In the same way that we saw the number of spikes per second produced by a neuron changes if its input current changes, at the simplest level of description all neural responses correspond to changes in their firing rates. These firing-rate changes are brought about by stimulus-dependent changes in the neurons' input currents.

The standard procedure for finding receptive fields is to present an animal with many different stimuli and measure the neuron's firing rate as a function of stimulus. In the simplest method one simply counts the number of spikes produced by a neuron while a stimulus is presented and divides by the duration of stimulus presentation to obtain the neuron's average firing rate.

When a parameter of the stimulus—such as spatial location of a visual stimulus, pitch of an auditory stimulus, frequency of a vibrational stimulus—can be altered in a graded manner, it is straightforward to plot the neuron's firing rate as a function of that parameter. The resulting graph is a plot of the neuron's *tuning curve*. The receptive field corresponds to the range of stimuli for which the firing rate is significantly different from its rate in the absence of stimulus. The tuning curve then describes the properties or characteristics of the neuron's receptive field.

> **Tuning curve**   A plot of a neuron's firing rate in response to a range of stimuli.

In figure 3.1, some examples of receptive field are shown. In figure 3.1A the receptive field is a spatial area, indicating where on the skin surface a touch can produce a response in a particular neuron. This is an example of a standard use of receptive field, indicating the shape and size of the region of locations that produce a response. Indeed, the first use of the term "receptive field" by Sherrington in 1906 referred to the regions of body surface that produced a reflexive movement mediated by neural activity.[1]

For visual stimuli, the location is typically referred to in visual space, i.e., angular coordinates centered on the direction of eye gaze. For auditory stimuli, the location more often refers to location on the frequency spectrum (i.e., the pitch) of a stimulus rather than its physical location. These differences arise because the initial sensory neurons of vision on the retina respond to light from a particular location relative to eye position,[2] while the initial sensory neurons of audition in the cochlea respond to sound of a particular frequency.[3]

> **Sensory neuron**   A cell that responds to a particular physical change impacting an organism and responds with a change in its membrane potential that can be transmitted to other neurons.

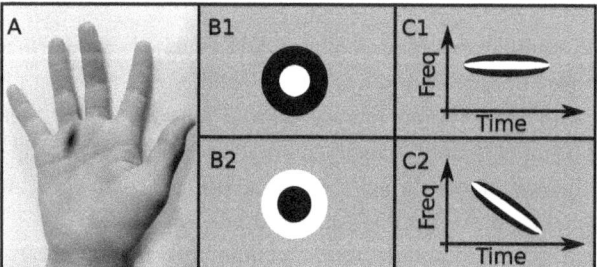

Figure 3.1
Sketches of receptive fields and their general structure. (A) A neuron in primary somatosensory cortex is most active when a particular area of skin is touched (marked dark). (B) A ganglion cell in the retina responds best to a center-surround arrangement of visual stimulus, either light center with dark surround (upper panel, "On-center") or a dark center with a light surround (lower panel, "Off-center"). White, light; black, darkness. (C) The spectrotemporal receptive field of a neuron in auditory cortex can indicate a response to a particular, fixed, frequency of a tone (upper panel) or to a sweep of frequency (a downward sweep in the lower panel). The response is accentuated if there is no sound intensity at surrounding frequencies. White, high intensity; dark, low intensity.

A more complex use of receptive field is shown in figure 3.1B, indicating the *type* of stimulus that produces a response as well as its location. In this example, the basic structure of the receptive field shows where a positive neural response is produced by a lighter than background stimulus and by a darker than background stimulus, in white and black respectively.

Center-surround   A feature of many receptive fields in which the sign of the stimulus needed to produce an increase in a neuron's firing rate switches as the parameters of a stimulus are shifted in either direction from the set of parameters that produce maximum response in the neuron.

The center-surround organization of figures 3.1B1 and 3.1B2 is likely due to lateral inhibition, which is a common feature of neural circuits.[4-8] Lateral inhibition causes a neuron to be inhibited by neurons with different, but perhaps similar, receptive fields. Therefore, a neuron receiving lateral inhibition responds most strongly when it receives direct input from the "center" of its receptive field *and* when those neurons, which would respond most to its "surround," do not receive input and so remain inactive.

Figure 3.1C includes a further complexity, namely the dynamics of a stimulus that most readily causes a neural response. In particular, the lower panel, c2, indicates the receptive field of a neuron that responds most vigorously when a tone starts at a high frequency then sweeps down to a lower frequency over a brief period of time. In tutorial 3.1 you will produce and characterize a similar spatiotemporal receptive field.

Spatiotemporal receptive field   A plot of the combined parameter-dependence and time-dependence of a stimulus that will produce a response in a neuron.

As we consider neurons increasingly further removed from sensory receptors in the periphery, there is a trend of increasing size and complexity of receptive fields. Moreover, neural responses become more and more dependent on context and the history of prior stimuli. Therefore, the concept of receptive field is applicable only for neurons near the periphery, where there is a chance that the ensemble of potential stimuli can be suitably tested. For example, if a neuron responds most to a dark vertical bar at a particular location of the visual scene, it is possible to discover this by placing bars of different orientations in different positions until an increase in firing rate is detected. For such simple receptive fields, a bar partially within the receptive field or a bar at a near-optimal orientation would also produce a neural response, so one can home in on the receptive field and find the stimuli that produce the strongest response.

However, if a neuron happens to be most active whenever I look down my street toward my home, the odds of this response being found in a systematic stimulus-focused manner are miniscule (given a billion possible homes and a myriad of possible features other than a home that could be tried). Rather, knowledge of what is important to the subject would need to play a role.[9]

Interestingly, fortuitousness played a role for Hubel and Wiesel in the early measurements of receptive fields in the primary visual cortex of cats.[10, 11] The accidental movement of a notepad across a projector screen caused a dramatic burst of neural activity in the anaesthetized animals. The two scientists later received the Nobel Prize in 1981 for their work, in which they described such responses that were sensitive to the orientation of moving edges or slits.

Primary sensory cortex An area of a cortex that receives the most direct input, in terms of shortest path-length, from sensory neurons of a particular modality.

Sensory modality Specifies one of the particular senses (vision, taste, etc.).

Neurons a few synapses or more from the periphery—such as those in the cerebral cortex—are not driven primarily by direct sensory input. Even neurons in primary sensory cortex receive far more connections from other neurons in the cortex than from neurons preceding it in the sensory pathway. While feedforward sensory responses dominate studies of neural responses in anesthetized animals, it is only because the stronger feedback input to cortical cells is silenced by anesthesia. Therefore, care is needed when interpreting the results of experiments using methods—such as anesthesia—that simplify the behavior of a neural circuit. Such simplification may provide the interpretable, constraining data that is necessary in the early stages of discovery to put together the building blocks of a circuit. However, the simplified situation may provide only a small piece of any description of that neuron's role in natural behavior.

3.1.2   Time-Varying Responses and the Peristimulus Time Histogram (PSTH)

In vivo neural responses are highly variable—if an identical stimulus is presented multiple times (i.e., on multiple trials), both the timing of spikes with respect to stimulus onset and the total number of spikes change from trial to trial. Therefore, it is typical to average across trials simply to get a better estimate of a neuron's mean response to a stimulus. Moreover, the averaging of spike trains across trials can allow us to uncover the dynamics of a neuron's response to a stimulus, not just the mean number of spikes. The peristimulus time histogram (PSTH) is the principal method for such averaging.

The PSTH estimates the dynamics of the response by aligning spike trains to the time of stimulus onset, then averaging across trials to obtain an estimate of the mean firing rate as a function of time. We might ask whether the mean response across trials is at all relevant beyond being descriptive, since the spikes on each individual trial can impact only the animal's behavior on that trial. One argument that extends the relevance of across-trial averaging is that we are measuring one neuron out of many, perhaps hundreds, of a population with a similar trial-averaged response. The trial-averaged response could be reproduced on each individual trial by the population's mean response—and it is the population's activity that impacts other neurons and the animal's behavior. If neurons were uncorrelated on a trial-by-trial basis, this would be a good argument, but as we will discuss in chapter 9, more sophisticated methods should be applied when correlations are important.

Two distinct methods can be used to combine across trials the temporal information associated with each spike. In the most straightforward method, time is binned into small windows, and then the number of spikes in each time-window is counted and averaged across trials (see figure 3.2B, C). The resulting PSTH is a series of adjacent rectangles whose height indicates the mean firing rate with an accuracy that depends on the number of spikes contributing to each time bin. Therefore, there is a trade-off, with larger time bins containing more spikes so producing a more accurate estimate of the mean rate, but at the same time the larger time bin has less sensitivity to time-dependent variation of the neuron's activity.

In the standard binning method, all information regarding the spikes' relative position in the bin is lost. One might think that a spike falling near the boundary of a bin could contribute equally to the next bin to avoid such loss of information. Alternative methods aim to overcome such information loss. In a variant of the straightforward method, a sliding window can be used (figure 3.2D), so that each spike enters a series of time bins with centers ranging from a half of the bin width before the spike to a half of the bin width after the spike.

A final alternative is to smooth each spike with a Gaussian filter (see figure 3.2E for the result and figure 3.3 for an introduction to linear filters). In this method, each spike is replaced with a Gaussian function of unit area, centered on the spike time. Such a filter is linear, as the contributions of all spikes are added together linearly. The time resolution can be quite fine as the Gaussian is a continuous function. The width of the Gaussian filter plays a similar role to bin width in the straightforward method—the wider the Gaussian, the less noisy the resulting PSTH, but the lower the temporal resolution.

3.1.3   Neurons as Linear Filters and the Linear-Nonlinear Model

Linear filter   A system whose response is equal to a weighted sum of its inputs.

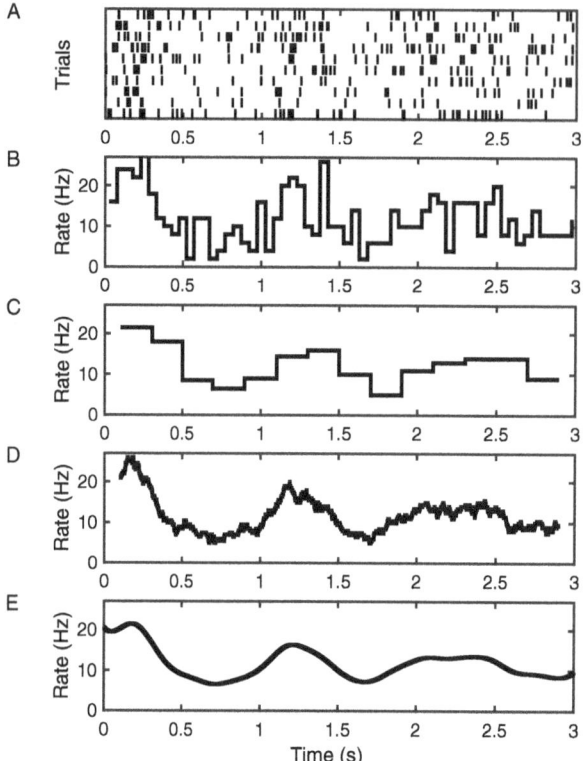

Figure 3.2
PSTH for an example of oscillating activity. (A) Each of the 10 rows contains the series of spike times from a single neuron in one trial. Spikes are produced as an inhomogeneous Poisson process (see text or code) with the firing rate in Hertz as $r(t) = 12 + 6\sin(2\pi t)$ where t is time in seconds. (B) Spikes are accumulated across trials in consecutive time bins of width 50 ms. (C) Spikes are accumulated across trials in consecutive time bins of width 200 ms. (D) Spikes are accumulated across trials in a sliding bin of width 100 ms. (E) Each spike is filtered (smoothed) with a Gaussian of width 100 ms to produce a continuous, smooth function. In this example, because the rate is itself defined as a smooth function, the smoothed spike train produces the best approximation to the underlying rate. This figure was created by the available online code `PSTH_osc.m`.

Linear-nonlinear model   A system that takes a linear filter of its inputs, then produces its output as a nonlinear function of the filtered inputs.

When receptive field structure is indicated in the manner of figure 3.1B, C, the suggestion is that the neuron is acting as a linear filter. To understand the response of a linear filter, we can consider figure 3.3, which shows the response of a neuron with a center-surround (On-center) receptive field to two different stimuli. The neuron's response is defined by the filter function—or

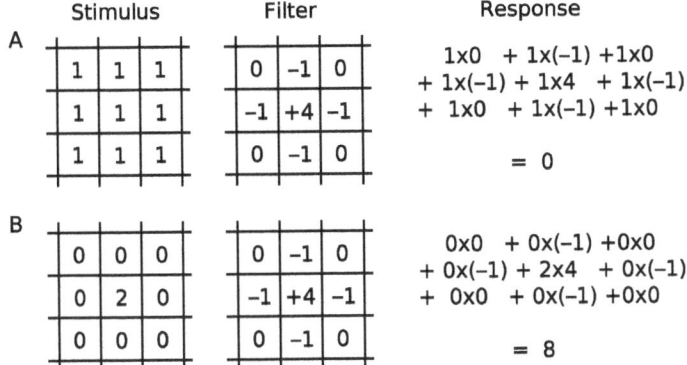

Figure 3.3
Response of a center-surround linear filter or kernel. The response of many neurons in primary sensory areas can be modeled as a linear filter of the stimulus. In the simplified example shown, the neuron filters a spatial stimulus by multiplying the stimulus amplitude at each spatial location (left) by the corresponding value of the filter—also called the kernel—(center) to produce a response (right). (A) For the center-surround filter shown (middle) a spatially constant stimulus (left) generates no response (right) because the positive contribution at the center of the filter is canceled by the negative contributions of the surround. (B) Such a filter produces a strong response to a spatially localized stimulus.

kernel—shown in the middle column. The value of the stimulus at each point within the receptive field is multiplied by the value of the kernel at that location. The contributions obtained in this manner from each point within the receptive field are then summed together to produce the neuron's response. The filter is called linear because the neuron's response is a linear combination—i.e., a weighted sum—of all the stimulus values. In the example shown, the linear combination is obtained as a sum over individual discrete stimulus values—more generally, the kernel would be a continuous function and the sum would be replaced by an integral.

Kernel   A function that describes the weight (i. e., relative importance) given to each component of the inputs.

Just as a linear filter can be used to approximate a neuron's response to the spatial properties of a stimulus, so too can it be used to approximate a neuron's time-varying response to a time-varying stimulus. We will use methods that assume such linear responses in the next section and in tutorial 3.1.

As we have seen in chapter 2, neural responses, even in the simplest of models, are not linear. In fact, since a firing rate cannot be negative, it must be a nonlinear function of input for any cell, simply because of the threshold for response (the threshold nonlinearity). However, a straightforward extrapolation of the linear filter model, namely the "linear-nonlinear model" has proven to be of great value in describing neural responses.[12–14] In the linear-nonlinear model, a weighted sum of the stimulus over the neuron's receptive field, as in figures 3.1B, C and 3.3, produces the inputs to a model neuron. The model neuron's response is then a nonlinear transformation—equivalent to sending the input through an f-I curve—of that linearly weighted input.

3.1.4 Spike-Triggered Average

A technique that can be used to characterize simple receptive fields is the spike-triggered average.[15, 16] With this technique, a neuron is recorded while it responds to a multitude of different stimuli. Each time the neuron produces a spike, the stimulus that caused the spike is stored. All of the stored spike-causing stimuli are then averaged together. The process can be carried out for spatial receptive fields, in which case each stimulus is maintained long enough to ascribe each spike to the correct spatially patterned stimulus—for example, in visual processing it takes a few tens of milliseconds for a stimulus to produce a spike in V1).

More commonly, the temporal properties of neural responses are also considered. The sequences of stimuli preceding each spike for 100–200 ms are then stored. As a useful control, those stimuli following each spike can also be stored. When the stored sequences of temporal stimuli are averaged together, they should be aligned to the time of each spike. The same stimulus can contribute at multiple time points in such a time-dependent spike-triggered average. For example, if two spikes occur, one 50 ms after the stimulus and the other 100 ms after the stimulus, then that stimulus contributes to the –50 ms and to the –100 ms bins of the spike-triggered average (see figure 3.4).

3.1.5 White-Noise Stimuli for Receptive Field Generation

As suggested earlier in this chapter, when characterizing a neuron by its receptive field, great care is needed when designing the set of stimuli. At the simplest level, it is important to realize that any correlation included in the stimuli used to discover a receptive field will be found to be present in that receptive field.

In a possibly exaggerated story from the early days of neural networks, one network was trained to identify tanks in a pattern recognition test. The network was astoundingly successful, beyond what seemed reasonable given the subtle features, such as the end of a tank's gun turret, present in some of the pictures. On further analysis, it was found that pictures of tanks were taken on one day, whereas pictures without tanks were taken on a separate day. Differences reflecting changes in the weather, such as brightness of the sky, could be used to distinguish tank from non-tank. Thus, accidentally, one stimulus feature—presence of a tank—was completely correlated with another stimulus feature—the weather pattern. If the output of the neural network is thought of as a neuron, then its trained response to sky brightness appeared like a response to a tank because of the unfortunate correlation.

> **White noise**   A type of random variation in time such that the variation at one point in time is not at all correlated with the variation at any prior point in time.

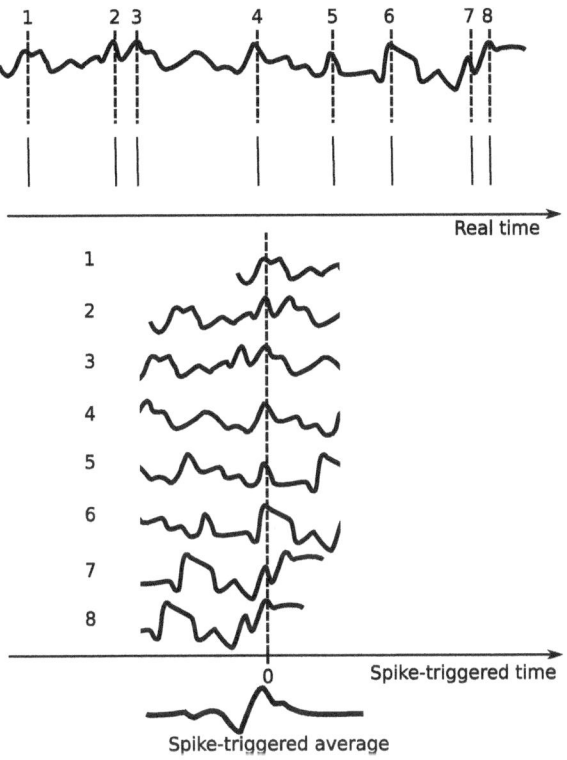

Figure 3.4
Producing a spike-triggered average. A time-varying signal (solid continuous line, top) produces a sequence of eight spikes (vertical solid lines below the stimulus). The stimulus is then realigned to the time of each spike (vertical dashed lines top and middle) to produce eight samples of the stimulus, one sample around the time of each spike. These samples are averaged to produce the spike-triggered average (STA, bottom). The STA preceding the neuron's spike (set to the time of zero) is the average stimulus to cause a spike. Since the spike cannot cause the stimulus, the period after the spike time over which the STA decreases to zero is an indication of the timescale of correlations within the stimulus. Any appreciable decay time—indicating stimulus correlations that do not rapidly decay to zero—is a confound if this method is used to estimate properties of the neuron.

The type of stimulus that avoids any such confounding correlations is white noise.[17, 18] Production of white noise stimuli is limited by the resolution of the projecting apparatus—light intensity is always correlated on the length scale of a pixel and on the time scale of a frame. However, the "white noise" stimuli used to measure receptive fields are typically discretized at a much coarser level than these limits, with correlations over the range of many pixels, so that there is sufficient chance of generating a response. Use of stimuli with correlations limits the resolution of any receptive field produced but ensures the stimuli produce a sufficient number of responses before averaging, as explained in what follows.

Consider a neuron that responds best to illumination covering a circle of radius one degree of the visual field. If a stimulus contains randomly chosen luminance over each pixel covering one-thousandth of a degree, the number of independently chosen luminances within the neuron's receptive field would be on the order of one million. These one million independent values would have to be correlated in the right way to generate a significant response. The likelihood of sufficient correlation by chance would be so small that the number of such stimuli needed to map the receptive field would be prohibitively large.

Indeed, one can estimate how likely it is that any one combination of a million independent stimuli would produce 1 percent greater total input to a neuron. For example, if each of a million pixels is equally likely to be "on" or "off," the mean number "on" is 500,000. If the cell responds when 505,000 are "on" (a 1 percent increase in input) the chances of this occurring in any one image turn out to be 2×10^{-45}, an astronomically small probability! (This number is arrived at by using the binomial distribution to get the standard deviation as $\sqrt{Np(1-p)} = \sqrt{10^6 \times 1/2 \times 1/2} = 500$, then using the normal distribution to find the chances of being ten standard deviations or more from the mean). If such stimuli were presented at the rate of one every 10 ms, then there would be negligible chance of seeing a response even if the experiment lasted for the lifetime of the universe!

The problem here is mathematically equivalent to the combinatorial problem when using stimulus-driven statistics to determine the response of a neuron with a complex receptive field—the probability of finding a stimulus that can drive a cell using random features alone becomes astronomically small. More progress can be made by considering the types of stimuli that are relevant to an organism based on its lifetime of experiences. Alternatively, in many studies, the relevant stimuli are imposed on the animal over the course of several months of repetitive experiments. This latter procedure enhances the likelihood of the scientist finding a neuron that responds to the experimentally relevant stimulus—though at the risk of producing responses that are not found in an animal's natural setting.

3.1.6 Spatiotemporal Receptive Fields

Receptive fields were first described for tactile stimuli, to indicate which regions of an animal's skin when touched would produce a response in a given neuron (figure 3.1A). Today, visual and auditory stimuli are the most commonly used to measure receptive fields (figure 3.1B, C). In all modalities, the dynamics of a stimulus is important. Auditory stimuli are inherently dynamic, even after transformation to the spectrum of pitch (frequency) versus time, because we rarely hear a constant pitch. For visual stimuli, any change over time indicates movement, which is of great importance to an animal. In fact, objects may be invisible within a scene until they move as a coherent whole and immediately pop out to our attention. Thus, it should not be a surprise that many neural responses depend on the dynamics of a stimulus.

We have already observed neural responses that depend on stimulus dynamics when we considered spike-rate adaptation in chapter 2. At its extreme, the combined effect of several such adaptation processes can render a neuron insensitive to a constant stimulus and only responsive to stimulus changes. Such an effect would appear in the spike-triggered average as a biphasic response—the neuron is most likely to spike when a "positive" stimulus follows a period of below-average stimulus. We will investigate an example of this in tutorial 3.1.

When the spatial and temporal components are combined, the resulting spatiotemporal structure of the receptive field of a neuron can be plotted as a colormap, or in grayscale (as in figure 3.5), or as a contour plot in two dimensions.[19]

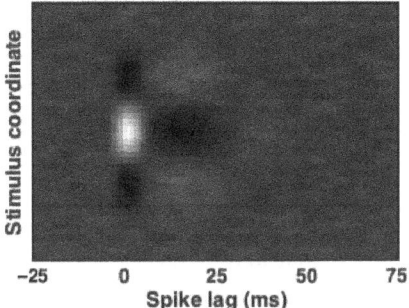

Figure 3.5
Model of a spatiotemporal receptive field of a V1 simple cell. A single spatial dimension is plotted (y-axis), corresponding to the location of a bar of optimal orientation, combined with a time dimension (x-axis). The center-surround spatial response combines with a biphasic temporal response. Light indicates high positive stimulus, dark indicates negative stimulus. This figure was generated by the online code `spatial_AELIF_randomI.m`.

Table 3.1
Structure of code for tutorial 3.1, part A

Main Code
1. Generate stimulus vector.
Simulate vector of spike times.
2. Downsample stimulus and spike-time vectors using the function expandbin.
3. Send downsampled vectors to the function STA.
4. Plot the spike-triggered average returned from STA.

function expandbin
1. Define new_vector with length new_dt/old_dt times the length of old_vector.
2. Find the mean of sets of points of length new_dt/old_dt in old_vector and add them to a single point in new_vector.

function STA
1. Define a time-window vector.
2. Initialize to zero the sta vector.
3. Loop through all spikes adding to the sta vector the stimulus offset by the time of each spike.
4. Divide the sta vector by the number of spikes.

3.2   Tutorial 3.1: Generating Receptive Fields with Spike-Triggered Averages

Neuroscience goal: Analyze receptive fields using the spike-triggered average.

Computational goals: Use a function; collapse or expand data into sets with different widths for time bins.

In this tutorial you will simulate an AELIF neuron, following the dynamics described in tutorial 2.3 Q.2, and using the following parameters: $E_L = -60$ mV, $V_{th} = -50$ mV, $V_{reset} = -80$ mV, $\Delta_{th} = 2$ mV, $G_L = 8$ nS, $C_m = 100$ pF, $a = 10$ nS, $b = 0.5$ nA and $\tau_{SRA} = 50$ ms. In both parts of this tutorial, the goal is to use the spike-triggered average to assess what stimulus is best at producing spikes in the model neuron.

Part A: Time-Varying Stimulus

In part A (see table 3.1), we will treat the stimulus as a single input current that varies over time and assess what temporal variation of the stimulus is most likely to lead to a spike in the model neuron. To achieve this, we will use a stimulus that is held constant for 5 ms, then randomly replaced with a new stimulus.

1.   The first goal is to simulate the neuron and record the spike times.
  a.   Produce a vector of 40,000 values for the applied current, with each value chosen randomly from a uniform distribution between −0.5 nA and +0.5 nA. These values will be used to produce successive 5 ms blocks with the current fixed in each 5 ms block.
  b.   Create a time vector in steps of 0.02 ms up to a total time of $40,000 \times 5$ ms and an applied current vector of the same size as the time vector. The current vector should contain the 40,000 randomly generated values from 1a), with each value repeated for 250 timesteps, so that the applied current is constant for 5 ms before changing to a new value.
  c.   Simulate the AELIF neuron with the applied current vector from 1b using the preceding parameters and the equations for tutorial 2.3, Q.2. Use a simulation timestep, $\Delta t = 0.02$ ms. Ensure that all spikes are recorded in a vector of the size of the time vector with a 1 in each time bin that a spike occurs and 0s otherwise.

2.   To improve the efficiency of the analysis we will *downsample* the stimulus and spike vector (see figure 3.6). This is best achieved by writing a function that we will call `expandbin`. The function will take as inputs an initial vector, an initial bin width, and a final bin width. It will return an output vector of smaller size than the initial vector, with the size reduced by a factor equal to the ratio of the bin widths. In particular, in this tutorial, we are simulating the neuron with a time bin of 0.02 ms but are only interested in changes on a timescale of 1 ms or more (and our applied current only changes on a 5 ms timescale). So, we want to replace

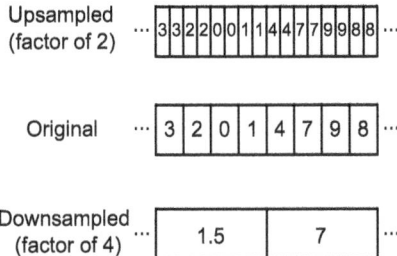

Figure 3.6
Downsampling and upsampling data. The original data stream (center) is a series of entries as a function of time. It can be downsampled by averaging across bins (bottom), reducing time resolution and using fewer time bins, or it can be upsampled (top) to create more time bins.

our stimulus vector and spike vector with vectors that contain a value for each 1 ms of time rather than for each 0.02 ms. This requires downsampling by a factor of 50.

a.   To achieve the necessary downsampling:

(i)   Create a function, `expandbin`, with inputs and outputs as follows:

`function [new_vector] = expandbin(old_vector, old_dt, new_dt)`

(ii)   Within the function (that will be saved in the file "`expandbin.m`") find the length of `old_vector` then the scaling ratio of `new_dt` divided by `old_dt` calculated to the nearest integer.

(iii) Calculate the size of `new_vector` by dividing the size of `old_vector` by the scaling ratio. Define `new_vector` as an array of zeros of this size.

(iv)   For successive elements of `new_vector` calculate the mean of successive blocks of entries of `old_vector`. The length of each block of entries is the scaling factor. In this tutorial, the mean of a block of 50 of the 0.02 ms time bins in `old_vector` will be placed into a 1 ms time bin of `new_vector`.

b.   Send to the function, `expandbin`, which you have just created, the applied current vector used to update the membrane potential in the simulations with its timestep of `dt` (note that while your values of applied current change every 5 ms, the vector used in the simulations should have a value every `dt` and the latter is used here), along with a variable `new_dt` defined as 1 ms. The function will return downsampled versions of this vector. Repeat by sending the simulated vector of spike times to the function, `expandbin`. Since spikes are defined as binary, 0 or 1, you should then change all the reduced spike values in the returned vector back to 1.

3. You will produce and plot the spike-triggered average derived from the two downsampled vectors (the stimulus and the spike times). You will create and use another function, STA, to achieve this.

    a. To create such a function for generating spike-triggered averages:

        (i)   Create the function, STA, with outputs and inputs in the following format:

```
function [sta, tcorr] = STA(Iapp, spikes, dt, tminus, …tplus)
```

               where the inputs are respectively: applied current vector; vector of spike times, with 1 at the time of every spike; the timestep used in these vectors; the time of the stimulus before a spike to begin recording; and the time of the stimulus after a spike to stop recording. Note that the vectors Iapp and spikes must be the same size.

        (ii)   It is useful in a function to supply default values that are used if a variable is not passed to it. If only three variables are passed to the function STA, then the last two will not exist within the function. In this case, tminus and tplus define the time window over which the spike-triggered average is calculated. Within the function, they can be created if they do not already exist using a conditional statement and the in-built function exist:

```
if (~exist('tminus'))
```

               Note that the tilde, ~, means "not," so the line of code following the conditional is only executed if the variable tminus has not yet been defined.

               Include two such conditional statements to set default values for tminus and tplus of 75 ms and 25 ms respectively.

        (iii)   Produce integer variables nminus and nplus that correspond to the number of time bins before and after a spike over which the stimulus is recorded for averaging.

        (iv)   Define the time window, tcorr, as a vector spanning from -nminus*dt to nplus*dt in steps of dt. This will provide the x-coordinates when the spike-triggered average is plotted.

        (v)   Initialize the variable sta to be zero in a vector the same size as tcorr. This will accumulate values to provide the y-coordinates when the spike-triggered average is plotted.

        (vi)   Use find to find the time bins that contain spikes.

        (vii) Loop over all the spike-times found in (vi), for each spike defining a window that begins a number of bins, nminus, before the spike and ends a number of bins, nplus, after the spike. Ensure the window neither begins earlier than the stimulus nor ends after the stimulus is over (you can ignore the spike in such cases).

        (viii) Within the loop, add to the vector sta the values of the stimulus vector within the window you have just defined for the spike (as in figure 3.4).

(ix)   After the loop, to calculate the average stimulus, just divide `sta` by the total number of spikes.

Note: It is easiest just to ignore those spikes too near the beginning or the end of the stimulus. However, if the time window for the STA is not much smaller than the stimulus duration it is better if they are not ignored (to avoid loss of a large fraction of the data). In that case, care must be done when averaging the stimulus: values in each time bin of `sta` should be divided only by the number of spikes that can contribute to that time bin, rather than by the total number of spikes. The value of the stimulus before and after the end of the trial is assumed to be zero, so, for example, a spike very early in the trial would otherwise contribute a lot of unwanted zeros to the average stimulus at earlier times.

b.   Now send the downsampled applied current vector and spike train to the function `STA`, which will return the spike-triggered average and the time window.

c.   Plot the spike-triggered average as a function of the time window around the stimulus. To adopt conventional coordinates, the x-axis should be the negative of the time window, so as to represent the time lag of the spike after the stimulus that produced it. With this transformation, the STA at positive values of x denotes a causal relationship.

d.   Comment on how any specific parameters of the AELIF model that you have simulated helped to determine any features of the spike-triggered average in c.

Part B: Spatiotemporal Stimulus (Optional Challenge)

We now extend the tutorial to include a stimulus that varies in both space and time. We will also assume that the neuron's input current is a linear combination of the values of the stimulus at each spatial location. (A linear combination is a weighted sum where the weights can be negative.) Therefore, the neuron receives a time-varying input current as in part A, but at each point in time the value of the input current depends on many values of the spatially varying stimulus. It is worth noting that while we discuss the additional coordinate of the stimulus as being spatial, corresponding to a visual stimulus, the mathematics and the associated computer code are identical if the stimulus coordinate is the frequency of an auditory stimulus.

1a.   Generate an array of 40 rows by 40,000 columns to contain the stimulus, $S(x,t)$, with each entry randomly chosen from a uniform distribution between –0.5 nA and 0.5 nA.

1b.   Assume the row number corresponds to a spatial coordinate, x, (with $x_{max} = 40$) and that the input current to the cell, $I_{app}(t)$, is given by the weighted sum of the stimulus at each spatial coordinate according to:

$$I_{app}(t) = \sum_{x=1}^{x_{max}} W(x)S(x,t) \quad \text{with} \quad W(x) = \cos\left[4\pi\left(\frac{x-x_0}{x_{max}}\right)\right]e^{-16\left(\frac{x-x_0}{x_{max}}\right)^2}$$

where $x_0 = 20.5$ is at the center of the spatial stimulus. Plot the input weight vector, $W(x)$ (which must be a 1×40 row vector) and generate the complete time-dependent input current as a row vector by direct matrix multiplication of the weight vector by the stimulus array.

1c. Simulate the AELIF cell following the instructions of part A, except with two of the parameters altered: $a = 40$nS, $b = 1$ nA; and with the new input current. As usual, record the times of spikes as 1 in an array that otherwise contains zeros.

1d. Downsample the spikes array and "upsample" the stimulus array to produce new arrays with time-bins of size 1 ms. Care must be taken with the stimulus array, since you must either carry out the upsampling one row at a time or adapt the function, `expandbin`, you created in part A to handle arrays of greater than one dimension. While downsampling involves taking the mean of values from many bins to enter into one new bin, upsampling involves taking the value from one old bin and repeatedly entering it into many new bins.

2. Generate a new function, `STA_spatial`, by adapting the function `STA`. The new function will take a two-dimensional array (like $S(x,t)$) as input and return a two-dimensional array as the spike-triggered average, sta.

   Hint: a few changes are needed to make the conversion, but you might want to use a command like `[Nspace, Nt] = size(stim_array)` to extract the number of spatial and temporal bins and to be careful when using `mean` to take the mean of a block of columns (since time varies from column to column) when downsampling rather than the mean of a block of rows.

3. Use your newly created function to calculate the spatiotemporal receptive field. Use the command `imagesc(fliplr(sta))` to visualize the receptive field in the conventional manner. (The command `fliplr` reverses the order of columns, i.e., flips the array from left to right).

4a. Plot rows 12, 20, and 28 of the receptive field. How do they compare?

4b. Plot columns 25, 50, and 75 of the receptive field. How do they compare?

In vitro   Literally "in glassware," meaning an experiment carried out with the system maintained within laboratory equipment rather than remaining in the living animal, as occurs in vivo.

3.3   Spike-Train Statistics

If a constant or regularly varying input current is provided to a neuron, either in vitro, or in a model (in silico), any resulting activity of the neuron is itself regular except in rare cases. Yet the activity of neurons in the brains of wakeful animals is highly irregular. The traditional solution in neural models is to add noise—in the sense of random fluctuations—to the input current or conductance. However, in a real circuit, these random fluctuations in a neuron's input are largely due to the irregularity of spike times of other neurons receiving similar random fluctuations in their inputs. A constraint arises then, when understanding the behavior of neural circuits, because the observed irregularity in spike trains must be able to account for the noisy inputs to cells that can generate such irregularity. Therefore, it can be as important to measure the irregularity of spike trains—i.e., at least the second-order spike-train statistics—as it is to measure a neuron's mean response.

> **Second-order statistics**   Measures of data, such as variance and standard deviation, which depend on the square (second power) of the values of each data point.

3.3.1   Coefficient of Variation (CV) of Interspike Intervals
The coefficient of variation of any set of values is the standard deviation (a second-order statistic) divided by the mean of the values. In the case of measured interspike intervals (ISIs), a highly regular spike train has ISIs of nearly the same value, so the set of ISIs has very small standard deviation. Therefore, the CV of a regular spike train is close to zero. Spike trains of cortical neurons measured in wakeful mammals typically have CVs near one or greater than one.

> **Coefficient of variation (CV)**   Standard deviation divided by mean of a set of data.

The CV is directly related to the histogram of the ISIs. The full histogram of ISIs contains more information than the CV alone—which is one number that depends simply on the histogram's width and the position of its mean—so it can be better to plot or fit with a curve the histogram showing the full distribution of ISIs. For irregular spike trains, the distribution of ISIs can be shaped like an exponential decay, albeit with a small dip near an ISI of zero corresponding to the refractory period. In fact, the exponential ISI distribution has a CV of one and is a feature of the Poisson process (to be described later), so the Poisson process is sometimes used to model individual spike trains.

> **Poisson process**   Probabilistic production of events, such as spikes, at any point in time, with equal probability per unit time.

Since the CV depends on all of the spike intervals measured over a period of time, a high CV does not imply a highly irregular spike train (although an irregular spike train will produce a high

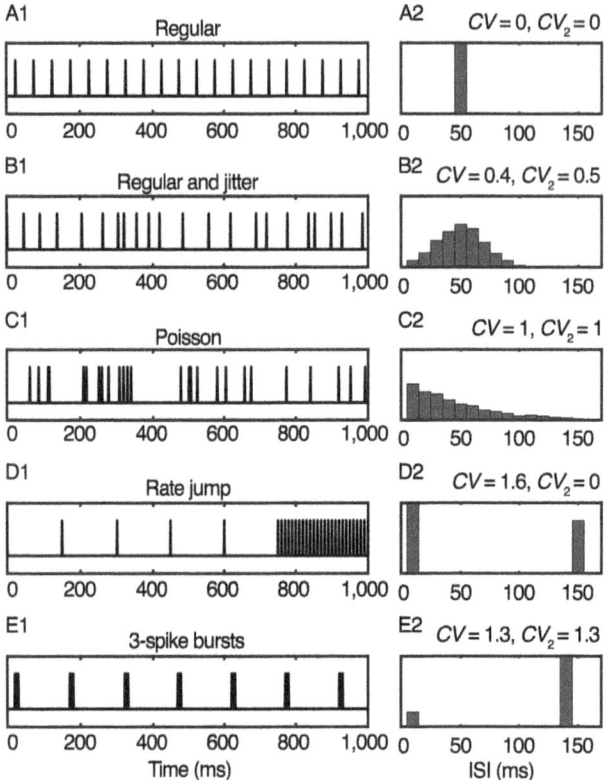

Figure 3.7
Spike trains, ISI distributions, and statistics. (A) A regular spike train with a fixed ISI of 50 ms. (B) A spike train with ISIs arising randomly from a normal distribution with mean 50 ms and standard deviation of 20 ms. (C) A Poisson spike train with mean ISI of 50 ms. (D) A regular spike train with ISIs of 150 ms switches to a regular spike train with ISIs of 10 ms. (E) A spike train with bursts of three spikes, with 150 ms between bursts and 4 or 5 ms between spikes within a burst. (Left) A1–E1. Example of 1 s of each spike train. (Right) A2–E2. Histograms show the distribution of ISIs generated by a 300-s simulation. *CV* and *CV*$_2$ of each distribution is given above each example. This figure was created by the online code `spikes_cv_isis.m`.

CV; see figure 3.7). A high CV, meaning a large range of interspike intervals relative to their mean, can arise because the neuron's firing rate is changing over time, either slowly, or rapidly but periodically as it does for regularly bursting neurons. Therefore, an improved measure of irregularity was proposed, based on the similarity between neighboring ISIs. Since only two ISIs were compared at a time, the measure is called CV$_2$.[20] If there is no time-dependence, so that the value of one ISI has no correlation with the value of the previous one—i.e., each ISI is a random sample from the complete distribution of ISIs—then CV$_2$ is the same as the CV. However, if the neuron's firing rate varies slowly, on a timescale longer than typical ISIs, then CV$_2$ will be smaller than the CV and CV$_2$ will better correspond to the level of irregularity.

Of course, CV_2 is itself a poor measure of irregularity when, for example, a neuron fires regular doublets of spikes with large separations between two closely spaced spikes. Such doublet firing is an example of bursting, producing an ISI distribution that is bimodal (i.e., contains two peaks; see figure 3.7E). For bursting neurons, the ISIs within bursts should be analyzed separately from the times between bursts (i.e., the interburst intervals).

Bursting neuron   A neuron producing sets of closely separated spikes with longer intervals in between those sets.

3.3.2 Fano Factor

The Fano factor is another measure of variability, in this case one that considers the total number of spikes in a given time interval instead of the interspike intervals. It is defined as the variance of the number of spikes, $N(T)$, in a time bin, T, divided by the mean:

$$F(T) = \frac{\mathrm{Var}[N(T)]}{\overline{N(T)}} = \frac{\overline{N^2(T)} - \left[\overline{N(T)}\right]^2}{\overline{N(T)}} \tag{3.1}$$

where the overline denotes the mean of a quantity.

Fano factor   Variance divided by the mean of multiple measures of the number of a quantity.

As with the CV, the Fano factor is increased by fluctuations that cause irregular timings of spikes and by variation in the underlying firing rate. The Fano factor can be calculated from a single series of consecutive time bins, in which case one number is produced that may depend on the chosen size of time bins, so it can be plotted as a function of this size. Alternatively, time bins can be aligned across trials to a specific time in the trial. In the latter case the Fano factor can vary both with bin size and as a function of time during the trial. Such time-dependence of the Fano factor can indicate when, in a task, neural activity is particularly constrained and well defined, so it can suggest "resets" or the presence of fixed activity states that may be important steps during information processing.

A Poisson process with constant firing rate (called a homogeneous Poisson process, described in the next subsection) has a Fano factor of one, independent of the size of time bin used (see appendix A). If the firing rate varies within each trial but in an identical manner on each trial, then the Fano factor in each time bin can still be calculated by comparing time bins across trials with the same underlying firing rate. In this case, a Poisson process (now called inhomogeneous) still has a Fano factor of 1, even though the CV of ISIs would be greater than one because of the within-trial variation.

3.3.3 The Homogeneous Poisson Process: A Random Point Process for Artificial Spike Trains

At times, we may need to simulate a series of spikes that has the gross features of an in vivo spike train—a given mean rate that may vary in time and a similar irregularity of spike times—and not wish to take the time to adjust parameters in a model neuron until it reproduces these features. A good reason to need such simulated spike trains is to compare the results of analyses on such control or "dummy" datasets with the results of identical analyses carried out on datasets acquired in vivo.

A simulated spike train is an example of a point process, which means a process whose contents are entirely represented by a set of discrete points—in this case the spike times. The point process is in continuous time, as the discrete spike times are instantaneous values that could arise at anywhere in a range of values of the continuous variable, time. Here we focus on the Poisson process, which is an example of such a continuous-time point process, in part because its statistics are similar to those of neurons measured in vivo, and in part because it is the simplest such process. Therefore, it is a good springboard for those who want to delve more deeply into point processes at a later date.

A Poisson process is defined by the probability of an emission in any small time interval, δt. The probability is proportional to the length of the small time interval, so it can be written as $r \cdot \delta t$. For a homogeneous Poisson process the proportionality constant, r, which corresponds to the firing rate of a neuron, is fixed across time. The single value of r is the only parameter needed to fully define the homogeneous Poisson process.

A key result (derived in appendix A) for a Poisson process is the probability of a given number, N, of spikes in a time interval, T. For a Poisson process the entire probability distribution, $P[N]$, depends on a single parameter, $\lambda = rT$, via

$$P[N] = \frac{\lambda^N e^{-\lambda}}{N!}. \tag{3.2}$$

Equation 3.2 defines the Poisson distribution, with $\lambda = rT$ being the expectation value for N —the expected mean number of spikes from many examples of the process. We see that equating r with the firing rate of a neuron is justified by this result as the mean number of spikes in a time window is expected to be rate × time.

Poisson distribution   The probability of a given number of events, N, produced by a Poisson process in a fixed time interval, as a function of N.

The distribution defined in equation 3.2 may appear to be in conflict with the definition of a Poisson process, which is based on the probability of a single spike in a small time interval being $r \cdot \delta t$, and includes no exponential term. However, the definition of a Poisson process assumes the small time-interval, δt, is in the limit that the product $r \cdot \delta t$ is much less than 1. In this case $e^{-r\delta t} \approx 1$ so that equation 3.2 with $N = 1$ can be simplified from $P[1] = \left((r \cdot \delta t)^1 e^{-r\delta t} \right) / 1!$ to produce $P[1] = r \cdot \delta t$ as required.

From the Poisson distribution, we can produce the distribution of interspike intervals for a Poisson process. To calculate this, we need to find the probability that following any spike the next spike is in a small time interval, from T until $T + \delta t$ later. The dependence of this probability on T is the ISI distribution. It is evaluated by multiplying together the probability that there are no spikes in the interval of size T, by the probability, $r \cdot \delta t$, of a spike in the following small time interval. Substituting $N = 0$ into equation 3.2 yields $P[0] = e^{-\lambda} = e^{-rT}$. Therefore, the probability of the interspike interval being between T and $T + \delta t$ for a Poisson process is $r \cdot \delta t \cdot e^{-rT}$, or:

$$P(T)\delta t = r \cdot \delta t \cdot e^{-rT}. \tag{3.3}$$

A Poisson process of rate r can be simulated most easily by choosing time-bins of width, Δt, much smaller than $1/r$, so that the probability of more than one spike in the interval is negligible, in which case $P[0] \approx 1 - r\Delta t$ and $P[1] \approx r\Delta t$ (respectively the probability of no spikes or a single spike in each time-bin). Such a spike train can be implemented in a single line of code, e.g., in MATLAB:

```
spikes = rand(size(tvec)) < rate*dt;
```

where `tvec` is the vector of time-points, and `rate*dt` is the same as $r\Delta t$.

Alternatively, and in a method that can be generalized to other processes with different ISI distributions, successive ISIs can be randomly chosen from the ISI distribution. To achieve this, the cumulative sum of the ISI distribution should be plotted and scaled between 0 and 1 on the y-axis. A random number between 0 and 1 is then chosen as the y-coordinate. The corresponding x-coordinate is read from the graph and used as the ISI. For the Poisson process, with an exponential distribution of ISIs, the cumulative sum becomes an integral, which is also an exponential distribution. The resulting ISI chosen becomes:

```
ISI = -log(rand())/rate;
```

If `rand()` is replaced by `rand(1,Nspikes)` then a set of `Nspikes` spike times is generated from the set of ISIs as

```
spiketimes = -cumsum(log(rand(1,Nspikes)))/rate;
```

3.3.4 Comments on Analyses and Use of Dummy Data

Data are analyzed in order to test whether they support one hypothesis or another. In some cases, the data expected by one (or both) of the hypotheses can be simulated. In such an ideal situation, the data generated for each hypothesis can be produced with the same degree of variability as the real data, by either additional noise or by parameter variation. Results of the analyses of these different samples of simulated data can be directly compared with the analysis of the real data. In principle, statistically significant results can be confirmed when the real analysis falls within the spread of results generated by the many random instantiations of one hypothesis and not the other.

Dummy data   A set of simulated data that can be analyzed in the same manner as the data from the real system being studied, usually as a control study.

Even if simulated data are produced in too simplistic a manner to be compared directly to the real data, one can, and should, still analyze simulated data to test the validity of any analysis. For example, if we see a change in the Fano factor in the real data, it is worth simulating many sets of spike trains with similar statistics to the real data, using a single underlying model. We can then measure how much the Fano factor can vary across trials when the underlying process is constant. If the observed variation in the real data falls into the range of simulated variation, we realize that such statistical noise could easily explain our observations and do not need to pursue other explanations. That is, while the change in Fano factor across our data would not rule out other explanations, it would not provide evidence for them.

Finally, even before carrying out an experiment designed to distinguish two hypotheses, it is valuable to simulate data in a simple manner (such as via an inhomogeneous Poisson process) according to each hypothesis. The spike trains produced by the simulated data should be analyzed in the same manner as planned for the real data. Following many such "simulated experiments" it will be clear whether the two hypotheses can be distinguished reliably by the planned methods. If the simulated datasets are not reliably distinguishable, then new analyses, or a different experiment, or more samples are needed.

3.4   Tutorial 3.2: Statistical Properties of Simulated Spike Trains

Neuroscience goal: Discover how different types of within-trial variability and across-trial variability impact measures such as the ISI distribution and Fano factor.
Computational goals: Analyze distributions of time differences and cumulative sums of events; gain more practice at altering the size of time windows for analyzing data.

Part A: AELIF Neuron with Noise
1.   Simulate an AELIF neuron with the following parameters: $E_L = -70$ mV, $V_{th} = -50$ mV, $V_{reset} = -80$ mV, $\Delta_{th} = 2$ mV, $G_L = 10$ nS, $C_m = 100$ pF, $a = 2$ nS, $b = 0$ nA and $\tau_{SRA} = 150$

ms. Use a timestep, Δt, of 0.01 ms and simulate a duration of 100 s. Set the input current, I_{app}, to have a mean of zero but with a different value on each time step selected from the normal distribution with a standard deviation of $\sigma/\sqrt{\Delta t}$. (Compare tutorial 1.1, Q.2).

a. Initially set $\sigma = 50 \, \mathrm{pA} \cdot \mathrm{s}^{0.5}$ (the exponent of 0.5 in the time units is canceled when divided by $\sqrt{\Delta t}$). Record the spike times and the set of interspike intervals (ISIs) by taking the difference between the spike times.

   (i) Plot the histogram of ISIs (use `histogram` with 25 bins for the values of ISI).

   (ii) Calculate the CV of the ISIs as the standard deviation divided by the mean.

   (iii) Calculate the number of spikes in each consecutive 100 ms window. Calculate the variance and mean of these numbers (you can use in-built functions) and use these results to calculate the Fano factor.

   (iv) Repeat iii, but use a loop that allows the window size for spike-counting to range from 10 ms up to 1 s. Plot Fano factor against window size.

b. Repeat all of the steps of 1a but with AELIF parameter $b = 1$ nA. Explain any differences in your results from 1a and comment on any dependence on the time-window of the Fano factor in iv. *Hint: How does the variance compare to the mean of a set of values that are only 1s or 0s?*

c. Repeat steps i–iii of 1a with AELIF parameter $b = 0$ nA, while reducing the noise to set $\sigma = 20 \, \mathrm{pA} \cdot \mathrm{s}^{0.5}$. On separate trials add to the input current constant terms of 0 nA, 0.1 nA, and 0.2 nA. Comment on how the results change with added input current.

Part B: Homogeneous and Inhomogeneous Poisson Process (Optional)

Note: You can use the online function `alt_poissrnd`, or an in-built function to generate a Poisson process, or write your own function for this part of the tutorial. If you write your own function you can approximate the Poisson process by assuming that a single spike occurs in each time bin with probability of rate multiplied by bin width and otherwise no spikes occur. Such an approximate Poisson process becomes the Poisson process by definition, when the bin width approaches zero. However, for finite bin widths the possibility of more than one spike per time bin must be included for an exact Poisson process. Since neurons never fire two spikes within 1 ms of each other, the approximate Poisson process provides a better approximation to spike trains of real cells than the exact Poisson process, so the approximate process can be used for this tutorial.

2a. Simulate a 100-s spike train as a homogeneous Poisson process (i.e., one of fixed rate) with a rate of 20 Hz, using time bins of size 0.1 ms. Proceed through parts i–iv of question 1a to assess the statistics of the Poisson spike train.

2b. Simulate a set of 1000 trials, each of 10 s duration, of the homogeneous Poisson process. Store the spikes as an array of 1s and 0s, with each row corresponding to a trial.

2c. Use the function `cumsum` to accumulate across columns (the second dimension of the array) the total spike count as a function of time, $N(t)$, in each trial.

2d. Now calculate the variance, $\mathrm{Var}[N(t)]$, and mean, $\overline{N(t)}$, across rows (the first dimension of the array) of the cumulative spike count. Plot the ratio to see the Fano factor as a function of elapsed time, $F(t)$.

2e. In an alternative method, use a fixed size of 200 ms to produce a series of successive time-windows. Count the numbers of spikes in each 200 ms time window independently in each trial. Then compare values across trials as in question 2d to calculate the Fano factor for each 200 ms time bin through the trial.

3. Simulate spike trains for 1000 trials of 10 s as an *inhomogeneous* Poisson process (one with a time-varying rate) that follows a firing rate, $r(t) = 25 + 20\sin(2\pi t)$ on each trial. Repeat 2c–2e.

Part C: Doubly Stochastic Point Processes (Challenge)

4a. Simulate spike trains for 1000 trials of 10 s as an inhomogeneous Poisson process according to a firing rate, $r(t)$, with an initial value of $r(0) = 5$ Hz and a final value of $r(10) = 25$ Hz. For each trial select a different random time point with uniform likelihood between 2.5 s and 7.5 s. At that point in the trial, change the rate from the initial rate to the final rate.

4b. Plot the mean and the variance of firing rate across trials as a function of time (i.e., the rate in each time point can now have a different value on different trials—for each time point calculate the mean and variance of these values).

4c. Repeat 2c–2e for this doubly stochastic point process (the process is doubly stochastic because underlying rate varies randomly and spikes are produced randomly according to that rate).

5a. Simulate spike trains for 1000 trials of 10 s as an inhomogeneous Poisson process according to a firing rate, $r(t)$, with an initial value of $r(0) = 25$ Hz that accumulates noise in the firing rate as a random walk according to the equation:

$$r(t + \Delta t) = r(t) + \sigma_r \sqrt{\Delta t}\, \tilde{w}_n$$

with $\sigma_r = 5$ Hz and where $\tilde{w}_n$ is a random number selected from a Gaussian (Normal) distribution with zero mean and unit variance (as produced by the randn function in MATLAB; see section 1.6.6). Notice that the term added can be first calculated as an array, then the function cumsum can be used to accumulate the changes in firing rate.

Use the max and min functions to ensure that no value in the array of firing rates is less than zero or greater than 50 Hz.

5b. Calculate the across-trial mean and standard deviation of the firing rate as a function of time using the values of $r(t)$ that you have generated.

5c. Repeat 2c–2e with the new doubly stochastic point process.

The online function fano.m *can be used to carry out the Fano factor analyses as needed.*

3.5 Receiver-Operating Characteristic (ROC)

Given the variability of neural firing, both in the presence and absence of a stimulus, one might wonder whether a single neuron's response can indicate with any reliability whether a stimulus is present or absent. A means of assessing the reliability of such signal detection—perhaps better termed the discriminability—is through the receiver-operating characteristic (ROC) to be described in the following subsections. ROC methods are valuable beyond neuroscience—for example, in medical diagnostic tests[21]—whenever a single, potentially unreliable measurement is used to determine which one of two conditions exists.

Receiver-operating characteristic (ROC) curve   A plot of the probability of true positives (being correct when a test gives a positive result) against the probability of false positives (being incorrect when a test gives a positive result) for different thresholds of a positive test result.

3.5.1 Producing the ROC Curve

A receiver-operating characteristic (ROC) is used when assessing how much evidence a given response provides in favor of one alternative hypothesis over another. In signal detection theory, one hypothesis would be the presence of a signal; the alternative would be the absence of that signal. For example, in medical tests the response indicates the presence or absence of a disease with different degrees of likelihood. Tests are rarely 100 percent accurate, so a threshold is set at a level based on the trade-off between avoiding false positives (better achieved with a high threshold for detection) versus increasing the power to capture all true positives correctly (better achieved with a low threshold for detection). The receiver-operating characteristic can be viewed as a curve, by plotting the probability of a true positive result (True +ve, or the hit rate, $\alpha(x)$) versus the probability of a false positive result (False +ve, $\beta(x)$) as the threshold for detection is varied (figure 3.8).

ROC analysis can also be used to assess how much the spike train of a single neuron (or the aggregate of a group of neurons) can contribute to an animal's ability to distinguish between two stimuli or distinguish the presence from the absence of a stimulus. Even if the firing rates are very different across the two conditions, the duration of the stimulus or the time across which any circuit in the animal's brain can accumulate spikes may be limited. Therefore, if spikes are emitted by the neuron as a noisy process, it is possible that fewer spikes are counted on some trials with a higher expected firing rate than are counted on some trials with a lower expected firing rate. That is, if the numbers of trials with different spike counts are accumulated and plotted as a histogram (figure 3.8A), then it is possible that the two distributions overlap. Such overlap makes it impossible to produce a threshold such that all trials of one condition produce a spike count that is higher than the threshold, while all trials of the other condition produce a spike count that is lower than the threshold.

A

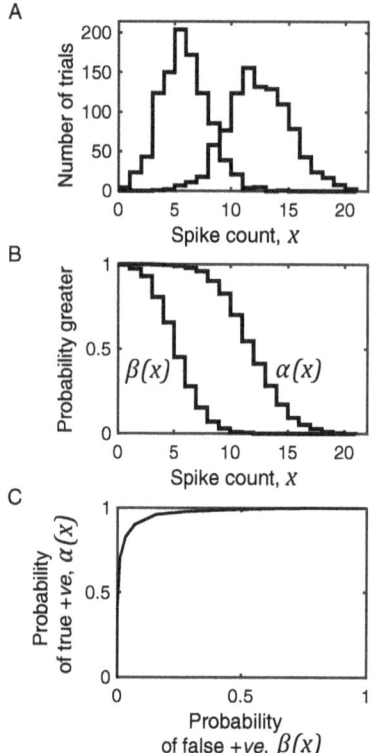

Figure 3.8

The ROC curve. (A) Histograms of the number of spikes produced by a noisy AELIF neuron across 1000 trials with a time window of 0.5 s and a stimulus of 0.1 nA (right histogram) or no stimulus (left histogram) combined with input noise with $\sigma = 20$ pA.s$^{0.5}$. (B) The cumulative sum of each histogram is calculated and divided by the total number of trials, then subtracted from 1 to obtain the fraction of trials with more spikes than a given spike count. (C) The y-values (α and β) from each of the two curves in B) at the same x-value are plotted against each other to produce the ROC curve as the x-value varies. That is, each point on the ROC curve can be produced by choosing a threshold at a particular spike count (x-value) such that a positive (+ ve) test result is produced if the spike count is above that threshold. The fraction of trials with above-threshold spike counts in the absence of stimulus is the false positive value on the x-axis (β). The fraction of trials with above-threshold spike counts in the presence of a stimulus is the true positive value (or hit rate) on the y-axis (α). Parameters for these data are those given in tutorial 3.3, question 1. This figure is produced by the online code AELIF_noise_ROC.m.

The overlap of the two distributions is reduced—and thus an observer's ability to discriminate the two conditions increased—if either the separation of the means of the distributions increases, or if the standard deviations of the distributions decrease. A measure of this overlap is the discriminability index, d', which is defined by $d' = \Delta(mean)/\sigma$ where $\Delta(mean)$ is the difference between the means of the two distributions and where $\sigma = \sqrt{\frac{1}{2}(\sigma_A^2 + \sigma_B^2)}$ with σ_A and σ_B being the standard deviations of each individual distribution. If the noise does not vary between the two distributions such that their variances are equal, then $\sigma = \sigma_A = \sigma_B$.

A little calculus (see appendix B) reveals that if the two distributions are Gaussian with equal variances, then the probability of an error, $P(\text{Error})$, (where an error is either a "miss" or a false positive), given an optimal positioning of the threshold (optimized to minimize probability of an error) is equal to

$$P(\text{Error}) = \frac{1}{2}P(\text{Miss}) + \frac{1}{2}P(\text{False} + \text{ve}) = \frac{1}{2}\text{erfc}\left[\frac{\Delta(mean)}{2\sqrt{2}\sigma}\right] = \frac{1}{2}\text{erfc}\left(\frac{d'}{2\sqrt{2}}\right), \tag{3.4}$$

where erfc is the complementary error function.

3.5.2   Optimal Position of the Threshold

If our goal is to minimize the probability of an error when a priori (before measurement) a stimulus is as equally likely to be present as absent, the position of the threshold for stimulus detection should be at the intersection of the two probability distributions (figure 3.9A).

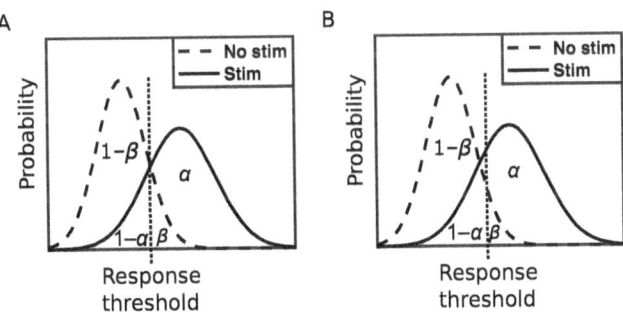

Figure 3.9
Optimal position of the threshold for stimulus detection. (A) If both possibilities are equally likely a priori, then the optimal threshold is at the intersection of the two probability distributions. (B) Raising the threshold above the optimum reduces the fraction of errors that are false positives (β) but reduces true positives (α) (i.e., increases the fraction of errors that are "misses") by a greater amount.

To see this, consider moving the threshold from the intersection to a position slightly to the right (figure 3.9B). In this case, the higher threshold would reduce the number of false positives (fraction of stimulus-absent trials to the right of the threshold) while increasing the number of misses (fraction of stimulus-present trials to the left of the threshold). However, the reduction in false positives is less than the increase in misses (the stimulus-present curve is above the stimulus-absent curve to the right of the intersection) so the net effect is to increase the fraction of errors.

Similarly, if the threshold were decreased from the intersection then the number of misses would decrease, but the number of false positives would increase by more, again generating more errors. So, in this case the minimum number of errors is at the intersection.

It is useful to notice that when the threshold is at the intersection, the rate of change of false positives with a change in threshold exactly matches the rate of change of true positives (hits) with a change in threshold. This translates to the ROC curve having a gradient of one, since the rate of change of β equals the rate of change of α.

The argument can be generalized to cases where the known prior likelihood of a stimulus differs from 1/2. For example, if the fraction of trials with a stimulus were 1/4 and we now want to choose an optimal threshold, we will be minimizing $P(\text{Error}) = \frac{1}{4} P(\text{Miss}) + \frac{3}{4} P(\text{False} + \text{ve}) = \frac{1}{4}(1-\alpha) + \frac{3}{4}\beta$. We can use the same argument as before if we plot the stimulus-absent probability distribution at three times the size of the stimulus-present probability distribution (since the number of stimulus-absent trials is three times the number of stimulus-present trials) and evaluate the new intersection. Given this scaling, the new intersection occurs at a threshold where the rate of change of true positives is three times the rate of change of false positives. On the ROC curve, this corresponds to a point where the gradient is three.

In general, a priori the probability of a stimulus being absent can be some number, which we will call k, times more likely than the probability of a stimulus being present. We then scale the stimulus-absent probability distribution up by k-fold and look at its intersection with the stimulus-present probability distribution. The threshold for stimulus detection is set at the intersection, which is where the stimulus-present probability is k times the unscaled stimulus-absent probability (for a proof see appendix B). This means the rate of change of true positives is k times the rate of change of false positives, so the optimal threshold appears on the ROC curve where the gradient is k.

Finally, in real-life situations the costs of a false positive and a miss are rarely equal. For a housefly, the cost of a false positive when responding to the possibility of a looming object such as a hand descending upon it is the waste of energy entailed in flying away. However, the cost of a false negative, or a miss, is likely death as the hand squashes it. In experimental trials the relative costs can be controlled by adjusting the amount of reward (e.g., the number of drops of fruit juice) an animal receives for the various outcomes of the trial. In such cases, one wants to weight the probability of each type of error by the cost, or the loss, associated with that error.

The loss associated with an error is the difference between the value to the animal of a correct response and the value to the animal of an incorrect response for that stimulus. To select an action optimally, a common unit, such as value, is needed to compare and weigh the potential outcomes against each other. Given such rational accounting, a missed reward counts as a loss just as does a punishment that could have been avoided. For example, in the unfortunate case of a harmful stimulus that cannot be avoided even if correctly identified, then there is in fact no loss associated with a failure to respond to it—the outcome is equally detrimental whether the test for the stimulus is a hit or a miss.

If the loss functions associated with the different stimulus-response possibilities are known, a rational goal for any animal is to minimize its overall loss. In the case of stimulus detection, we minimize expected loss, L, given by

$$L = L(\text{Miss}) P(\text{Stim}) P(\text{Miss}) + L(\text{False} + \text{ve})[1 - P(\text{Stim})] P(\text{False} + \text{ve}) \tag{3.5}$$

where $L(\text{Miss})$ and $L(\text{False} + \text{ve})$ are the losses associated with a miss and a false positive respectively.

Loss ratio   The cost of missing a stimulus divided by the cost of incorrectly identifying an absent stimulus as present (a false positive).

We see that minimization of the loss requires almost the same calculation as minimization of the probability of error, except with the prior probability of a stimulus, $P(\text{Stim})$, replaced by the product of that probability with its associated loss. Similarly, the prior probability of the absence of a stimulus, $[1 - P(\text{Stim})]$ should be weighted by its associated loss. Therefore, following the previous arguments for optimal positioning of the threshold (and the associated calculations of appendix B) the threshold for stimulus detection should be set so that the ratio of the probabilities for response with stimulus-present to response with stimulus-absent—i.e., the gradient of the ROC curve (figure 3.8C)—is given by multiplying the loss ratio by the ratio of prior stimulus probability

$$\frac{d\alpha}{d\beta} = \frac{L(\text{False} + \text{ve})[1 - P(\text{Stim})]}{L(\text{Miss}) P(\text{Stim})}. \tag{3.6}$$

In summary, a change in the loss ratio should have the same impact on behavior as a change in the prior stimulus probability. For example, making the reward for a true positive three times as great as the reward for a correct rejection (true negative) should have an equivalent effect to presenting the stimulus on three times the number of trials as omitting it. In experiments either technique can be used to alter the threshold used in binary response tasks.

3.5.3 Uncovering the Underlying Distributions from Binary Responses: Recollection versus Familiarity

Figure 3.10 indicates how ROC analysis can be used to gain information about the underlying probability distributions used for making different responses. It can be particularly useful to know if behavior arises from a bimodal probability distribution, because bimodality is an indication of two distinct underlying processes. For example, the underlying pitch of adult speakers has a bimodal distribution arising from differences between male and female speakers—and that the pitch is bimodal allows us to tell quite reliably the gender of someone speaking.

Here we will consider the usefulness of ROC analysis for revealing whether a particular cognitive function is based on one or two distinct underlying processes.[22] If some trials initiate one process while other trials initiate a different process, then the overall probability distribution of responses would be a combination of the two distinct underlying probability distributions. Such a combined distribution may be bimodal (though is not necessarily so).

Recollection   Ability to recall the occurrence of a stimulus, usually by connecting it in the context of surrounding stimuli (an episodic memory).

Recognition   Synonymous with familiarity, knowing that a stimulus has been encountered previously without recall of the context.

An ongoing debate in psychology, with considerable evidence for two distinct processes based on ROC analysis, is whether recollection involves a process that is distinct from recognition memory or familiarity.[23-25] Recollection occurs when we recall some of the details of an event, so that we can be quite certain that the stimulus was observed (although false memories are common). Recollection is suggestive of a binary, all-or-none, process—we recall doing something or we do not. The associated neural activity is likely to be bimodal, with a lack of recollection—meaning no memory of an event—being equivalent to the absence of the event we are asked to recollect (figure 3.10A1).

Recognition occurs when we think we have seen the stimulus before but do not recall the event of seeing the stimulus—we are not certain the way we are when we manage to recall an event. A recognized event would suggest neural activity that is somewhat distinct but not completely distinct from that which would arise in the absence of the event.

If the processes of recollection and recognition are the same, one might expect a broad distribution of responses when trying to remember a real event (as in figure 3.10A3), in which case the demarcation between recognition and recollection just appears when activity is sufficiently

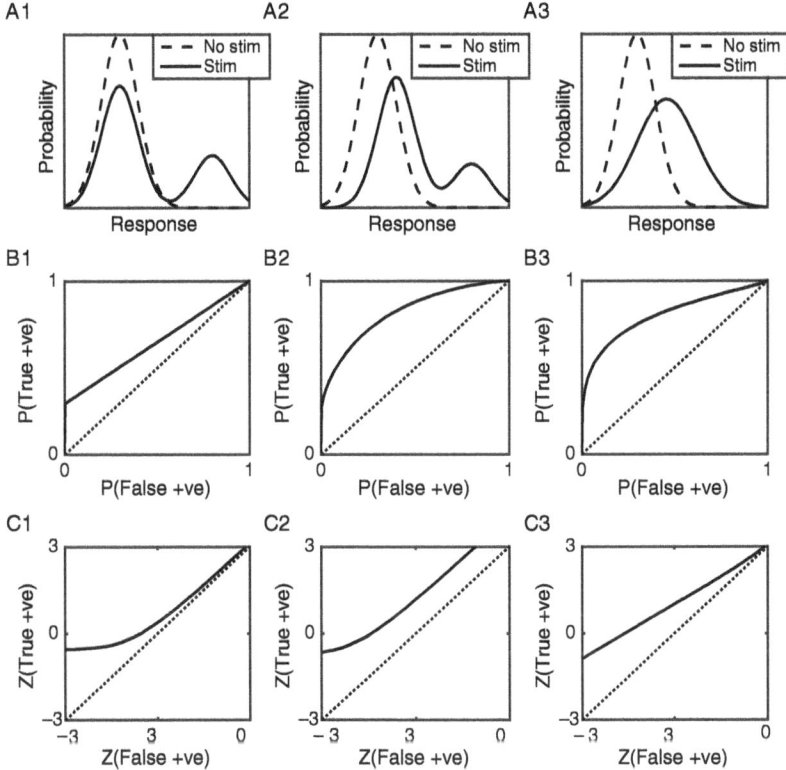

Figure 3.10

ROC analysis to reveal the stimulus-response distribution. (A) Three alternative stimulus-response distributions are depicted. (A1) The response to the stimulus is bimodal, with a strong response on a fraction of trials (recall), but otherwise no distinction from the absence of a stimulus (guessing). (A2) The response is also bimodal, but the weak response (recognition) differs from the no-stimulus condition. (A3) The stimulus response arises from a single Gaussian distribution with increased mean and variance compared to the no-stimulus condition. (B) ROC curves produced from the distributions in 3.10A (solid) with the chance diagonal line (dotted). (B1) The fraction of trials with high response produces a nonzero y-intercept. Other trials arise from the same distribution as the no-stimulus condition, so the curve is linear with gradient equal to the ratio of the sizes of the lower distributions. (B2) The fraction of high-response trials still produces a positive y-intercept, but now the low-response trials are sampled from a distribution with higher mean than the no-stimulus condition, so the curve is concave. (B3) A single broad stimulus-distribution produces a similarly shaped ROC curve to 3.10B2, making it nontrivial to distinguish distributions 3.10A2 and 3.10A3 from ROC curves alone. (C) ROC curves plotted on z-score coordinates, where z-score is number of standard deviations above or below the mean assuming a single Gaussian distribution. (C1, C2) Sampling from two Gaussians for response to the stimulus leads to a curved plot. (C3) When both stimulus-present and stimulus-absent conditions each produce a single Gaussian distribution in 3.10A, then a plot of the z-scores against each other is linear. The gradient is the ratio of the standard deviations. Code to produce these figures is available online as ROC_distributions.m.

distinct from the activity in the absence of the event. However, if recollection relies on a different process from mere recognition, one might expect to see a bimodal distribution (one with two peaks, as in figure 3.10A2) in some of the neural responses.

Given the difficulty of acquiring the underlying neural responses in humans, attempts are made to infer the underlying distributions from aspects of the behavior, such as response speed. Alternatively, an experimenter can use confidence in the reliability of a memory as a readout of neural response and use confidence as a substitute for response threshold to produce an ROC curve (all choices with a given confidence or greater are considered "above threshold"). The confidence in a memory can be requested, or stimuli can be grouped by difficulty, or attempts to measure the confidence can be made by altering the relative losses associated with false positives or misses.[26]

Different distributions of confidence produce different ROC curves. If a subject were only either 100 percent certain (and then correct) or simply guessing, the resulting ROC curve would be linear with a positive y-intercept (figure 3.10B1). In this case, the intercept would indicate the fraction of trials on which the subject is certain. If a subject's confidence were based on two distinct processes that produced either recall or recognition, then the corresponding neural response distributions would be distinct from each other and distinct from the stimulus-absent response distribution (figure 3.10A2). The resulting ROC curve would be nonlinear (figure 3.10B2). However, such a curve can appear very similar to the curve produced when the stimulus produces just a single, broad response distribution arising from a single memory process (figure 3.10A3).

Z-score   For a data point, the number of standard deviations away from the mean of the distribution of the data.

The probabilities of a false positive and a true positive can be converted to z-scores,[27] where the z-score is the number of standard deviations away from the mean assuming a Gaussian distribution. Thus, a cumulative probability of one half corresponds to a z-score of zero, since half of the area lies above the mean of a Gaussian distribution. The details of such a conversion are provided in appendix B with the results reproduced in figure 3.10C. In this case, the significant nonlinearity produced by bimodal probability distributions (figures 3.10C1, C2) disappears when the stimulus distribution is a broad Gaussian (figure 3.10C3), allowing for a distinction between a single process (one broad peak) and multiple underlying processes (producing two peaks).

In practice, great care is needed in producing such curves, since the significant deviations from linearity in figure 3.10C2 lie below z-scores of -2 for the false positive responses. Only approximately 2.5 percent of responses lie in the range of 2 or more standard deviations above the mean, so a lot of data must be collected before one can be confident in the position of the curve in this region.

More generally, whenever such curves are produced following multistep data analysis, it is recommended that results be compared with those arising from identical analysis of dummy data. A scientist can sample from the inferred response distributions to produce a dummy dataset with the same number of data points as the original data. The process can be repeated with a large number of dummy datasets, so a range of results can be obtained and significance of the original data can be assessed accordingly.

3.6   Tutorial 3.3: Receiver-Operating Characteristic of a Noisy Neuron

Neuroscience goal: Practice in how a neuron's distribution of spike counts and the ROC curve depends on noise level, stimulus duration, and stimulus-responsiveness.

Computational goal: Manipulation and comparison of distributions of data.

In this tutorial, we assume a neuron receives one type of input current when a stimulus is present and a second type of input current in the absence of a stimulus. The two types of input current can differ in their mean value and/or their standard deviation. We will assume the number of spikes over a certain time interval can be detected and used to indicate the presence or absence of the stimulus. We will assess how well these two cases can be distinguished by setting different values of the threshold for stimulus detection.

For all questions, you will simulate an AELIF neuron, following the dynamics described in tutorial 2.3 Q.2, using the parameters: $E_L = -70$ mV, $V_{th} = -50$ mV, $V_{reset} = -80$ mV, $\Delta_{th} = 2$ mV, $G_L = 10$ nS, $C_m = 100$ pF, $a = 2$ nS, $b = 0$ nA and $\tau_{SRA} = 150$ ms.

1. Set the mean input current in trials with no stimulus to be 0 and in trials with a stimulus present to be 0.1 nA. For each time step, add a random number taken from the normal distribution of zero mean and standard deviation of $\sigma / \sqrt{\Delta t}$ (see section 1.6.6) where Δt is the timestep of 0.01 ms. Set $\sigma = 20 \text{pA.s}^{0.5}$ in both conditions (stimulus-absent and stimulus-present).

    a. Simulate each condition for 1000 trials of duration 0.5 s.

    b. Count and store the number of spikes in each trial of each condition.

    c. Calculate the mean firing rate of each condition.

    d. Use the `histcounts` function (or your own function) to count the number of trials with a given number of spikes in each condition separately. To produce identically sized vectors for the two conditions, use a range from 0 to the maximum number of spikes found in a trial in either condition for the possible values of number of spikes per trial.

    e. Plot the histogram of the number of spikes per trial for each condition (use a staircase plot with the function `stairs`) (as in figure 3.8A).

    f. Calculate the fraction of trials with greater than a given spike count by using the cumulative sum (`cumsum`) of the histograms, normalizing by (i.e., dividing by) the total number of trials and subtracting from one—or instead of subtracting from one you can

use the command `cumsum(histogram_output,'reverse')`. Plot this monotonically decreasing fraction for each condition (as in figure 3.8B).

g. For each value of spike count as a threshold, plot the fraction of trials with greater
    than that spike count in the stimulus-on condition as "True positives," α, on the y-axis
    against the fraction of trials with greater than that spike count in the stimulus-off con-
    dition as "False positives," β, on the x-axis (as in figure 3.8C). This is the receiver-
    operating characteristic curve.

2. Repeat Q1, but with a stimulus duration of 0.2 s instead of 0.5 s. For each of the three figures,
    explain any differences you observe from the results of Q1.

3. Repeat Q1 with a stimulus duration of 0.5 s, but with the following alterations:

    Mean applied current is 0.5 nA in the presence of a stimulus (and still 0 nA in the absence of
    a stimulus). The noise is reduced in the presence of a stimulus ($\sigma = 5\text{pA.s}^{0.5}$) and increased
    in the absence of a stimulus ($\sigma = 50\text{pA.s}^{0.5}$). Comment on and explain any difference in the
    results due to these altered stimulus and noise properties. Just by counting spikes in a trial,
    could you ever be nearly certain that a stimulus is present? Could you ever be nearly certain
    that a stimulus is absent?

Questions for Chapter 3

1a. You see a dip in a spike-triggered average about 50 ms before the spike time, a peak close to
    the spike time, and zero after the spike time. Explain what this can mean.

1b. You see a peak in a spike-triggered average near the spike time that lasts for 20 ms after the
    spike time. Explain what this can mean.

2. Sketch the spatiotemporal receptive field of a neuron that responds most to a drifting grating
    with a spatial period of 5 degrees, moving from left (negative-x) to right (positive-x) at a
    speed of 20 degrees per second.

3. Rank the following spike processes from smallest to largest CV of ISIs, then from smallest
    to largest CV_2 of ISIs:

    a. A neuron with regular spikes every 50 ms.

    b. A Poisson process with mean rate of 20 Hz.

    c. A regularly bursting neuron with 5 spikes separated by 5 ms within a burst, then a pause
        of 500 ms before the next burst.

    d. A regularly spiking neuron whose rate gradually increases from 0.01 Hz to 100 Hz over
        an hour.

4. Sketch the ROC curve of a neuron, which responds to a positive stimulus with any number
    from 11 to 20 spikes, with equal likelihood, and otherwise produces 1–15 spikes with equal
    likelihood. What is the area under the curve?

3.7. Appendix A: The Poisson Process

The homogeneous Poisson process is a point process that describes a series of events, which occur at discrete instants of time, with the probability of an event at any one point in time being identical to the probability at any other point in time. The process is a Markov process, which means that the probability is not dependent on history—the series of outcomes of fair coin flips is a Markov process, because whether we have just flipped a series of heads has no impact on whether the next flip is a head or a tail.

3.7.1   The Poisson Distribution

In this section, we will derive the Poisson distribution (equation 3.2), which is the probability of a given number of events in a certain time interval produced by a Poisson process. We have already seen that the formula for the Poisson distribution can be used to show that the distribution of times between events (the interevent interval distribution, or for neurons the interspike interval, ISI, distribution) is an exponential decay.

The definition of a Poisson process is one in which the probability of an event in a very small interval, δt, is a constant, $r\delta t$. We consider a much longer interval, T, comprising N_T small intervals ($T = N_T \delta t$) and ask the probability of a given number, N, of events. We assume that the small interval, δt, is small enough that $r\delta t \ll 1$, and therefore—since most time-bins are empty—that $N \ll N_T$.

We will calculate the probability of N events in an interval based on the probability of $N-1$ events in the interval then use the fact that the probabilities must sum to one when all possible values of N are combined. For example, we define the probability of no events in the long interval, T, as $P(0)$ (which we need to calculate). The probability of a single event in a particular interval and no events in other intervals is related to $P(0)$, but has an additional factor, because one time bin that had no spike now contains a spike—a factor of $(1 - r \cdot \delta t)$ that contributed to $P(0)$ is replaced by $r \cdot \delta t$ when contributing to $P(1)$.

Therefore, the probability of an event in one particular small time bin and no other time bin is $P(0) \cdot r\delta t / (1 - r\delta t)$.

There are N_T such small intervals and an event in any one of them could give rise to a single spike in the total interval. Summing these N_T different ways of getting one spike yields $P(1) = N_T r\delta t P(0) / (1 - r\delta t) = rTP(0) / (1 - r\delta t)$.

We can assume the factor $(1 - r\delta t) \approx 1$ (because $r\delta t \ll 1$) and see that $P(1) = rTP(0)$.

We can make a similar argument in general for $P(N)$ as any arrangement of N spikes in an interval is given by an arrangement of $N-1$ spikes that contributes to $P(N-1)$, but with one time bin that did not contain a spike now containing a spike. As with comparing $P(1)$ with $P(0)$, this difference, when accounting for any time-bin the extra spike can be in, leads to a factor of $(N_T - N)r\delta t = rT$ (where we have used $N \ll N_T$, so the result is true as $\delta t \mapsto 0$). Notice that with point processes, each event has a duration of 0, so however many events there are in a time interval, the total time in between events that contains no events is equal to the time interval itself.

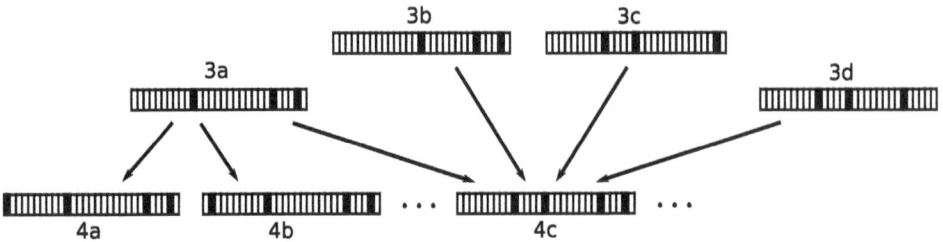

Figure 3.11
Relating $P(N)$ to $P(N-1)$. Any combination of events that contributes to the probability of three events in an interval—such as the combination labeled 3a—is one event away from each of N_T combinations of events that contribute to the probability of four events in an interval, like those labeled 4a, 4b, and 4c. However, each of those four-event combinations can be arrived at four different ways—for example, the combination labeled 4c is reached by adding one event to any of the combinations labeled 3a, 3b, 3c, or 3d. Therefore, the actual number of four-event combinations is $N_T / 4$ times the number of three-event combinations. Since addition of an extra event occurs with probability $r\delta t / (1 - r\delta t)$, we find—by multiplying the relative probabilities of a single combination by the relative numbers of combinations—that the probability of four events, $P(4)$, is related to the probability of three events, $P(3)$, by the formula $P(4) = P(3) \cdot r\delta t N_T / [4(1 - r\delta t)]$. When considering a general whole number of events, N, this formula becomes $P(N) = P(N-1) \cdot r\delta t N_T / [N(1 - r\delta t)]$ via similar logic.

The story does not end there, for a little care is needed (see figure 3.11). Any one of the N spikes in an arrangement that contributes to $P(N)$ could be omitted to produce an arrangement that contributes to $P(N-1)$. That is, if we took all arrangements of spikes that contribute to $P(N-1)$ with a spike added to any time bin, we would find each arrangement that contributes to $P(N)$ appearing N times. To combat such overcounting, we must divide by N when evaluating $P(N)$ from $P(N-1)$.

Therefore, the general formula is

$$P(N) = P(N-1)\frac{rT}{N} \quad \text{(for } N > 0\text{)}. \tag{3.7}$$

Using equation 3.7 to iterate up from $P(0)$, we see that

$$P(1) = P(0)rT \quad P(2) = P(1)\frac{rT}{2} = P(0)\frac{(rT)^2}{2!} \quad P(3) = P(2)\frac{rT}{3} = P(0)\frac{(rT)^2}{3!}$$

and so on. This can be written more generally as

$$P(N) = P(0)\frac{(rT)^N}{N!}. \tag{3.8}$$

All that remains, is to find $P(0)$ from the normalization

$$\sum_{N=0}^{\infty} P(N) = 1, \tag{3.9}$$

which formalizes the certainty that in any interval there will be a particular number of events that is a nonnegative integer.

Finally, recognizing that the exponential function is defined by

$$e^x = \sum_{N=0}^{\infty} \frac{x^N}{N!}$$

we can evaluate $P(0)$ by combining equations 3.8 and 3.9 to yield

$$\sum_{N=0}^{\infty} P(N) = P(0) \sum_{N=0}^{\infty} \frac{(rT)^N}{N!} = P(0)e^{rT} = 1,$$

Hence $P(0) = e^{-rT}$, and in general the Poisson distribution is given by

$$P(N) = e^{-rT} \frac{(rT)^N}{N!}. \tag{3.10}$$

3.7.2 Expected Value of the Mean of a Poisson Process

The expected value of any distribution is given by multiplying the value of each possible outcome by the probability of that outcome, then summing these values together. This leads to the expected value, $\langle N \rangle$, for the number of events in an interval T of a Poisson process with rate parameter r as:

$$\langle N \rangle = \sum_{N=0}^{\infty} N \cdot P(N) = e^{-rT} \sum_{N=0}^{\infty} N \cdot \frac{(rT)^N}{N!}$$

$$= rTe^{-rT} \sum_{N=1}^{\infty} \frac{(rT)^{N-1}}{(N-1)!} = rTe^{-rT} \sum_{N'=0}^{\infty} \frac{(rT)^{N'}}{(N')!} \tag{3.11}$$

$$\left(\text{now use the formula } \sum_{n=0}^{\infty} \frac{x^n}{n!} = e^x \right) = rTe^{-rT}e^{rT} = rT.$$

This result confirms that r is the mean rate of events for a Poisson process.

3.7.3 Fano Factor of the Poisson Process

The Fano factor of a process is the variance divided by the mean of the number events. To evaluate the variance, we need to calculate the expected value of the square of the number of events, $\langle N^2 \rangle$. Using a similar procedure to the one followed when calculating the mean, and rewriting N^2 as $N(N-1)+N$ to simplify the calculation, we find:

$$\langle N^2 \rangle = \sum_{N=0}^{\infty} N^2 \cdot P(N) = \sum_{N=0}^{\infty} [N(N-1) + N] \cdot P(N)$$

$$= \sum_{N=0}^{\infty} N(N-1) \cdot P(N) + N = e^{-rT} \sum_{N=0}^{\infty} N(N-1) \cdot \frac{(rT)^N}{N!} + N \qquad (3.12)$$

$$= e^{-rT} \sum_{N=2}^{\infty} \frac{(rT)^N}{(N-2)!} + N = (rT)^2 e^{-rT} \sum_{N'=0}^{\infty} \frac{(rT)^{N'}}{N'!} + NN^2 = (rT)^2 + rT.$$

Therefore, the variance in N, obtained by combining equations 3.11 and 3.12, is:

$$\text{Var}(N) = \langle N^2 \rangle - \langle N \rangle^2 = (rT)^2 + rT - (rT)^2 = rT \qquad (3.13)$$

and the Fano factor, $F(N)$ is given as:

$$F(N) = \frac{\text{Var}(N)}{\langle N \rangle} = 1. \qquad (3.14)$$

This proves the result that a Poisson process has a Fano factor of 1.

3.7.4   The Coefficient of Variation (CV) of the ISI Distribution of a Poisson Process

The CV of a distribution is its standard deviation divided by its mean. The CV of an exponential distribution—such as the distribution of interspike intervals (ISIs) when spike emission follows a Poisson process—can be shown to be 1.

For such a Poisson process, equation 3.3 shows that the probability of an ISI being between T and $T + \delta T$ is given by

$$P(T)\delta T = r\delta T e^{-rT} \qquad (3.15)$$

The mean ISI is calculated using integration by parts:

$$\langle T \rangle = \int_0^{\infty} TP(T)dT = \int_0^{\infty} Tre^{-rT}dT = [-Te^{-rT}]_0^{\infty} - \int_0^{\infty} -e^{-rT}dT = 0 + \left[-\frac{e^{-rT}}{r} \right]_0^{\infty} = \frac{1}{r}. \qquad (3.16)$$

Similarly, the mean of the ISI-squared is given by:

$$\langle T^2 \rangle = \int_0^{\infty} T^2 P(T)dT = \int_0^{\infty} T^2 re^{-rT}dT = [-T^2 e^{-rT}]_0^{\infty} - \int_0^{\infty} -2Te^{-rT}dT = 0 + \frac{2}{r}T = \frac{2}{r^2}. \qquad (3.17)$$

Hence, combining equations 3.16 and 3.17, the standard deviation of the ISI distribution is:

$$\sigma_T = \sqrt{\langle T^2 \rangle - \langle T \rangle^2} = \sqrt{\frac{2}{r^2} - \left(\frac{1}{r} \right)^2} = \frac{1}{r}. \qquad (3.18)$$

Therefore, the standard deviation is equal to the mean of an exponential probability distribution (from equations 3.16 and 3.18) and its coefficient of variation, the ratio of standard deviation to the mean, is equal to one.

3.7.5 Selecting from a Probability Distribution: Generating ISIs for the Poisson Process

If we know the probability distribution of ISIs, we can generate an artificial spike train with that distribution. If one ISI has no impact on successive ISIs, then the sequence of ISIs is a Markov process, meaning each ISI is independent of history. In such cases the artificial spike train can be statistically equivalent to the original spike train, and so be useful as a control in any statistical analyses.

The procedure for selecting numbers according to a given distribution is sketched in figure 3.12. The goal is to select a number with probability proportional to the height of the ISI distribution. If we correctly normalize the ISI distribution to be a probability distribution with area 1 (that is, the integral under the curve is 1), then selecting an ISI is equivalent to randomly selecting a point uniformly within this area and reading off the x-coordinate. For very complicated functions, this can be achieved by selecting random pairs of (x,y) coordinates, testing whether the pair falls under the curve, and recording the first value that is under the curve and therefore within the distribution. Clearly, any x-coordinate at which the probability distribution is greater will have more likelihood of being chosen, as a randomly chosen y-coordinate is more likely to fall under the curve.

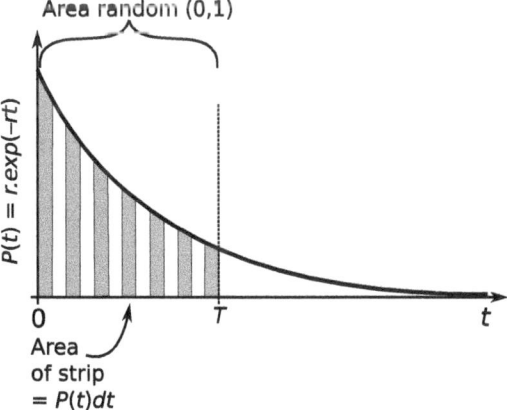

Figure 3.12
Random selection from a known distribution. The area to the left of a randomly chosen interval, T, is the sum of the areas of strips of height $P(t)$ and width dt. If a point is chosen at random with equal probability from within the total area of the distribution, its likelihood to fall on a given strip is equal to the area of that strip, which is $P(t)dt$, so proportional to the corresponding probability. Such selection can be made by choosing a random number from a uniform distribution between zero and 1 and requiring either the area to the left, or the area to the right of the chosen interval, T, is equal to that random number.

For distributions that can be integrated, the process can be made computationally more efficient. For example, when the ISI distribution is exponential, as is the case for the Poisson process, then we can use the result that the integral of an exponential is also an exponential, $\int_0^T re^{-rt}dt = 1 - e^{-rT}$ and $\int_T^\infty re^{-rt}dt = e^{-rT}$. Selecting a point at random within the area of the distribution can then be achieved by randomly choosing the area that it to the left of that point or to the right of that point. For the exponential distribution of ISIs that allows us to choose an ISI of length, T, by requiring that e^{-rT} is a uniformly distributed random number between 0 and 1. A little rearrangement leads to:

$$T = -\frac{1}{r}\ln[random(0,1)] \tag{3.19}$$

where $random(0,1)$ means select a random number from a uniform distribution between 0 and 1, i.e., use rand() in MATLAB.

3.8.   Appendix B: Stimulus Discriminability

We consider a test for the presence of a stimulus based on whether a neuron's response is above a given threshold. The goal is to set the level of the threshold so as to minimize the probability of an error. An error can be a miss (a false negative) meaning the stimulus was present but went undetected ($1 - \alpha$ in figure 3.9), or can be a false positive meaning the response was above threshold, indicating the presence of a stimulus when in fact the stimulus was absent (β in figure 3.9).

We assume that the test produces a mean response of μ_A when the stimulus is absent and a mean response of μ_B (where $\mu_B > \mu_A$) when the stimulus is present. In each case responses are distributed about those means as Gaussians with standard deviations σ_A and σ_B respectively. If we define the threshold for stimulus detection as θ then the probability of a false positive, which is the probability the response is above-threshold when the stimulus is absent, becomes:

$$P(r > \theta \mid A) = \frac{1}{\sqrt{2\pi\sigma_A^2}}\int_\theta^\infty e^{-\frac{(x-\mu_A)^2}{2\sigma_A^2}}\, dx. \tag{3.20}$$

Similarly, the probability of a false negative, i.e., a miss, which is the probability the response is below-threshold when the stimulus is present, becomes:

$$P(r < \theta \mid B) = \frac{1}{\sqrt{2\pi\sigma_B^2}}\int_{-\infty}^\theta e^{-\frac{(x-\mu_B)^2}{2\sigma_B^2}}\, dx. \tag{3.21}$$

3.8.1   Optimal Value of Threshold

The overall probability of an error is then:

$$P(\text{Error}) = P(r > \theta \mid A)P(A) + P(r < \theta \mid B)P(B), \tag{3.22}$$

where $P(A)$ and $P(B)$ are respectively the a priori probabilities of the stimulus being absent or present (a priori meaning the probability prior to any information regarding the neuron's response).

The first term of equation 3.22 decreases as we raise the threshold while the second term increases when we raise the threshold. As discussed in section 3.5.2, the optimal threshold is where the rate of change from the two terms cancels out. In calculus, this arises because the minimum of any function (such as $P(\text{Error})$ as a function of a variable threshold) corresponds to a point where its gradient is zero. That is, if look at the derivative, $dP(\text{Error})/d\theta$, and find a value of θ where the derivative is zero it may be the optimal value. (It could also be the value with highest error, so one should compare the error at each value of zero derivative and select the value with lowest error so long as the error is below 1/2, which is the error obtained in the limit of very high or very low threshold.)

If we evaluate the derivative of equation 3.22, we obtain

$$\frac{dP(\text{Error})}{d\theta} = P(A)\frac{1}{\sqrt{2\pi\sigma_A{}^2}}\frac{d}{d\theta}\int_\theta^\infty e^{-\frac{(x-\mu_A)^2}{2\sigma_A{}^2}}\,dx + P(B)\frac{1}{\sqrt{2\pi\sigma_B{}^2}}\frac{d}{d\theta}\int_{-\infty}^\theta e^{-\frac{(x-\mu_B)^2}{2\sigma_B{}^2}}\,dx$$

$$= P(A)\frac{-1}{\sqrt{2\pi\sigma_A{}^2}}e^{-\frac{(\theta-\mu_A)^2}{2\sigma_A{}^2}} + P(B)\frac{1}{\sqrt{2\pi\sigma_B{}^2}}e^{-\frac{(\theta-\mu_B)^2}{2\sigma_B{}^2}} \tag{3.23}$$

so $dP(\text{Error})/d\theta = 0$ occurs at a value of θ defined by:

$$P(A)\frac{1}{\sqrt{2\pi\sigma_A{}^2}}e^{-\frac{(\theta-\mu_A)^2}{2\sigma_A{}^2}} = P(B)\frac{1}{\sqrt{2\pi\sigma_B{}^2}}e^{-\frac{(\theta-\mu_B)^2}{2\sigma_B{}^2}}, \tag{3.24}$$

which can be rewritten as

$$\frac{\frac{1}{\sqrt{2\pi\sigma_B{}^2}}e^{-\frac{(\theta-\mu_B)^2}{2\sigma_B{}^2}}}{\frac{1}{\sqrt{2\pi\sigma_A{}^2}}e^{-\frac{(\theta-\mu_A)^2}{2\sigma_A{}^2}}} = \frac{P(A)}{P(B)}. \tag{3.25}$$

In words this means that the ratio of the values of the probability distributions (left-hand side) match the inverse ratio of their prior probabilities (right-hand side).

3.8.2 Calculating the Probability of an Error

The position of the optimal threshold can be written more simply in the special case where $P(A)\sigma_A = P(B)\sigma_B$, as for example would happen if the two distributions have equal variance and a stimulus is present on half of the trials.

In this case, equation 3.25 shows that the optimal threshold is defined by

$$e^{-\frac{(\theta-\mu_B)^2}{2\sigma_B^2}} = e^{-\frac{(\theta-\mu_A)^2}{2\sigma_A^2}} \tag{3.26}$$

so

$$\frac{(\theta-\mu_B)^2}{2\sigma_B^2} = \frac{(\theta-\mu_A)^2}{2\sigma_A^2} \tag{3.27}$$

and because the threshold lies between the two means, $\mu_A < \theta < \mu_B$ this rearranges to give $\sigma_A(\theta-\mu_B) = \sigma_B(\theta-\mu_A)$. Writing the difference between the means as $\Delta = \mu_B - \mu_A$ and rearrangement of equation 3.27 allows us to write the threshold as either $\theta = \mu_A + \sigma_A\Delta/(\sigma_A + \sigma_B)$ or $\theta = \mu_B - (\sigma_B\Delta/(\sigma_A + \sigma_B))$.

Given this value for the threshold, the probability of error becomes, by substituting $x' = x - \mu_A$ in the first term and $x' = \mu_B - x$ in the second term:

$$P(\text{Error}) = P(A)\frac{1}{\sqrt{2\pi\sigma_A^2}} \int_{\frac{\sigma_A\Delta}{\sigma_A+\sigma_B}}^{\infty} e^{-\frac{x'^2}{2\sigma_A^2}} dx' + P(B)\frac{1}{\sqrt{2\pi\sigma_B^2}} \int_{\frac{\sigma_B\Delta}{\sigma_A+\sigma_B}}^{\infty} e^{-\frac{x'^2}{2\sigma_B^2}} dx'$$

$$= P(A)\frac{1}{\sqrt{\pi}} \int_{\frac{\Delta}{\sqrt{2}(\sigma_A+\sigma_B)}}^{\infty} e^{-x^2} dx + P(B)\frac{1}{\sqrt{\pi}} \int_{\frac{\Delta}{\sqrt{2}(\sigma_A+\sigma_B)}}^{\infty} e^{-x^2} dx \tag{3.28}$$

$$= \frac{1}{2}\text{erfc}\left[\frac{\Delta}{\sqrt{2}(\sigma_A+\sigma_B)}\right]$$

where we have used $P(A)+P(B)=1$ in the final line and used the complementary error function, $\text{erfc}(z)$, which is defined in terms of the Gaussian integral as $\text{erfc}(z) = \frac{2}{\sqrt{\pi}}\int_z^{\infty} e^{-x^2} dx$.

In the special case where $P(A) = P(B)$ and $\sigma_A = \sigma_B$ then $P(\text{Error}) = \frac{1}{2}\text{erfc}\left[\frac{\Delta}{2\sqrt{2}\sigma_A}\right] = \frac{1}{2}\text{erfc}\left[\frac{d'}{2}\right]$ as used in equation 3.4.

3.8.3 Generating a Z-Score from a Probability

ROC curves are usually plotted as the probability of a true positive (a hit) against the probability of a false positive. They can also be plotted as the z-score of true positives against the z-score of false-positives. In the latter case, a Gaussian distribution of responses is assumed to convert the probability that a response is greater than a given value to the number of standard deviations above or below the distribution's mean corresponding to that value. The complementary error function is needed because the integral from a value to infinity of a Gaussian probability distribution is simply the fraction of the area higher than that value, which is the probability of a response greater than that value.

For a Gaussian distribution of standard deviation, σ, the area below a value, r_0, is given from the definition of a Gaussian by

$$P(r < r_0) = \frac{1}{\sqrt{2\pi\sigma^2}} \int_{-\infty}^{r_0} e^{-\frac{x^2}{2\sigma^2}} dx \qquad (3.29)$$

which becomes

$$P(r < r_0) = \frac{1}{\sqrt{\pi}} \int_{-\infty}^{\frac{r_0}{\sqrt{2}\sigma}} e^{-x^2} dx = 1 - \frac{1}{2}\mathrm{erfc}\left[\frac{r_0}{\sqrt{2}\sigma}\right]. \qquad (3.30)$$

So, to calculate the probability, p, that the response, r, is greater than a number of standard deviations, z, from the mean, we write $z = (r_0/\sigma)$,

$$p = P(r < \sigma z) = 1 - \frac{1}{2}\mathrm{erfc}\left[\frac{z}{\sqrt{2}}\right]. \qquad (3.31)$$

From equation 3.31 the inverse transformation is needed to calculate z from the probability, p:

$$z = \sqrt{2}\,\mathrm{inverfc}[2(1-p)]. \qquad (3.32)$$

where inverfc is the inverse of the complementary error function.

4

Conductance-Based Models

Each neuron is a miniature analog computer—analog, because its behavior depends on the membrane potential, which is a continuous variable. The information processed by a neuron is via charge flow across its membrane, which causes changes in its membrane potential, our variable of interest. The charge flow depends on the conductance of ion channels, which in many cases varies with the membrane potential. The ensuing feedback loops, in which the membrane potential impacts channel conductance, which in turn impacts membrane potential, give rise to a wealth of exciting and useful dynamical behaviors, such as voltage spikes, oscillations, and bistability. This chapter focuses on such voltage-dependent single-cell phenomena.

4.1 Introduction to the Hodgkin-Huxley Model

In 1952, Alan Hodgkin and Andrew Huxley published a series of five papers[1-5] (the first with Bernard Katz) culminating in their model for the generation and propagation of the action potential by active sodium and potassium channels. Their work combined computational neuroscience with meticulously analyzed electrophysiological data acquired from the axon of the giant squid. Their model, now known as the Hodgkin-Huxley model, will be described in detail in this chapter, because it is the basis for all conductance-based models. In 1963 Hodgkin and Huxley (jointly with John Eccles) received the Nobel Prize for their work, which began the field of computational neuroscience at a time when computers were rarely available—indeed, Huxley spent a few weeks carrying out the necessary calculations on a Brunsviga, a hand-operated mechanical calculating machine (figure 4.1).

Electrophysiological data   Refers to measurements of the membrane potential or transmembrane current of neurons.

Figure 4.1
A mechanical calculating machine, the Brunsviga, of the type used by Huxley. For each mathematical operation,
Huxley would need to set levers in place and turn a crank to produce an answer. The huge amount of effort required
to simulate the model equations gave Hodgkin and Huxley the impetus to use sophisticated mathematical methods to
minimize the calculation time. Photo from www.vintagecalculators.com, used with permission.

4.1.1 Positive versus Negative Feedback

While the impressive calculations of Hodgkin and Huxley led to a prediction of the speed of
propagation of a spike of voltage called the action potential along the giant axon of the squid, we
will just model the dynamics of the voltage spike (not its spatial dependence) in this book. The
key to the dynamics of a single spike (figure 4.2) is very fast (submillisecond) positive feedback,
followed by slow negative feedback. In the mathematical sense, positive or negative says nothing about whether it is good or bad—in fact, in most scientific situations positive feedback is a
problem (think explosion) while negative feedback is essential (stabilization and control).

In terms of the action potential, positive feedback means that as the membrane potential
increases, a biophysical process acts to further increase the membrane potential. In this case, it
is the voltage-dependent activation of sodium channels: Channels open as the membrane potential rises and their opening admits an influx of positive sodium ions to the cell, which raises the
membrane potential further. In principle, sodium activation can raise the membrane potential to
a level as high as the sodium reversal potential, the point at which the sodium current would fall
to zero however large the sodium conductance. However, typically before that point is reached,
negative feedback comes into play.

Channel activation   A membrane potential-dependent process necessary for the opening of a
channel, usually by depolarization.

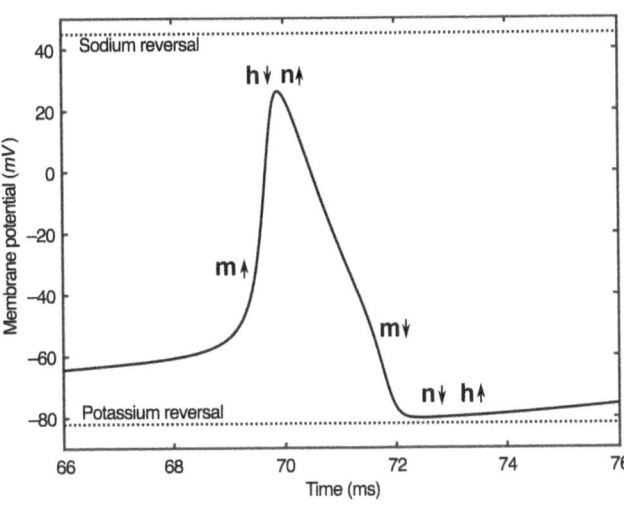

Figure 4.2
Stages of the action potential. (A) Sodium activation, m↑, causes a rapid upswing of the membrane potential, V, by opening sodium channels. (B) Slower sodium inactivation, h↓, closes sodium channels and potassium activation, n↑, opens potassium channels at high-V to bring about the fall in membrane potential. (C) Deactivation of sodium channels, m↓, as V falls below threshold. (D) Slower deinactivation of sodium channels, h↑, and deactivation of potassium channels, n↓, at hyperpolarized V allows V to rise again ready for a new spike. This figure is produced by the online code HH_old_base.m.

Channel deactivation   The opposite of channel activation, usually by hyperpolarization.

Channel inactivation   A process that occurs with depolarization to prevent channels from opening.

Channel deinactivation   The opposite of inactivation, a process that is necessary for a channel to open, with the opposite voltage-dependence of activation.

Negative feedback arises through three mechanisms. The first, a reduction in sodium current as its reversal is approached, is not that important in shaping the action potential—sodium reversal ensures that a spike has a maximum height, but it cannot bring the membrane potential back to baseline. The spike's downswing depends on sodium inactivation and potassium activation.

On the one hand, sodium channels inactivate with depolarization of the membrane potential. Ions cannot flow through an inactivated channel. Inactivation is slower than activation, so that as the membrane potential is shifted from low to high, the channels are activated (allowing current flow) for a short period of time before they inactivate (preventing current flow). This means the sodium current is a transient current—it is not maintained at high levels if the membrane potential is held fixed, even far below sodium's reversal potential. Inactivation of the sodium channels alone would allow the membrane potential to decay back to rest at the leak potential, but slowly over a time on the order of 10 ms that corresponds to the neuron's base time constant ($\tau_m = C_m R_m$).

Transient current   A current through an ion channel that rises and decays when the membrane potential is changed but does not persist.

Delayed rectifier current   The outward potassium current that activates more slowly than the sodium current and acts to terminate a spike by returning the membrane potential back to baseline.

However, real action potentials are much sharper, with a much more rapid return of the membrane potential to its resting level thanks to the opening of potassium channels. Like sodium channels, the potassium channels activate with increasing membrane potential. They are slower to activate though, permitting the initial peak in membrane potential before they start dominating and producing an outward current. The greater the peak potassium conductance, the more rapidly charge flows back out of the neuron and the narrower the voltage spike. These potassium channels are known as delayed rectifier channels. They are called delayed because their slower responsiveness means their full effect is delayed compared to the more rapid sodium activation. The term "rectifier" is added because the channels only open when the membrane potential is above the reversal potential for potassium, allowing for an outward unidirectional or "rectifying" flow of current. Unlike sodium channels, the potassium delayed-rectifier channels do not inactivate, so they produce a persistent conductance—if the voltage is clamped at a depolarized level, the conductance (and hence the outward current) is sustained.

Although these spike-producing processes were uncovered in the axon of the giant squid, the biophysical processes are the same as those producing the action potentials in our own brains. The main difference when looking at spikes in mammalian cortex, for example, is the spike width being submillisecond due to faster channel kinetics.

4.1.2 Voltage Clamp versus Current Clamp

When studying neurons, one often controls the applied current and observes the dynamics of the membrane potential in what is called current-clamp mode. Alternatively, one can control the membrane potential and observe the voltage-dependent dynamics of currents and conductances in voltage-clamp mode. The voltage-clamp approach is useful, or even essential, in electrophysiological experiments that attempt to uncover the behavior of a particular conductance. In particular, it underlay the process by which Hodgkin and Huxley deduced the dynamics of the gating variables in their original model.

In real neurons, some skill is needed to ensure the membrane potential is clamped to the same value throughout the cell in voltage-clamp mode. However, in simulations one need only preassign the value of the membrane potential at each time point and remove the membrane potential update step from the code. When building up a simulation of a neuron, it is worth testing the behavior of each added conductance in voltage-clamp mode, to ensure it acts as it should, before simulating the full dynamics of the membrane potential.

Voltage clamp   Measurements of the current flowing into or out of a neuron when its membrane potential is controlled in an experiment.

Current clamp   Measurements of the changes in the membrane potential of a neuron when its input current is controlled in an experiment.

4.2   Simulation of the Hodgkin-Huxley Model

The Hodgkin-Huxley model is based on four variables: membrane potential, V_m; sodium activation variable, m; sodium inactivation variable, h; and potassium activation variable, n. Therefore, four coupled differential equations are simulated to produce the dynamics of the model. The activation and inactivation variables are together called gating variables, because they characterize whether the corresponding channels are open or shut (i.e., able to pass ionic current). Each gating variable has a voltage-dependent steady-state value that it would reach if the membrane potential were fixed. Each also has a voltage-dependent time constant, which determines the rate at which it approaches that steady state.

Gating variable   A variable between 0 and 1, representing the fraction of channels in a particular state. Multiplication together of all such variables for a type of channel indicates the fraction of the channels that are open and able to transmit current.

For many channel-types these voltage-dependent functions are obtained by direct fitting to empirical data. Hodgkin and Huxley went further by deriving rate constants and fitting functions to those rate constants. The connection between steady-state values and time constants with the underlying rate constants is best understood by consideration of a two-state system.

4.2.1   Two-State Systems

A simple, but very important system in biology is one of many identical components, each of which can be in one of two states. Here we call the states A and B with rate constants k_A for the switch from B to A and k_B for the switch from A to B:

$$
A \underset{k_A}{\overset{k_B}{\rightleftharpoons}} B \tag{4.1}
$$

To be concrete, we can assume A and B to be two conformational states of a particular protein. Then we will use lower-case a and b to represent the fraction of protein molecules in states A and B respectively. Clearly a and b (like the gating variables we are considering in the Hodgkin–Huxley model) can vary in the range from 0 to 1. Also, in a two-state system, all protein molecules are in either state A or state B, so we have the constraint

$$a + b = 1$$

or equivalently

$$b = 1 - a.$$

The rate constants give the rate of change of a and b from first-order reaction kinetics in chemistry as

$$
\frac{da}{dt} = k_A b - k_B a = k_A(1-a) - k_B a \tag{4.2}
$$

$$
\frac{db}{dt} = -k_A b + k_B a = -k_A b + k_B(1-b). \tag{4.3}
$$

We will just consider equation 4.2, because equation 4.3 is essentially identical, and if we know the dynamics of a we can just set $b = 1 - a$ to obtain the dynamics of b.

First, we can find the steady state of a (a_∞) by solving $da/dt = 0$ in equation 4.2, which leads to

$$
k_A(1 - a_\infty) - k_B a_\infty = 0 \tag{4.4}
$$

and after rearranging,

$$a_\infty = \frac{k_A}{k_A + k_B}. \tag{4.5}$$

Equation 4.5 produces the expected behavior, with the steady state of a ranging from near zero when rate of transition away from A is much greater than rate of transition to A (i.e., when $k_B \gg k_A$, $a_\infty \to 0$) or to near one when rate of transition to A is much greater than rate of transition away from A (i.e., when $k_A \gg k_B$, $a_\infty \to 1$).

The time constant, τ_a, for the dynamical process can be evaluated by rewriting the first order linear differential equation in the form

$$\frac{da}{dt} = \frac{a_\infty - a}{\tau_a} \tag{4.6}$$

(see section 1.4.1).

To extract the time constant, one notices that the prefactor of a in equation 4.6 is $-1/\tau_a$ so equating this with the prefactor of a in equation 4.2 yields

$$\frac{-1}{\tau_a} = -k_A - k_B,$$

which leads to

$$\tau_a = \frac{1}{k_A + k_B}. \tag{4.7}$$

Thus, given rate constants, k_A and k_B, one can calculate steady state, a_∞, and time constant, τ_a. Similarly, given the steady state and time constant, the rate constants can be evaluated from equations 4.5 and 4.7 as:

$$k_A = \frac{a_\infty}{\tau_a} \text{ and } k_B = \frac{1 - a_\infty}{\tau_a}. \tag{4.8}$$

In the Hodgkin-Huxley model, the two rate constants for each gating variable are voltage-dependent.

4.2.2 Full Set of Dynamical Equations for the Hodgkin-Huxley Model

The membrane potential of the Hodgkin-Huxley model follows the dynamical equation

$$C_m \frac{dV_m}{dt} = G_L (E_L - V_m) + G_{Na}^{(max)} m^3 h (E_{Na} - V_m) + G_K^{(max)} n^4 (E_K - V_m) + I_{app} \tag{4.9}$$

which is like the dynamical equation for the leaky integrate-and-fire model (equation 2.9), but with the two extra (middle) terms on the right-hand side, one for the sodium conductance and the other for the potassium conductance. Those conductance terms include the gating variables, m for sodium activation, h for sodium inactivation, and n for potassium activation. The exponents for each gating variable were obtained by fitting to the time-dependence of data long before the biophysical interpretation of the gating variables was established. Intriguingly the potassium delayed-rectifier channel contains four subunits: If each subunit opens independently with a probability of n then the expected fraction of channels with all four subunits in the open state is n^4, providing a plausible account for that term in the total potassium conductance. However, in general the channel proteins have many more states, and these equations are best considered as an empirical fit.

Each gating variable follows its own voltage-dependent dynamical equation, in which we use the symbol α as the rate constant for increase of that gating variable and the symbol β as the rate constant for its decrease. The three equations then have identical forms (see the prior section):

$$\frac{dm}{dt} = \alpha_m (1-m) - \beta_m m \tag{4.10}$$

$$\frac{dh}{dt} = \alpha_h (1-h) - \beta_h h \tag{4.11}$$

$$\frac{dn}{dt} = \alpha_n (1-n) - \beta_n n \tag{4.12}$$

but the rate constants have different voltage dependences. In particular, for sodium activation, α_m increases whereas β_m decreases with depolarization; for potassium activation, α_n and β_n have similar voltage-dependences but are smaller, leading to slower dynamics than sodium activation; for sodium inactivation, α_h decreases whereas β_h increases with depolarization, leading to the opposite dependence on the membrane potential for the activation and inactivation gating variables (figures 4.3 and 4.4).

When the voltage-dependence of the rate constants are known, as in table 4.2, then the set of equations 4.9–4.12 can be simulated together (figures 4.5 and 4.6) to show the membrane potential's response to any time-dependent input current, I_{app}. The following section describes the different behaviors produced by such simulations.

4.2.3 Dynamical Behavior of the Hodgkin-Huxley Model: A Type-II Neuron

Type-II neuron   A neuron with a discontinuous jump in its firing rate curve and which can be either inactive or active in a history-dependent manner for some values of input current.

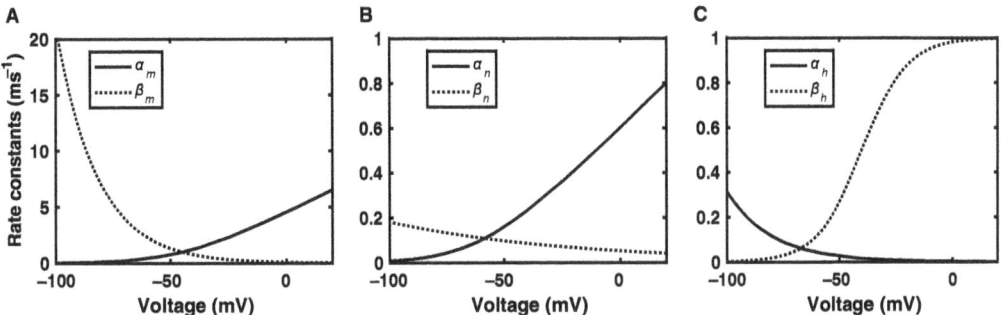

Figure 4.3
Rate constants for the gating variables of the Hodgkin-Huxley model. (A) Rates of sodium activation, α_m, and deactivation, β_m. (B) Rates of potassium activation, α_n, and deactivation, β_n. (C) Rates of sodium deinactivation, α_h, and inactivation, β_h. Note the twentyfold increase in the y-axis scale for the activation variable of sodium, which responds most rapidly. Equations are provided in table 4.2. This figure was produced by the online code HHab_plotnewV.m.

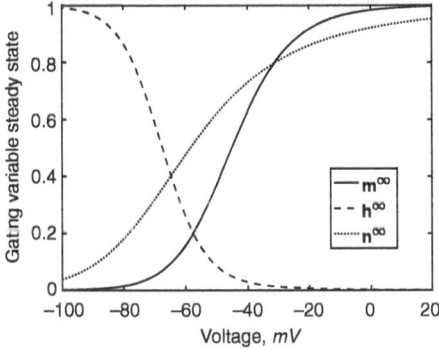

Figure 4.4
Steady state of the gating variables in the Hodgkin-Huxley model. The activation variables approach 1 at positive membrane potentials, whereas the inactivation variable approaches one at very negative membrane potentials. The small region where sodium activation, m, and inactivation, h, are both greater than zero provides a voltage "window" in which a sodium current may be sustained. Hence the sodium current can be called a "window current." This figure was produced by the online code HHab_plotnewV.m.

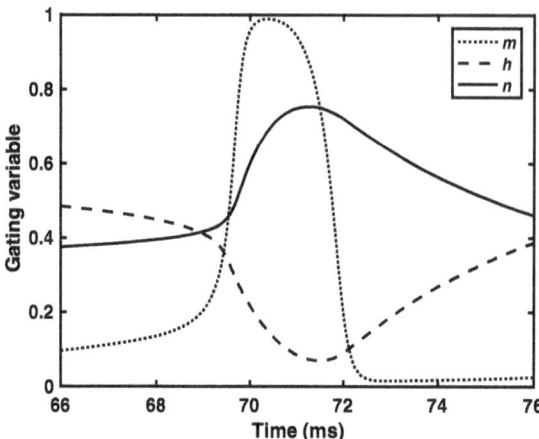

Figure 4.5
Dynamics of gating variables during a single spike. In this example (as in figure 4.2), the peak of the spike is at 70 ms, almost simultaneous with the peak of sodium activation, m (dotted). The sodium inactivation variable, h (dashed) reaches a minimum and potassium activation, n (solid) reaches a maximum over 1 ms later. The neuron is not able to spike again until h and m have increased and n has decreased back to near baseline. This figure is produced by the online code HH_old_base.m.

In this section, we consider those behaviors of the Hodgkin-Huxley model that are specific to type-II neurons. All of these behaviors are symptomatic of subthreshold oscillations: (1) a jump from a single spike, or unsustained firing, to a high firing rate; (2) bistability (two possible firing rates for the same input current); (3) spike generation via hyperpolarization (also called anode break); and (4) resonance in response to oscillating input currents of different frequencies.

Subthreshold oscillations   Oscillations in the membrane potential, of insufficient amplitude to produce a spike.

Bistability   The existence of two stable states of activity, such as spiking and quiescence, for the same value of all inputs and other parameters.

Anode break   A spike in the membrane potential that is produced following release from hyperpolarization.

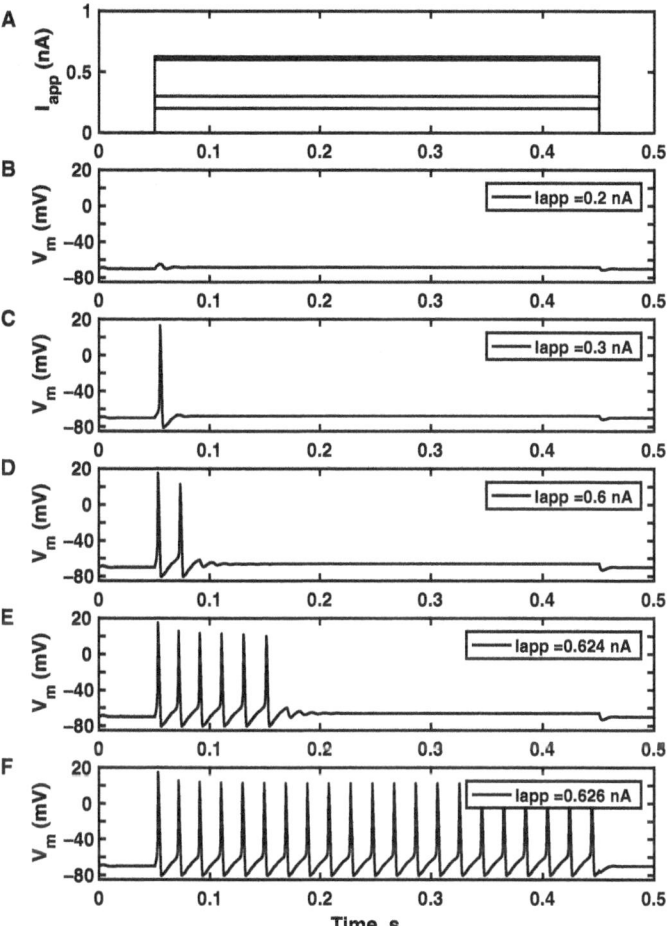

Figure 4.6
Type-II behavior of the Hodgkin-Huxley model. (A) Current steps that are applied in panels (B–E). (B) A 0.2 nA applied current produces a decaying subthreshold oscillation in the membrane potential. (C) A 0.3 nA applied current produces a single spike followed by decaying sub-threshold oscillations. (D) A 0.6 nA applied current produces two spikes, the second of lower amplitude than the first, followed by more slowly decaying subthreshold oscillations. (E) A 0.624 nA applied current produces a succession of six spikes of near constant interspike interval, but not sustained firing. Subthreshold oscillations follow the last spike. (F) An applied current of 0.626 nA is just sufficient to produce sustained spiking, but at a high frequency of over 50 Hz. This figure is produced by the online code HH_manyItrials.m.

Resonance An enhancement in the response of a system when it is periodically stimulated at a particular frequency, called its resonant frequency.

If the Hodgkin-Huxley model is simulated with step currents of different amplitudes (figure 4.6), a level of input is reached at which a single spike is produced, followed by decaying sub-threshold oscillations. At a higher level of input continuous spiking arises, but the frequency of spiking is never lower than that of the sub-threshold oscillations. Therefore, the f-I curve contains a jump up from zero to a significant firing rate at the current threshold (figure 4.7). In fact, the vertical line in the figure, which guides the eye, should strictly be omitted, since it passes

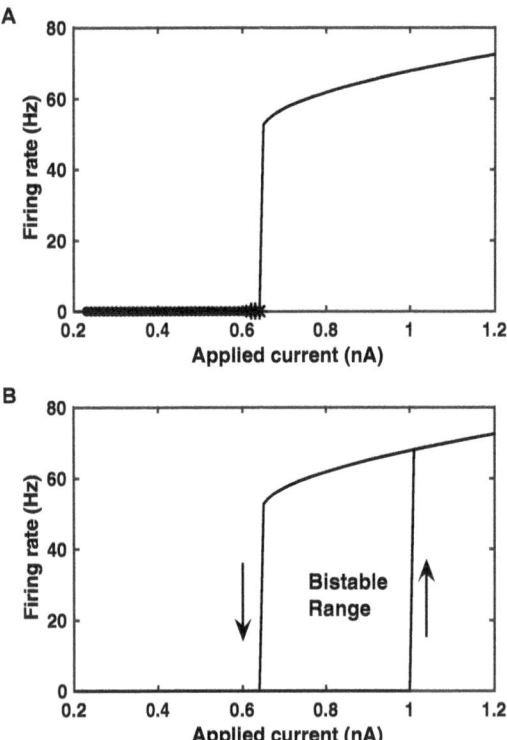

Figure 4.7
Firing rate curve of Hodgkin-Huxley model. (A) Firing rate in response to current steps from a baseline of 0 nA, as used in figure 4.6. Small solid circles indicate a single spike is produced; crosses indicate multiple spikes are produced without sustained firing. (B) Firing rate in response to a protocol of current steps that successively increase from the previous value without reset, then successively decrease. For a wide range of currents the system is bistable, with the neuron remaining quiescent (zero firing rate) if initially quiescent, or sustaining firing if spiking is already initiated. The code used to generate this figure can be found online as HH_new_f_I_bi.m.

through values of firing rate that can never be attained by the neuron. Such a discontinuity in the firing rate response defines the type-II class of neurons.

Moreover, one must take care in plotting an f-I curve, since the neuron's spiking behavior at a given level of applied current is history-dependent (figure 4.7, lower panel). For example, if the applied current is gradually increased a much higher applied current is needed for the neuron to produce an action potential than if the current were stepped up instantaneously from zero.

To understand the history-dependence, we first recall that the neuron can produce a spike because the positive feedback of sodium activation is much more rapid than the negative feedback of sodium inactivation and potassium activation. However, if the membrane potential is raised incrementally, then the negative feedback terms have time to become stronger so that the membrane potential (and hence the applied current) must be higher before sodium activation can dominate and "take off."

Conversely, once the neuron is producing spikes, a lower value of applied current is needed to maintain further spikes. This is because, following each spike, sodium channels are more deinactivated (higher h) and potassium channels are more deactivated (lower n) than the steady state level produced by gradual increment of the applied current. Both sodium deinactivation and potassium deactivation make it easier for the neuron to produce a new spike by sodium activation. That is, whether a spike is produced at a given level of applied current or depolarization depends on the state of gating variables—in particular of sodium inactivation, h, and potassium activation, n—not just on the membrane potential alone.

A second example of the need to consider more than the membrane potential is the demonstration of the "anode break" action potential (figure 4.8). In this case, the Hodgkin-Huxley model neuron produces a single spike when the applied current is returned to zero following a period of hyperpolarization. Such a feature of the Hodgkin-Huxley model and other type-II neurons is impossible in an integrator model of a neuron, such as the leaky integrate-and-fire. However, when we think of the neuron as an oscillator, the behavior becomes more understandable.

For example, instead of pushing a swing to get it going, one can pull backward then release the swing. Mathematically the same is occurring when a negative current is applied to the neuron, which is then released. The key for the neuron is the process of deinactivation of sodium—the inactivation variable, h, approaches its maximum, 1, upon hyperpolarization, making a spike easier—combined with deactivation of potassium—the potassium activation variable, n, approaches zero upon hyperpolarization, reducing the potassium current that counteracts any sodium spike.

These behaviors show that the notion of a threshold for spiking as used in the simplest leaky integrate-and-fire models is purely a simplification and convenience. It should not be taken literally in more realistic models, yet alone in real neurons. Whether a model neuron produces a spike depends on the combination of all of the variables in the model, so when the variables change at different rates, production of a spike is history-dependent. Since electrophysiological experiments do not allow for measurement of gating variables, it is common to plot the membrane

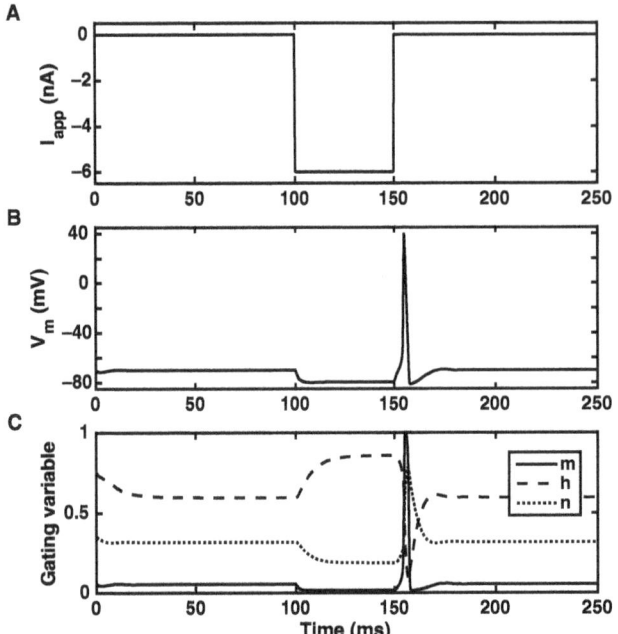

Figure 4.8
Anode break, a spike from hyperpolarization. (A) Applied current is negative (outward and hyperpolarizing) for a
period of 50 ms. (B) When the applied current is released and returns abruptly to its baseline value of zero an action
potential is produced. (C) A plot of the gating variables reveals that the period of hyperpolarization increased the inactivation variable, h, in a process called deinactivation. In combination with a reduced potassium conductance due to
the reduction in potassium activation, n, the deinactivation dramatically reduces the threshold for a sodium spike. This
figure was produced by the online code `HH_old_anode_break_3plot.m`.

potential, V_m, against its time derivative, dV_m / dt, to gain at least one extra factor in determining
whether an action potential should be produced.

The subthreshold oscillations underlying much of the type-II behavior can be observed following a step current that is just insufficient to produce an action potential (figure 4.9). In the
Hodgkin-Huxley model these oscillations are strongly damped, meaning they decay rapidly. It
is possible to observe sustained subthreshold oscillations in real neurons and in other models,
though usually at a lower frequency due to the presence of other currents. Complete removal of
the subthreshold oscillations transforms a type-II neuron into a type-I neuron, which possesses
none of the features discussed here. In the next section, we will add an extra potassium current to
produce the Connors-Stevens model, which exhibits such type-I behavior.

Type-I neuron   A neuron whose firing rate increases from zero without any rate jump as input
current incrementally increases above threshold.

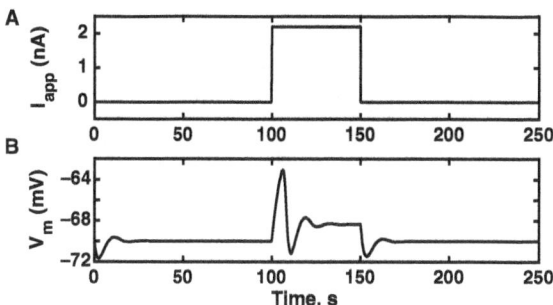

Figure 4.9
Subthreshold oscillations following step changes in current. (A) A subthreshold step of current is applied for 50 ms. (B) The membrane potential responds with a brief period of damped oscillations. This figure was produced by the online code: HH_old_sub_thresh.m.

A final property of intrinsic oscillators is their resonant response to nonuniform applied currents. Resonance is a ubiquitous feature of oscillating systems. Resonance means that the system responds most strongly to inputs that match its natural frequency. We all know that when pushing a swing we must time successive pushes to get it higher and that if we push at the wrong time we can slow the swing down and stop it.

In neurons, resonance can be observed in the response to a frequency sweep of oscillating applied current[6] known as the ZAP protocol (ZAP for impedance amplitude profile, where Z is the symbol in physics for impedance, which is proportional to resistance but includes phase). Figure 4.10 demonstrates the results of such a protocol applied to the Hodgkin-Huxley model using two different amplitudes for the peak of the oscillating applied current. Whereas the applied current oscillations have constant amplitude, the neuron's response peaks at intermediate frequencies—approximately at the frequency of the subthreshold oscillations. In figure 4.10C, the stronger response in a particular frequency range leads the model neuron to produce spikes only when the input frequency is near the neuron's natural frequency—its resonant frequency.

In tutorial 4.1 we will see how the time spacing of short current pulses can determine whether a spike is initiated and conversely how sustained spiking can be ended with a positive current pulse at just the right time.

4.3   Tutorial 4.1: The Hodgkin-Huxley Model as an Oscillator

Neuroscience goal: Gain appreciation of the Type-II properties of the Hodgkin-Huxley model.

Computational goal: Careful transposition of mathematical formulae and parameters into computer code; practice of simulations with several interdependent variables.

In this tutorial, you will simulate a full four-variable model similar to the original Hodgkin-Huxley model (though using modern units). The four variables, sodium activation, m, sodium inactivation, h, potassium activation, n, and membrane potential, V, will all be updated on each

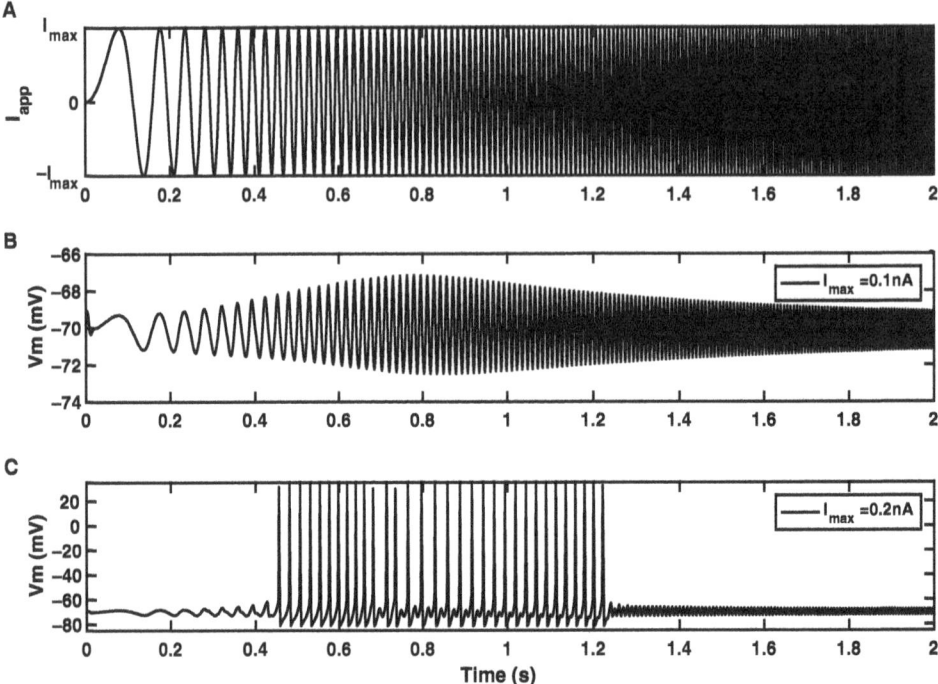

Figure 4.10
Resonant response to a frequency sweep or "ZAP" current. (A) The injected current is given by a sine wave of linearly increasing frequency, $I_{app}(t) = I_{max} \sin[\pi f(t) \times t]$ where $f(t) = f_0 + (f_{max} - f_0)t / t_{max}$, $f_0 = 0$ Hz, $f_{max} = 160$ Hz and $t_{max} = 2$ s. (B) With $I_{max} = 0.1$ nA, the membrane potential oscillates in response at the same frequency as the applied current, with an amplitude that peaks near 0.8 s when the frequency is 64 Hz. (C) With $I_{max} = 0.2$ nA, the membrane potential responds with action potentials when the oscillation frequency is within the resonant range. Notice that at the resonant frequency for subthreshold oscillations (at a time near 0.8 s) the neuron spikes on alternate current peaks because it takes the neuron longer to recover from the large deviation of a spike (bottom) than the smaller subthreshold oscillation (center). This figure was produced by the online code `HH_new_zap.m`.

timestep, since they depend on each other. Initial conditions should be 0 for all gating variables and E_L for the membrane potential, unless otherwise stated.

a. Set up a simulation of 0.35 s duration of the Hodgkin-Huxley model as follows:

$$C_m \frac{dV_m}{dt} = G_L(E_L - V_m) + G_{Na}^{(max)} m^3 h(E_{Na} - V_m) + G_K^{(max)} n^4 (E_k - V_m) + I_{app}$$

$$\frac{dm}{dt} = \alpha_m(1-m) - \beta_m m$$

$$\frac{dh}{dt} = \alpha_h(1-h) - \beta_h h$$

$$\frac{dn}{dt} = \alpha_n(1-n) - \beta_n n$$

Table 4.1
Parameter values for H-H model (figures 4.2–4.10 and tutorial 4.1)

Parameter	Symbol	Value *
Leak conductance	G_{Leak}	30 nS
Maximum sodium conductance	$G_{Na}^{(max)}$	12 μS
Maximum delayed rectifier conductance	$G_K^{(max)}$	3.6 μS
Sodium reversal potential	E_{Na}	45 mV
Potassium reversal potential	E_K	–82 mV
Leak reversal potential	E_L	–60 mV
Membrane capacitance	C_m	100 pF
Applied current	I_{app}	Variable

*Assumes a total membrane area of 0.1 mm^2.

Table 4.2
Gating variables of the H-H model (figures 4.2–4.10 and tutorial 4.1)

Gating variable	Steady state	Time Constant	Rate constants
m	$\dfrac{\alpha_m}{\alpha_m + \beta_m}$	$\dfrac{1}{\alpha_m + \beta_m}$	$\alpha_m = \dfrac{10^5(-V_m - 0.045)}{\exp[100(-V_m - 0.045)] - 1}$
			$\beta_m = 4 \times 10^3 \exp\left[\dfrac{(-V_m - 0.070)}{0.018}\right]$
h	$\dfrac{\alpha_h}{\alpha_h + \beta_h}$	$\dfrac{1}{\alpha_h + \beta_h}$	$\alpha_h = 70 \exp[50(-V_m - 0.070)]$
			$\beta_h = \dfrac{10^3}{1 + \exp[100(-V_m - 0.040)]}$
n	$\dfrac{\alpha_n}{\alpha_n + \beta_n}$	$\dfrac{1}{\alpha_n + \beta_n}$	$\alpha_n = \dfrac{10^4(-V_m - 0.060)}{\exp[100(-V_m - 0.060)] - 1}$
			$\beta_n = 125 \exp\left[\dfrac{(-V_m - 0.070)}{0.08}\right]$

where the parameters are given in table 4.1 and the rate constants are the instantaneous functions of membrane potential given in table 4.2.

Initially set the applied current to zero and check the membrane potential stabilizes at −70.2 mV.

b. Produce a vector for the applied current that has a baseline of zero and steps up to 0.22 nA for a duration of 100 ms beginning at a time of 100 ms. Plot the applied current on an upper graph and the membrane potential's response on a lower graph of the same figure. (You should see subthreshold oscillations but no spikes.)

c. Alter your code so that the applied current is a series of 10 pulses, each of 5 ms duration and 0.22 nA amplitude. Create a parameter that defines the delay from the onset of one pulse to the onset of the next pulse. Adjust the delay from a minimum of 5 ms up to 25 ms and plot applied current and membrane potential as in (b) for two or three examples of pulse separations that can generate spikes. Be careful (especially if using the Euler method) to repeat your simulation with your time step reduced by a factor of 10 to ensure you see no change in the response, as a sign you have sufficient accuracy in the simulation. (A timestep of as low as 0.02 µs may be necessary.) Describe and explain your findings.

d. Now set the baseline current to be 0.6 nA. Set the initial conditions as $V_m(0) = -0.065 \mathrm{V}$, $m(0) = 0.05$, $h(0) = 0.5$, $n(0) = 0.35$. (Note that the "0" in parenthesis indicates a time of $t = 0$, which corresponds to element number 1 in an array.) Apply a series of 10 inhibitory pulses to bring the applied current to zero for a duration of 5 ms, with pulse onsets 20 ms apart. Plot the applied current and the membrane potential's response as in (b, c). Describe and explain what you observe.

e. Now set the baseline current to 0.65 nA. Set the initial conditions as $V_m(0) = -0.065$ V, $m(0) = 0.05$, $h(0) = 0.5$, $n(0) = 0.35$. Increase the excitatory current to 1 nA for a 5 ms pulse at the time point of 100 ms. Plot the applied current and resulting behavior in the same manner as (b–d). Describe what occurs and explain your observation.

f. Repeat (e) with the baseline current of 0.7 nA, but set the initial conditions as $V_m(0) = -0.065$ V, $m(0) = 0$, $h(0) = 0$, $n(0) = 0$. As in (e), increase the excitatory current to 1 nA for a 5 ms pulse at the time point of 100 ms. Plot the applied current and resulting behavior in the same manner as (b–e). Describe what occurs and explain your observation, in particular by comparing with your results to part (e).

4.4 The Connor-Stevens Model: A Type-I Model

A-current, I_A The A-current is an outward (hyperpolarizing) potassium current, which activates at low membrane potentials and inactivates at high membrane potentials. It does not fully deactivate at resting potentials, where it can provide a sufficient outward current to counteract rebound spikes.

Table 4.3
Parameter values for the Connor-Stevens model (figure 4.11)

Parameter	Symbol	Value *
Leak conductance	G_{Leak}	30 nS
Maximum sodium conductance	$G_{Na}^{(max)}$	12 μS
Maximum delayed rectifier conductance	$G_K^{(max)}$	2 μS
Maximum A-type conductance	$G_A^{(max)}$	4.77 μS
Sodium reversal potential	E_{Na}	55 mV
Potassium reversal potential	E_K	–72 mV
Reversal of A-type channels	E_A	–75 mV
Leak reversal potential	E_L	–17 mV
Membrane capacitance	C	100 pF
Applied current	I_{app}	Variable

*Assumes a total membrane area of 0.01 mm^2.

The Connor-Stevens model[7] is a variant of the Hodgkin-Huxley model with altered parameters and an additional potassium current called the A-current. The model is designed to reproduce the responses of typical cells in mammalian cortex, which are able to produce spikes at low firing rates, unlike the giant squid neuron that underlies the Hodgkin-Huxley model. The extra potassium current has both activation (a) and inactivation (b) variables (figure 4.11), so the Connor-Stevens model is a six-variable model that looks very similar to the Hodgkin-Huxley model:

$$C_m \frac{dV_m}{dt} = G_L (E_L - V_m) + G_{Na}^{(max)} m^3 h (E_{Na} - V_m) + G_K^{(max)} n^4 (E_K - V_m)$$
$$+ G_A^{(max)} a^3 b (E_A - V_m) + I_{app} \tag{4.13}$$

$$\frac{dm}{dt} = \alpha_m (1 - m) - \beta_m m \tag{4.14}$$

$$\frac{dh}{dt} = \alpha_h (1 - h) - \beta_h h \tag{4.15}$$

$$\frac{dn}{dt} = \alpha_n (1 - n) - \beta_n n \tag{4.16}$$

$$\frac{da}{dt} = \frac{a_\infty - a}{\tau_a} \tag{4.17}$$

$$\frac{db}{dt} = \frac{b_\infty - b}{\tau_b} \tag{4.18}$$

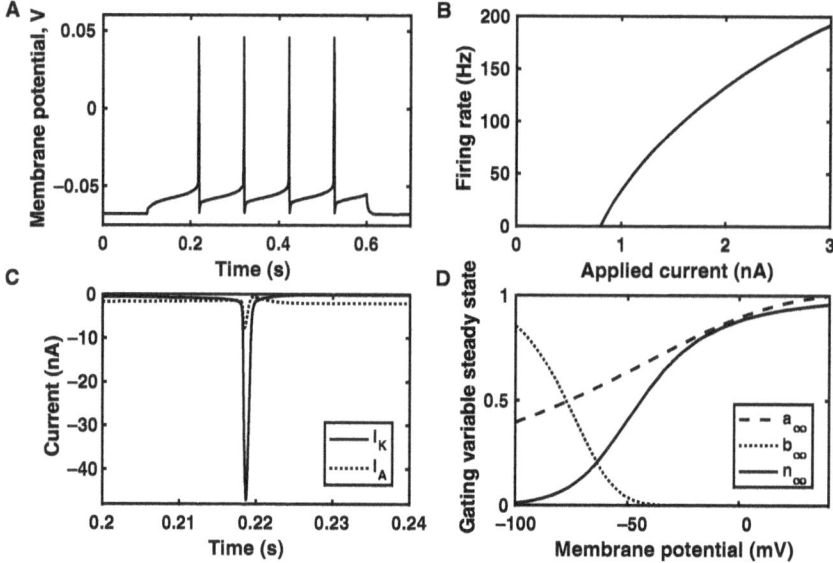

Figure 4.11
The Connor-Stevens model. (A) The spike train with an applied current of 850 pA between $t = 0.1$ s and $t = 0.6$ s. The delay to first spike is typical of a type-I model. (B) Type-I behavior is further exemplified by the f-I curve, which increases from zero without any discontinuity of firing-rate. (C) The A-current (I_A, dotted) becomes stronger than the delayed rectifier current (I_K, solid) shortly after a spike. Deinactivation of the outward A-current (b↑) counteracts deinactivation of the inward sodium current (h↑), preventing the rebound that causes oscillations in the membrane potential and type-II behavior. (D) Steady states of the gating variables show that at low membrane potential the conductance of the A current can remain high (both a_∞ and b_∞ are nonzero at $V = -70$ mV). Hyperpolarization can even cause the steady-state of the A-type conductance to increase via deinactivation (b_∞ rises) while the conductance of the delayed rectifier current falls dramatically via deactivation (n_∞ falls). This figure is produced by the online code CS_figure.m.

In equations 4.17 and 4.18, the gating variables for the A-current are just described in terms of their voltage-dependent steady states (a_∞, b_∞) and voltage-dependent time constants (τ_a, τ_b) without an attempt to derive the corresponding rate constants. The standard parameters for the model are given in table 4.3 (where a neuron of surface area 0.01 mm^2 is assumed). The membrane potential-dependences of the gating variables are given in table 4.4.

These functions are incorporated in the online codes. The sodium and potassium gating variables are qualitatively similar to those of the Hodgkin-Huxley model. The steady states of the A-current gating variables are shown in figure 4.11. Their time constants are sigmoid functions, dropping to the value given by the first term at high membrane potential and increasing by the amount on the numerator of the second term as the membrane potential is decreased to hyperpolarized levels.

The attentive reader may wonder why the reversal potential of A-type channels differs from that of potassium channels, because we showed in section 2.1 that the reversal potential depends on the type of ion flowing through a channel—and the A-type current is a potassium current.

Table 4.4
Gating variables of the Connor-Stevens model (figure 4.11)

Gating variable	Steady state	Time Constant	Voltage-dependent variables
m	$\dfrac{\alpha_m}{\alpha_m + \beta_m}$	$\dfrac{1}{\alpha_m + \beta_m}$	$\alpha_m = \dfrac{3.8 \times 10^5 (V + 0.0297)}{1 - \exp[-100(V + 0.0297)]}$ $\beta_m = 1.52 \times 10^4 \exp[-55.6(V + 0.0547)]$
h	$\dfrac{\alpha_h}{\alpha_h + \beta_h}$	$\dfrac{1}{\alpha_h + \beta_h}$	$\alpha_h = 266\exp[-50(V + 0.048)]$ $\beta_h = \dfrac{3800}{1 + \exp[-100(V + 0.018)]}$
n	$\dfrac{\alpha_n}{\alpha_n + \beta_n}$	$\dfrac{1}{\alpha_n + \beta_n}$	$\alpha_n = \dfrac{2 \times 10^4 (V + 0.0457)}{1 - \exp[-100(V + 0.0457)]}$ $\beta_n = 250\exp[-12.5(V + 0.0557)]$
a	a_∞	τ_a	$\tau_a = 3.632 \times 10^{-4} + \dfrac{1.158 \times 10^{-3}}{1 + \exp[49.7(V + 0.05596)]}$ $a_\infty = \left\{ \dfrac{0.0761\exp[31.4(V + 0.09422)]}{1 + \exp[34.6(V + 0.00117)]} \right\}^{1/3}$
b	b_∞	τ_b	$b_\infty = \left\{ \dfrac{1}{1 + \exp[68.8(V + 0.0533)]} \right\}^4$ $\tau_b = 1.24 \times 10^{-3} + \dfrac{2.678 \times 10^{-3}}{1 + \exp[62.4(V + 0.050)]}.$

There are two reasons for such differences. First, ion channels are not 100 percent specific—a smaller flow of other ions, at a level that is channel-specific, does shift the reversal potential for a given channel-type. Second, when optimizing a model of a neuron, it may be beneficial to be a little bit off in one parameter in order to make the model's behavior more realistic overall. In that sense, the "incorrect" parameter is accounting for many other variables not included explicitly in the model.

Like the sodium current, the A-current is a window current because it has both activation and inactivation variables. The "window" of membrane potential at which the A-current is active is slightly lower than that of the sodium channel, so that after a spike the A-current is slightly active (figure 4.11C). Since the A-current is hyperpolarizing, its postspike conductance acts to prevent a rebound spike that is typical of type-II behavior.

Signal processing pathways   Chains of biochemical reactions, often involving activation of proteins via phosphorylation, or the reverse, which mediate a cell's response to an incoming signal.

Endoplasmic reticulum An extensive subcompartment with relatively high calcium concentration, which pervades the cell, connecting to the nucleus and separated from the cytosol by a membrane with active channels.

Mitochondria Small subcompartments with relatively high calcium concentration that reside throughout the cytosol, most notable for producing ATP, the cell's energy supply.

Cytosol The aqueous part of the cytoplasm, the fluid in the main cell body, within which the nucleus, mitochondria, the endoplasmic reticulum, and other organelles reside.

4.5 Calcium Currents and Bursting

Calcium ions play a crucial role in a vast number of signal processing pathways in biology. They are involved in fertilization,[8] the cell cycle,[9] transcriptional regulation,[10, 11] muscle contraction,[12] and have a number of functions specific to neurons including synaptic transmission[13] and synaptic plasticity[14, 15] that we will cover in this book. Table 2.1 indicates that calcium is maintained at very low concentrations in the cytosol—the bulk of the intracellular fluid—compared to the extracellular medium. This might be a consequence of the relative insolubility in water of calcium salts (such as calcium phosphate or calcium carbonate) whose precipitation could be harmful to cell function. As well as in the extracellular medium, higher calcium concentrations can be found in internal stores, such as the endoplasmic reticulum and mitochondria.[16] Transient release of calcium from such stores can lead to quite exquisite and complex calcium dynamics within cells, including neurons.[17]

Given the positive charge and low concentration of calcium ions in the cytosol, the opening of calcium channels in the cell membrane produces an inward current like the sodium current. Many different types of calcium channel exist in neurons, with their different functions determined by a combination of their location and their dynamics. As with the sodium and potassium channels that we have considered already, the dynamics of a calcium conductance depends on a combination of the time constants for activation and inactivation and on the voltage-dependence of the steady states of these gating variables. For example, rapidly activating high-threshold calcium channels located near axon terminals are able to respond to action potentials in a timely manner and mediate the calcium influx needed to initiate the release of vesicles of neurotransmitter[18, 19] (see chapter 5).

In the remainder of this chapter we will consider calcium channels with sufficient spatial spread that they can impact the membrane potential of the entire cell, leading to positive feedback in the same manner as a sodium spike. Indeed, we will see that calcium spikes are possible via a positive feedback process equivalent to the one that produces sodium spikes. The main difference being that a calcium spike follows slower dynamics than does a sodium spike, lasting tens to hundreds of milliseconds rather than one millisecond or less. If we consider that a calcium spike could be represented by a broad peak of applied current in our prior models, we see that the opening of calcium channels can lead to a burst of high frequency sodium spikes, which lasts until the calcium channels close.

> **Calcium spike**   A broad positive spike in the membrane potential that can last hundreds of milliseconds, wrought by positive feedback of voltage-gated calcium channels, which provide an inward current.

> **T-type calcium channel**   A voltage-gated calcium channel that deinactivates at membrane potentials below rest, such that it supplies inward current following periods of release from hyperpolarization.

> **Post-inhibitory rebound**   A positive rebound in the membrane potential following a period of inhibition (similar to "anode break"), which can result from the deinactivation of T-type calcium channels.

4.5.1   Thalamic Rebound and the T-Type Calcium Channel

T-type calcium channels are low-threshold calcium channels that respond to periods of hyperpolarization with deinactivation, which allows them to open more easily following release of inhibition—via a postinhibitory rebound—than without such inhibition. The channels are partially responsible for the pacemaking properties of heart tissue.[20] They are also found in thalamocortical "relay" neurons.[21, 22]

Thalamocortical neurons reside on the pathway between sensory receptors (as in the retina of the eye), which respond directly to external input, and neurons in sensory cortex, whose responses depend as much on the internal circuitry molded by prior experience. The behavior of the thalamocortical neurons is very different during sleep when we are less attentive to external inputs than during wakefulness. These differences in behavior can be attributed in part to the properties of the T-type calcium channels, which enable the neuron to switch between a depolarized tonic mode and a hyperpolarized bursting mode (see figure 4.12). In the tonic mode during wakefulness, the firing rate of the neuron is approximately a monotonic function of the inputs at that time. In the bursting mode during sleep, the neuron is mostly unresponsive to inputs, but produces intermittent bursts of high-frequency spikes.

Tonic mode   A state of neural activity with relatively regular spiking, with spike rate varying monotonically with input current.

4.6 Tutorial 4.2: Postinhibitory Rebound

Neuroscience goal: Understand the relationship of calcium conductance to bursting behavior and postinhibitory rebound.

Computational goal: Gain more practice writing and using a function, analyzing responses to variations of multiple parameters, and plotting functions of 2 parameters.

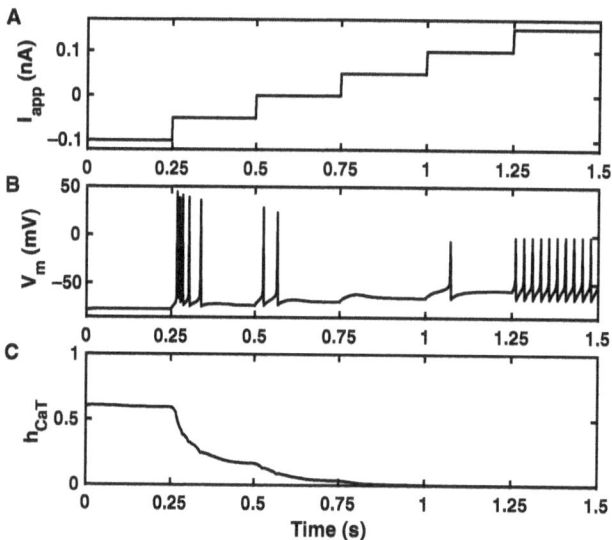

Figure 4.12
Deinactivation of T-type calcium channels produces postinhibitory rebound. (A) The current is stepped up every 250 ms from an initial inhibitory current of −100 pA in increments of +50 pA. (B) The membrane potential responds to the first step up to −50 pA with a burst of five spikes, but it does not respond with a spike following the third step from 0 pA to +50 pA. With higher currents the neuron responds with regular (tonic) spikes. (C) The inhibitory currents cause deinactivation of T-type calcium channels (a rise in h_{CaT}) allowing for the Ca^{2+} current to flow following hyperpolarization (left) but not depolarization (right). For parameters, see tutorial 4.2. This figure is produced by the online code PIR_steps.m.

Table 4.5
Parameter values for thalamic rebound model (figure 4.12 and tutorial 4.2)

Parameter	Symbol	Value *
Leak conductance	G_{Leak}	10 nS
Maximum sodium conductance	$G_{Na}^{(max)}$	3.6 μS
Maximum delayed rectifier conductance	$G_K^{(max)}$	1.6 μS
Maximum T-type calcium conductance	$G_T^{(max)}$	0.22 μS
Sodium reversal potential	E_{Na}	55 mV
Potassium reversal potential	E_K	−90 mV
Calcium reversal potential	E_{Ca}	120 mV
Leak reversal potential	E_L	−70 mV
Membrane capacitance	C_m	100 pF
Applied current	I_{app}	Variable

*Assumes a total membrane area of 0.01 mm^2.

Table 4.6
Gating variables of the thalamic rebound model (figure 4.12 and tutorial 4.2)

Gating variable	Steady state	Time Constant	Voltage-dependent variables
m	$\dfrac{\alpha_m}{\alpha_m + \beta_m}$	0	$\alpha_m = \dfrac{10^5(V+0.035)}{1-\exp[-100(V+0.035)]}$ $\beta_m = 4000 \exp\left[\dfrac{-(V+0.06)}{0.018}\right]$
h	$\dfrac{\alpha_h}{\alpha_h + \beta_h}$	$\dfrac{1}{\alpha_h + \beta_h}$	$\alpha_h = 350 \exp[-50(V+0.058)]$ $\beta_h = \dfrac{5000}{1+\exp[-100(V+0.028)]}$
n	$\dfrac{\alpha_n}{\alpha_n + \beta_n}$	$\dfrac{1}{\alpha_n + \beta_n}$	$\alpha_n = \dfrac{5\times10^4(V+0.034)}{1-\exp[-100(V+0.034)]}$ $\beta_n = 625 \exp[-12.5(V+0.044)]$
m_T	$m_{T,\infty}$	0	$m_{T,\infty} = \dfrac{1}{1+\exp\left[\dfrac{-(V+0.052)}{0.0074}\right]}$
h_T	$h_{T,\infty}$	τ_{h_T}	$h_{T,\infty} = \dfrac{1}{1+\exp[500(V+0.076)]}$ $\tau_{h_T} = 0.001 \exp[15(V+0.467)]$ if $V < -0.080$ $\tau_{h_T} = 0.028 + 0.001\exp\left[\dfrac{-(V+0.022)}{0.0105}\right]$ if $V \geq -0.080$

In this tutorial, you will produce a model thalamocortical neuron with a T-type calcium current. You will assess its response to single steps of applied current of variable size and baseline. You should simulate the neuron within a function rather than within a script, so that different levels of baseline current and step current can be manipulated in a separate file. You will find that storing separate aspects of the code in separate files makes for a neater, less cluttered set of codes that are easier to manipulate.

You will be simulating the following four dynamical equations of the model neuron:

$$C_m \frac{dV}{dt} = G_L(E_L - V) + G_{Na}^{(max)} m^3 h(E_{Na} - V) + G_K^{(max)} n^4 (E_K - V) + G_T^{(max)} m_T^2 h_T (E_{Ca} - V) + I_{app}$$

$$\frac{dh}{dt} = \alpha_h(1-h) - \beta_h h$$

$$\frac{dn}{dt} = \alpha_n(1-n) - \beta_n n$$

$$\frac{dh_T}{dt} = \frac{h_{T,\infty} - h_T}{\tau_{h_T}}$$

Parameters are given in table 4.5. The activation functions for the sodium and the T-type calcium current are not included as dynamical equations because they are assumed to respond instantaneously to changes in the membrane potential: $m = m_\infty = \alpha_m/(\alpha_m + \beta_m)$ and $m_T = m_{T,\infty}$. In this example, it is the long time constant for inactivation of the calcium current that will produce a slow timescale for burst duration. Membrane potential-dependence of all the gating variables are given in table 4.6.

a. Produce a function PIR to be saved in the file PIR.m, which simulates the model neuron described above for a period of 750 ms. You will find the voltage-dependence of the gating variables already encoded in the online file PIR_Vdependence.m, so you can copy expressions from that file if you prefer instead of typing them out.

A baseline current and current step will be supplied to the function as its two inputs. The applied current should be set to the baseline value from 0 to 250 ms, increased by the current step value in the midportion of the trial from 250 to 500 ms, and returned to the baseline value from 500 to 750 ms. The function should return two outputs: the membrane potential as a function of time and the time vector.

b. Produce a code that loops through a range of values for the baseline current (between −200 pA and +200 pA) and a range of values for the current step (between 0 and +100 pA) added to each baseline current. For each pair of the parameters (the baseline current and the current step), calculate the total number of spikes produced during the current step and, if two or more spikes are produced, calculate the minimum interspike interval. Use imagesc or another graphics tool for plotting functions of two variables (e.g., see also `surf` or `mesh` or `plot3`) to plot the total number of spikes on one graph and the minimum ISI on another graph, with baseline current on the x-axis and current step on the y-axis. Comment on your results and plot some examples of membrane potentials for each qualitatively distinct type of behavior.

4.7 Modeling Multiple Compartments

Up until now, we have ignored any impact of the spatial dependence of the membrane potential of a neuron. Such spatial dependence is important when the ion channel densities or the inputs a neuron receives vary across the neuron's membrane. Using a multicompartment model, we can take into account some of the heterogeneity in the density of ion channels in different parts of a neuron and begin to understand how a neuron might respond differently to input currents depending on which part of the neuron receives the input.

The idea of a multicompartmental model is to represent different sections of the neuron by individual compartments (figure 4.13), each of which is ascribed a single time-varying membrane potential. Resistors are included between the compartments to connect them in a similar

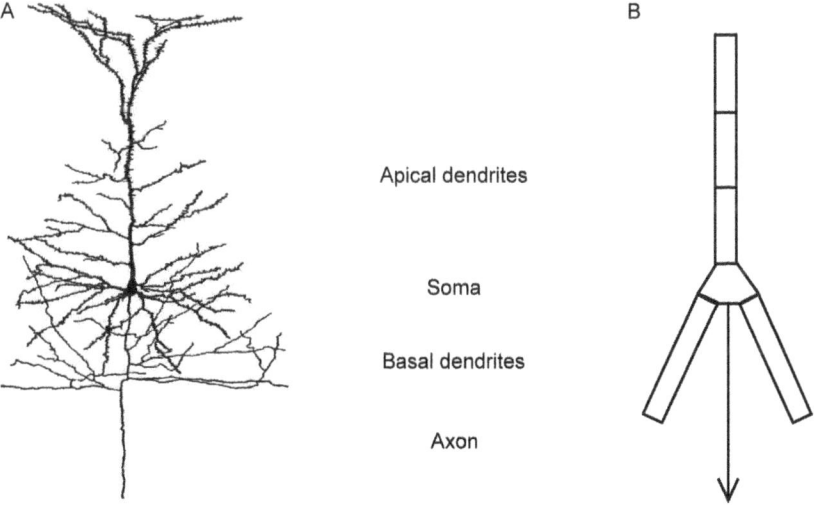

Figure 4.13
Simplifying a pyramidal cell with a multi-compartmental model. (A) The traced structure of a pyramidal cell with a long apical dendrite containing a large fraction of the cell's surface area (at the top) and other basal branches leaving the soma (at the bottom). (B) The dendritic structure is modeled as five separate compartments connected either to each other or to the soma's separate compartment. The axon is often not modeled, because once the action potential is generated in its initial segment (which is typically incorporated in the soma in a model) the axon simply transmits the spike with minimal attenuation or modulation. This image first appeared as figure 2.1 in the *Book of Genesis*, published under the GNU Free Documentation License, and was kindly provided by Dave Beeman, University of Colorado at Boulder.

morphology to that of the original cell. If the compartments are carefully chosen, the extensive spatial structure of a neuron can be simulated in a manner that captures the important characteristics of the original cell more fully than any of the single compartment, "point" neuron models that we have seen so far.

Several software packages exist for producing and simulating multicompartmental models. The most common examples include Neuron, Genesis, and Brian. For a more complete introduction to multicompartmental modeling, I recommend the *Book of Genesis*,[23] from which figure 4.13 is reproduced.

4.7.1   The Pinsky-Rinzel Model of an Intrinsic Burster

As well as those neurons that produce bursts of spikes transiently upon depolarization, many others produce bursts rhythmically without the need for any external current. Such intrinsic, rhythmic bursters (sometimes called pacemaker neurons) can drive a complete circuit in a periodic pattern, or act as a time signal to separate different phases of neural processing or coordinate information transfer.

Intrinsic burster   A neuron that fires bursts of rapid action potentials periodically, without external input.

Neurons can fire spikes intrinsically (i.e., without external input) if excitatory channels begin to activate at low membrane potentials, so that no depolarization is necessary. For an intrinsic burster, it is the calcium current that has such a low threshold for activation.

To simulate an intrinsically bursting neuron, we turn to a two-compartment model (figure 4.14), in which the membrane potential of a somatic compartment is treated as distinct from the membrane potential of a dendritic compartment. The particular two-compartment model that we will simulate is called the Pinsky-Rinzel model.[24] It should be noted that the Pinsky-Rinzel model is itself a great simplification of a 19-compartment model, the Traub model,[25] of a pyramidal neuron found in the hippocampus of the guinea pig. However, the two-compartment model reproduces the key features of the more intricate Traub model (see figures 4.15 and 4.16), and as we shall see it is plenty complicated enough (equations 4.19 and 4.20). Moreover, as an introduction to multicompartmental modeling, it contains the key features that can be built upon if more compartments are needed.

4.7.2   Simulating the Pinsky-Rinzel Model

The dynamical equations for the somatic (V_S) and dendritic (V_D) membrane potentials in the Pinsky-Rinzel model are:

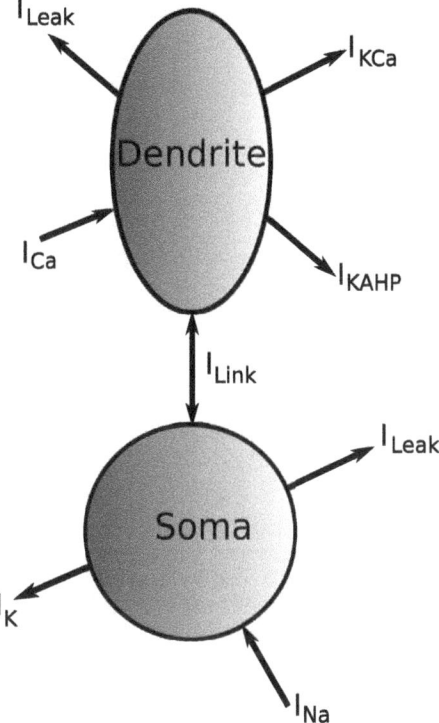

Figure 4.14
The Pinsky-Rinzel model is a two-compartmental model of a bursting neuron. The two compartments with their intrinsic currents are shown. The soma is similar to the standard Hodgkin-Huxley model with an inward fast sodium current (I_{Na}), an outward delayed rectifier potassium current (I_K), as well as a small leak current (I_{Leak}) that is mostly outward. The two compartments are connected by an electrical resistor, which conveys a link current (I_{Link}) that flows from the compartment with higher membrane potential to the one with lower membrane potential at a given time. The dendritic compartment receives inward calcium current (I_{Ca}), whose activation provides the positive feedback needed to generate a dendritic calcium spike. The outward calcium-dependent potassium current (I_{KCa}) is relatively fast and can terminate a dendritic spike, whereas the outward after-hyperpolarization current (I_{KAHP}) is slower and largely determines the interval between dendritic spikes. The dendritic compartment has its own small leak current (I_{Leak}).

$$C_S \frac{dV_S}{dt} = G_{Leak}^{(S)} (E_L - V_S) + G_{Na}^{(max)} m^2 h (E_{Na} - V_S) + G_K^{(max)} n^2 (E_K - V_S) + G_{Link} (V_D - V_S) + I_{app}^{(S)}$$

$$(4.19)$$

$$C_D \frac{dV_D}{dt} = G_{Leak}^{(D)} (E_L - V_D) + G_{Ca}^{(max)} m_{Ca}{}^2 (E_{Ca} - V_D) + G_{KCa}^{(max)} m_{KCa} X (E_K - V_D)$$

$$+ G_{KAHP}^{(max)} m_{KAHP} (E_K - V_D) - G_{Link} (V_D - V_S) + I_{app}^{(D)}$$

$$(4.20)$$

where C_S and C_D are the capacitances of each compartment, $G_{Leak}^{(S)}$ and $G_{Leak}^{(D)}$ the leak conductances, G_{Link} the conductance connecting compartments and $G_{Na}^{(max)}$, $G_K^{(max)}$, etc. are the maximal

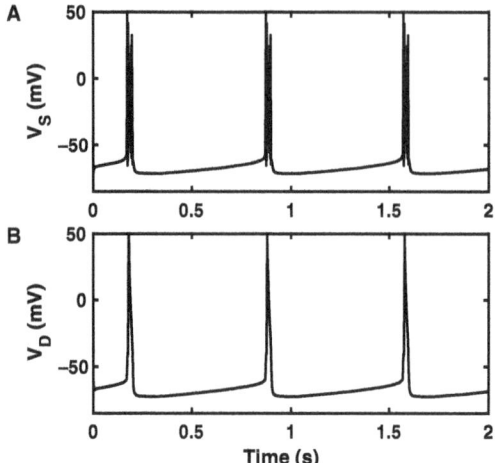

Figure 4.15
Intrinsic bursting in a variant of the PR model. (A) The somatic membrane potential produces bursts of high-frequency
sodium spikes with an interval of hundreds of ms between bursts. (B) The dendritic membrane potential produces
regular broad calcium spikes that are essential for the bursts in the soma. Parameters are given in table 4.7 and equations for rate constants of the gating variables in table 4.8. The code used to produce this figure can be found online as
PR_euler_final.m.

values of the corresponding active conductances. The gating variables vary with time according
to the somatic membrane potential (m, h, n), or dendritic membrane potential (m_{Ca}, m_{KCa}), or the
calcium concentration (X, m_{KAHP}). The dynamics of these gating variables is given in table 4.8.
Apart from the calcium-dependent term, X, which contributes to I_{KCa} (figure 4.14) and responds
instantaneously, the dynamics of all other gating variables are based on the usual rate constants,
α and β. The membrane-potential dependences (and for m_{KAHP}, the calcium-dependences) of
these rate constants are given in table 4.8 and are returned by the two functions PR_soma_
gating and PR_dend_gating, which can be found in the online material.

The model also includes the dynamics of calcium concentration, which depends on the calcium
current, I_{Ca}, via

$$\frac{d[Ca]}{dt} = -\frac{[Ca]}{\tau_{Ca}} + kI_{Ca},$$

where I_{Ca} is one of the dendritic currents given in equation 4.20: $I_{Ca} = G_{Ca}^{(max)} m_{Ca}^2 (E_{Ca} - V_D)$.
The time constant for decay of calcium is given by $\tau_{Ca} = 50$ ms here (an alteration of the original
Pinsky-Rinzel model). The constant, k, is a geometry-dependent factor required to convert the
total charge generated by the calcium current through the surface of the dendritic membrane
into a molar ionic concentration (with units of moles/liter, M) within the volume of the dendritic
membrane where the calcium-dependent potassium channels are located.

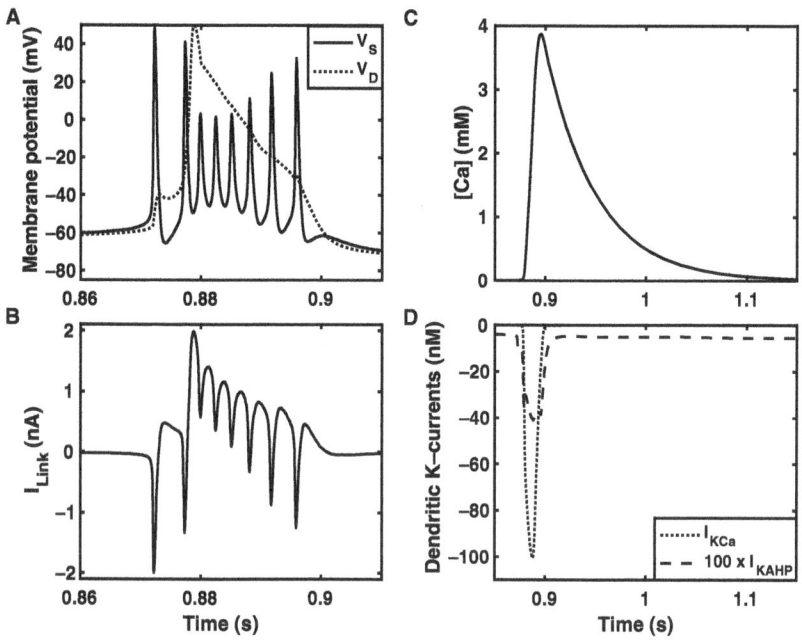

Figure 4.16
Dynamics of a single burst. (A) In this burst, eight sodium spikes (solid line) occur within 25 ms. The first sodium spike produces a subthreshold potential rise in the dendrite (dotted line). The second sodium spike initiates a full, broad calcium spike in the dendrite. During the calcium spike the sodium spikes are rapid and do not return fully to baseline. As the calcium spike decays the frequency of sodium spikes gradually decreases until the end of the burst. (B) I_{Link} is the current from the dendritic to somatic compartment. During a burst, there can be a "ping-pong" effect, a shuttling of current back and forth between the two compartments. In this example, the burst is initiated with a net flow of current from soma to dendrite ($I_{Link} < 0$), while during the burst the prolonged calcium spike in the dendrite produces a net current flow from dendrite to soma ($I_{Link} > 0$). (C) Intracellular calcium rises dramatically within a burst and decays between bursts. (Note the extended timescale compared to A and B.) (D) Hyperpolarizing potassium currents activate in the dendritic compartment during the burst. The faster, calcium-activated potassium current, I_{KCa}, which is also voltage-dependent, brings about the termination of the burst. The slower after-hyperpolarization current, I_{KAHP}, is also calcium-dependent, and after a burst must decay sufficiently before the next burst can commence. Parameters are those of figure 4.15. The code used to produce this figure can be found online as `PR_euler_final.m`.

4.7.3 A Note on Multicompartmental Modeling with Specific Conductances versus Absolute Conductances

Many models are produced using specific parameters such as specific conductance (i.e., conductance per unit area) and or specific capacitance (i.e., capacitance per unit area). The reason for their use is that such specific properties are more universal, being independent of cell size (section 2.3.1). In models with many compartments, a parameter such as the specific capacitance is constant.

If the mathematical equations are written and simulated using specific properties, then any current flow is given as current per unit surface area of the compartment into which it flows.

Table 4.7
Parameter values used for intrinsic bursting in figures 4.14 and 4.15

Parameter	Symbol	Value
Fractional area of Soma	A_S	1/3
Fractional area of Dendrite	$A_D = 1 - A_S$	2/3
Somatic leak conductance	$G_{Leak}^{(S)}$	$A_S \times 5$ nS
Dendritic leak conductance	$G_{Leak}^{(D)}$	$A_D \times 5$ nS
Maximum sodium conductance	$G_{Na}^{(max)}$	$A_S \times 3$ μS
Maximum delayed rectifier conductance	$G_K^{(max)}$	$A_S \times 2$ μS
Maximum calcium conductance	$G_{Ca}^{(max)}$	$A_D \times 2$ μS
Max. calcium-dependent potassium conductance	$G_{KCa}^{(max)}$	$A_D \times 2.5$ μS
Max. after-hyperpolarization conductance	$G_{KAHP}^{(max)}$	$A_D \times 40$ nS
Link conductance	G_{Link}	20 nS
Sodium reversal potential	E_{Na}	60 mV
Calcium reversal potential	E_{Ca}	80 mV
Potassium reversal potential	E_K	−75 mV
Leak reversal potential	E_L	−60 mV
Capacitance of soma	C_S	$A_S \times 100$ pF
Capacitance of dendrite	C_D	$A_D \times 100$ pF
Somatic applied current	$I_{app}^{(S)}$	0
Dendritic applied current	$I_{app}^{(D)}$	0
Calcium decay time constant	τ_{Ca}	50 ms
Conversion from charge to concentration	k	$\dfrac{2.5 \times 10^6}{A_D}$ MC^{-1}

Therefore, care is needed as while the sum of "link" currents between compartments of different surface areas is a conserved quantity—the current flowing out of one compartment flows into another—the current per unit area (specific current density) is not conserved when the two compartments have different areas. In particular, if current flows from a large compartment into a small compartment, the outflow is a small specific current density, whereas the inflow is much larger. In the equations, this is dealt with by dividing the link current into a compartment by the fractional area of each compartment.

On the other hand, if absolute conductance and capacitance are used in a model, then the value of these parameters for each compartment scales multiplicatively with the area of that compartment. In this case, currents are treated absolutely so that current from A to B is the negative of the current from B to A (like the link currents of equations 4.19 and 4.20).

Table 4.8
Gating variables of the Pinsky-Rinzel model

Gating variable	Steady state	Time Constant	Rate constants
m	$\dfrac{\alpha_m}{\alpha_m + \beta_m}$	$\dfrac{1}{\alpha_m + \beta_m}$	$\alpha_m = \dfrac{320 \times 10^3 (V_S + 0.0469)}{1 - \exp[-250(V_S + 0.0469)]}$
			$\beta_m = \dfrac{280 \times 10^3 (V_S + 0.0199)}{\exp[200(V_S + 0.0199)] - 1}$
h	$\dfrac{\alpha_h}{\alpha_h + \beta_h}$	$\dfrac{1}{\alpha_h + \beta_h}$	$\alpha_h = 128\exp\left[\dfrac{-(V_S + 0.043)}{0.018}\right]$
			$\beta_h = \dfrac{4000}{1 + \exp[-200(V_S + 0.020)]}$
n	$\dfrac{\alpha_n}{\alpha_n + \beta_n}$	$\dfrac{1}{\alpha_n + \beta_n}$	$\alpha_n = \dfrac{16 \times 10^3 (V_S + 0.0249)}{1 - \exp[-200(V_S + 0.0249)]}$
			$\beta_n = 250\exp[-25(V_S + 0.040)]$
m_{Ca}	$\dfrac{\alpha_{mCa}}{\alpha_{mCa} + \beta_{mCa}}$	$\dfrac{1}{\alpha_{mCa} + \beta_{mCa}}$	$\alpha_{mCa} = \dfrac{1600}{1 + \exp[-72(V_D - 0.005)]}$
			$\beta_{mCa} = \dfrac{2 \times 10^4 (V_D + 0.0089)}{\exp[200(V_D + 0.0089)] - 1}$
m_{KCa}	$\dfrac{\alpha_{mKCa}}{\alpha_{mKCa} + \beta_{mKCa}}$	$\dfrac{1}{\alpha_{mKCa} + \beta_{mKCa}}$	If $V_D > -0.010$: $\alpha_{mKCa} = 2000\exp\left[\dfrac{-(V_D + 0.0535)}{0.027}\right]; \beta_{mKCa} = 0$ If $V_D \leq -0.010$: $\alpha_{mKCa} = \dfrac{\exp\left(\dfrac{V_D + 0.050}{0.011} - \dfrac{V_D + 0.0535}{0.027}\right)}{0.018975}$ $\beta_{mKCa} = 2000\exp\left[\dfrac{-(V_D + 0.0535)}{0.027}\right] - \alpha_{mKCa}$
X	$\min(4000[Ca], 1)$	0	—
m_{KAHP}	$\dfrac{\alpha_{mKAHP}}{\alpha_{mKAHP} + \beta_{mKAHP}}$	$\dfrac{1}{\alpha_{mKAHP} + \beta_{mKAHP}}$	$\alpha_{mKAHP} = \min(20, 20000 \times [Ca]); \beta_{mKAHP} = 4$

4.7.4 Model Complexity

In general, the number of compartments we add to a model should depend on the question we are trying to answer. For example, the interplay between compartments is essential to the bursting behavior of the neuron modeled by Pinsky and Rinzel. Therefore, a minimum of two compartments is necessary if we are to understand bursting in this type of neuron, even though regular, intrinsic bursting can be generated in a single compartmental model. Alternatively, if we are concerned about whether the spreading of inputs over multiple subbranches of dendrites has the same impact on a cell as the clustering of inputs in a single subbranch, then a model with many more compartments, which includes those subbranches, is necessary.

While adding more compartments to a model can seem like adding more realism (as can adding more ion channels to a single compartment) and therefore be nothing but beneficial, the dangers are multifold:

1. With the addition of more unknown parameters, the model could behave almost in any manner desired, depending on the value of those parameters, and therefore become non-disprovable (which some find convenient!). Or to quote von Neumann, "with four free parameters I can make an elephant, with five I can waggle its trunk."

2. The more parameters there are, the harder it is to completely test the dependence on and relevance of each parameter, so it is unclear what features are important.

3. The more variables there are, the harder it is to gain any insight—from a dynamical systems perspective (chapter 7)—as to why the model acts the way it does. For example, it can be unclear how the different variables interact with and affect each other, in which case understanding the processes in question can become impossible.

In summary, to paraphrase Einstein, "we should make a model as simple as possible but no simpler."

4.8 Hyperpolarization-Activated Currents (I_h) and Pacemaker Control

---

Hyperpolarization-activated current, I_h   A mixed-cation inward current, without inactivation, activated by hyperpolarization and deactivated by depolarization.

Until now, all of the voltage-gated conductance terms that we have considered include an activation variable that increases with depolarization. While some channels also included an inactivation term, which became more positive at hyperpolarized potentials, they always required a subsequent depolarization to admit current. Here we consider the behavior of a hyperpolarization-activated cation conductance, which admits an inward, depolarizing I_h-current, when—as its name suggests—the cell is hyperpolarized.[26]

The I_h-conductance has no inactivation gating variable, just an activation variable, whose steady state increases monotonically with hyperpolarization. The channels allow a mixture of cations to flow through them, so the reversal potential, E_h, (−20 mV in our model) is not as high as that of typical excitatory channels (see table 4.9). We add I_h as a dendritic current to the Pinsky-Rinzel model (equation 4.20),

$$I_h = G_h^{(max)} m_h (E_h - V_D),$$ (4.21)

where the dynamics of the activation variable, m_h, follow the usual form,

$$\frac{dm_h}{dt} = \frac{m_h^\infty - m_h}{\tau_{m_h}}$$ (4.22)

with

$$m_h^\infty = \frac{1}{1 + \exp[(V_D + 0.070)/0.006]}$$ (4.23)

and

$$\tau_{m_h} = 0.272 + \frac{1.499}{1 + \exp[-(V_D + 0.0422)/0.00873]}.$$ (4.24)

These functions are plotted in figure 4.17.

The I_h-conductance provides negative feedback, because it increases as the membrane potential gets lower but acts to make the membrane potential higher. Unlike potassium activation and inactivation of sodium or calcium channels, both of which provide negative feedback to end a

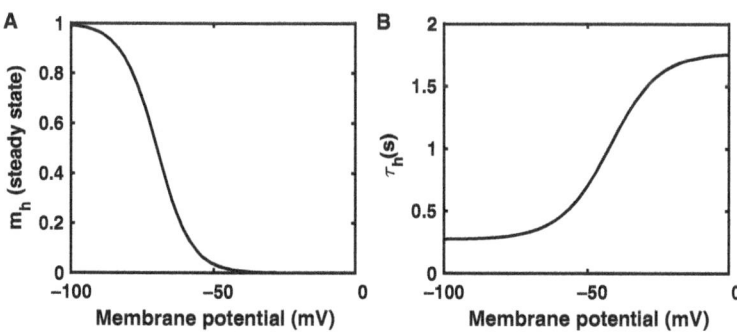

Figure 4.17
Ih gating variables. (A) Steady state of the activation variable, m_h^∞, plotted from equation 4.23, increases from 0 to 1 with hyperpolarization between −50 mV and −100 mV. (B) Time constant for the I_h-conductance is in the 100s of ms, which is greater than typical burst durations. This figure was produced by the online code IH_PR_loop.m.

spike, the I_h-conductance helps with regeneration following a spike. In this sense, it is like dein-activation of sodium or calcium channels. Therefore, a major effect of increased I_h-conductance is a reduction of the interspike interval, or, more importantly in bursting cells, a reduction of the inter-burst interval[27] (see figure 4.18). In this manner, the I_h-conductance may play a role in the homeostasis—i.e., in the long-term control of the period—of bursting neurons (section 8.6).

4.9   Dendritic Computation

Neurons receive inputs at many thousands of distinct sites across their dendrites or cell body. In chapter 5 we consider how to model these inputs, which arise due to activity in other neurons

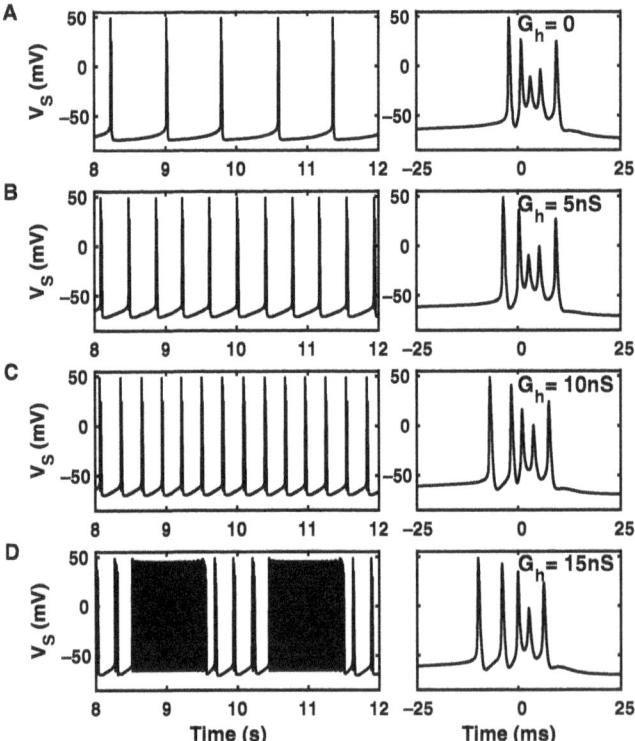

Figure 4.18
Impact of the hyperpolarization-activated current (I_h) on bursting. (A) The somatic membrane potential of a simulated bursting neuron without I_h. (B) I_h is added with $G_h = 50$ nS, causing more than doubling of the burst frequency with little change in burst waveform. (C) The burst contains an extra prespike with increased I_h with $G_h = 10$ nS. (D) For higher values of $G_h = 15$ nS, regular bursting is lost with intermittent periods of prolonged high-frequency spiking. (Left) 4 seconds of bursting activity shown. (Right) A 50 ms zoom-in of a burst, aligned to the upswing of dendritic membrane potential. This figure is produced using a modified Pinsky-Rinzel model (tables 4.8 and 4.9) with an additional I_h current produced using the method of Liu et al.[33] This figure is produced by the online code IH_PR_loop.m.

Table 4.9
Parameter values used for the burster with I_h current in figure 4.16

Parameter	Symbol	Value
Fractional area of soma	A_S	1/3
Fractional area of dendrite	$A_D = 1 - A_S$	2/3
Somatic leak conductance	$G_{Leak}^{(S)}$	$A_S \times$ 1 nS
Dendritic leak conductance	$G_{Leak}^{(D)}$	$A_D \times$ 1 nS
Maximum sodium conductance	$G_{Na}^{(max)}$	$A_S \times 3$ μS
Maximum delayed rectifier conductance	$G_K^{(max)}$	$A_S \times 2$ μS
Maximum calcium conductance	$G_{Ca}^{(max)}$	$A_D \times 2.5$ S
Max. calcium-dependent potassium conductance	$G_{KCa}^{(max)}$	$A_D \times 5$ μS
Max. after-hyperpolarization conductance	$G_{KAHP}^{(max)}$	$A_D \times 60$ nS
Max. hyperpolarization-activated conductance	$G_h^{(max)}$	0–200 nS
Link conductance	G_{Link}	25 nS
Sodium reversal potential	E_{Na}	60 mV
Calcium reversal potential	E_{Ca}	80 mV
Potassium reversal potential	E_K	–75 mV
I_h reversal potential	E_h	–20 mV
Leak reversal potential	E_L	–60 mV
Capacitance of soma	C_S	$A_S \times 100$ pF
Capacitance of dendrite	C_D	$A_D \times 100$ pF
Somatic applied current	$I_{app}^{(S)}$	0
Dendritic applied current	$I_{app}^{(D)}$	0
Calcium decay time constant	τ_{Ca}	50 ms
Conversion from charge to concentration	k	$(1 \times 10^6 / A_D) \mathrm{MC}^{-1}$

and appear at locations on the cell membrane called synapses where another neuron connects with the cell. In simple models, we just sum over individual inputs to obtain a total conductance or total current due to a particular kind of input. In these models the neuron performs a computation—which at a minimum is a nonlinear transformation—of its total input into a series of action potentials.

However, it is clear that in addition to the summed input, the relative location on the cell membrane of the different inputs is important.[28] In particular, if inputs are located near each other on the same dendrite they may generate a dendritic spike, in an example of supralinear summation.[29] The dendritic spike has a much stronger effect on the membrane potential of the soma than do those same inputs spread across multiple dendrites where they are locally too weak to generate a dendritic spike.

The generation of a dendritic spike can be thought of as a dendritic computation[30]—each dendrite acting as a local coincidence detector, producing a spike if it receives sufficient coincident excitatory inputs. In simplified models of complex, extensive neurons, as proposed by Bartlett Mel,[31, 32] the outputs of these dendritic computations provide the inputs to the soma. The soma itself performs a computation on these inputs, emitting an action potential only when it receives sufficient dendritic input.

Figure 4.19 shows the response of a four-compartment model neuron to different patterns of dendritic input. The model is based on the Pinsky-Rinzel model, but with the soma connected to three distinct dendritic compartments (for example, representing an apical dendrite, D1, and two basal dendrites, D2 and D3). If a current of 75 pA is applied simultaneously to each dendritic compartment (figure 4.19B1–2) the cell's response is subthreshold. However, if a current of 150 pA is applied with the same temporal pattern to just one of the dendritic compartments (figure 4.19C1–2) the resulting dendritic spikes produce regular spikes in the soma. This is noteworthy since the spike-producing total input to the cell in figure 4.19C is 150 pA, which is less than the subthreshold total input of 225 pA used in figure 4.19B. It is the higher local density of input that generates a neural response even when the cell's total input is lower.

4.10   Tutorial 4.3: A Two-Compartment Model of an Intrinsically Bursting Neuron

In this tutorial, you will simulate different variants of the two-compartment Pinsky-Rinzel model.

Neuroscience goal: Understand how the two compartments affect each other and how bursting neurons respond to applied current.

Computational goal: Gain further familiarity with function calls.

1. Save the files PR_soma_gating.m and PR_dend_gating.m, found in the online material, to the directory from which you will run your code for this tutorial.

2. Generate a vector of values for the membrane potential (between −0.085 V and 0.050 V) and a vector of values for the calcium concentration (between 0 and 2×10^{-3} M). Use the downloaded functions to calculate and plot all twelve of the rate constants for gating variables on a suitable number of figures. (Hint: type for example help PR_dend_gating to see how to send values to and return values from the function, or load the file PR_dend_gating.m to your editor to see how it works.)

3. Simulate the Pinsky-Rinzel model for 2 s, using a timestep of 2 μs, employing the same functions and parameters of tables 4.7 and 4.8, except set $k = (5 \times 10^6 \ \mathrm{MC}^{-1})/A_D$ and $G_{Link} = 50$ nS.

4. Detect somatic spikes in your model by triggering a spike when the somatic membrane potential increases past −10 mV and allowing a new spike to be triggered once a spike has ended only when the membrane potential decreases again to below −30 mV. (Or you can use the findpeaks function in MATLAB).

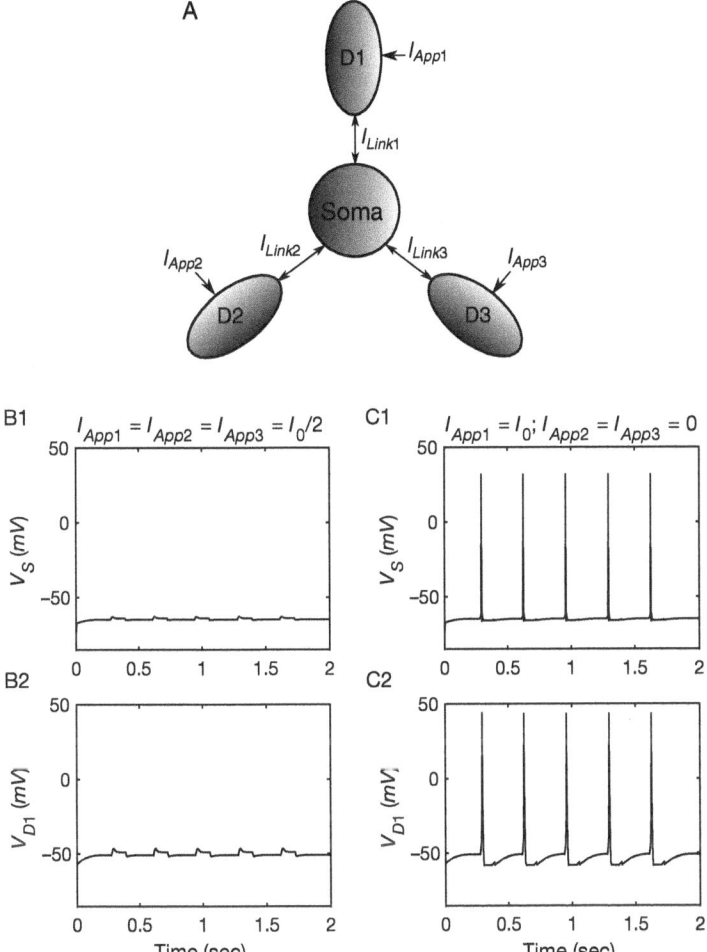

Figure 4.19
Neural activity depends on location of inputs. (A) The model neuron is based on the Pinsky-Rinzel model (figure 4.13) but with three identical dendritic compartments (D1, D2, and D3) instead of just one. (B1–2) Response of the membrane potential in the soma (top) and apical dendrite (bottom), when regular pulses of applied current are injected simultaneously to all three dendritic compartments. (C1–2) Responses of the membrane potential in the soma (top) and apical dendrite (bottom) when two-thirds of the total current injected to the neuron in B is injected into a single dendritic compartment (the apical compartment, D1). This figure is produced by the online code PR3branch.m.

5. Assess how the model's behavior depends on the strength of link between soma and dendrite by simulating the model with $G_{Link} = 0$ nS, $G_{Link} = 10$ nS, and $G_{Link} = 100$ nS. Plot appropriate graphs to demonstrate the differences in behavior with changes in G_{Link}. Explain any changes in behavior and provide further graphs as evidence for your explanation where necessary.

6. For the intrinsic bursting model of (3), assess how the spiking behavior changes when a constant current is applied to the dendrite—try currents of 50 pA, 100 pA, and 200 pA. Carry out a similar assessment when the current is applied to the soma. Comment on any differences in behavior that arise when a current is applied to the soma rather than to the dendrite.

Questions for Chapter 4

1a. Explain the positive feedback involved in producing an action potential in the Hodgkin-Huxley model.

1b. Describe how three different terms produce negative feedback in the Hodgkin-Huxley equations.

2   In the Hodgkin-Huxley model, the potassium activation variable, n, impacts the conductance as n^4, which is equivalent to the effect of four independent subunits, with each one in the state necessary for channel-opening with probability, n. Would the exponent change from 4, and if so, in what direction, if the subunits were not independent but had some degree of cooperativity to produce a positive correlation between their states? Explain your reasoning.

3   What is the key similarity between anode break in the Hodgkin-Huxley model and postinhibitory rebound in the thalamocortical model neuron? What differences are there between the two?

4   If a model of a bursting neuron has too strong a calcium conductance, it may no longer produce a burst of sodium spikes, even though the soma has regular periods of what seems to be above-threshold depolarization. Why is this?

5

Connections between Neurons

A single neuron can be a powerful computational device in its own right. Yet it is the huge number of interconnected neurons that gives the brains of higher organisms such as humans their immense capabilities. Estimates for the human brain suggest we possess more than 80 billion neurons, with more than a thousand times that number of connections. The pattern of those connections is essential for our ability to respond appropriately to the outside world, to recall memories, and to plan for the future. In chapter 8, we will consider how those connections change through learning and development via the process of synaptic plasticity. In the current chapter, we will consider the operation of individual connections, how one neuron can affect another through its connections, and how two or three interconnected neurons give rise to useful behavior in a small circuit.

5.1   The Synapse

Neurons connect with each other through *synapses* (from two Greek words meaning "fastened together"). Two types of synapse, chemical and electrical, are physically distinct in their operation. Chemical synapses can also be subdivided into excitatory and inhibitory synapses, based on their functional effect.

Synapse   A connection between two cells.

5.1.1   Electrical Synapses

Electrical synapses contain a direct, physical connection between two neurons called a gap junction. The junction contains small pores, through which ions can flow. In the simplest case the gap junction can be modeled as a resistor, such that a constant conductance is added between the two cells in the same manner as between adjoining compartments of a multicompartmental model. In other cases, the gap junction can be rectifying, meaning it allows current to pass in only one direction. In this case, the conductance is dependent on the voltage difference across the junction and is set to zero if the neuron with higher membrane potential would generate current flow in the disallowed direction.

Gap junction   A connection between two cells, across which ions can directly flow.

5.1.2   Chemical Synapses

Chemical synapses are the most common type in mammalian brains. Two neurons connected by a chemical synapse are separated by a small gap, the *synaptic cleft* (figure 5.1). Interaction between the two neurons is via the release of chemicals called *neurotransmitters*. The dominant method of interaction is unidirectional, from the axon terminal of a presynaptic neuron to the dendritic spine, dendritic shaft, or soma of a postsynaptic neuron.

Neurotransmitter   The general term for any chemical that diffuses across the space between two neurons, allowing communication between them.

Presynaptic neuron   Referring to a particular synapse, the presynaptic neuron is the neuron that releases neurotransmitter across the synapse when it produces an action potential.

Postsynaptic neuron   Referring to a particular synapse, the postsynaptic neuron is the neuron that responds to neurotransmitter following an action potential in the presynaptic neuron.

Vesicles   Tiny sphere-like compartments surrounded by membrane, which contain particular chemicals, such as neurotransmitter, to be transported or released outside the cell.

When the presynaptic neuron produces an action potential, the voltage spike is transmitted along the axon, which is typically very branched, generating a spike in the membrane potential at all of its axon terminals. The voltage spike causes high-threshold (N-type "neural" and P/Q-type "Purkinje") calcium channels in the vicinity of the axon terminal to open. The ensuing calcium

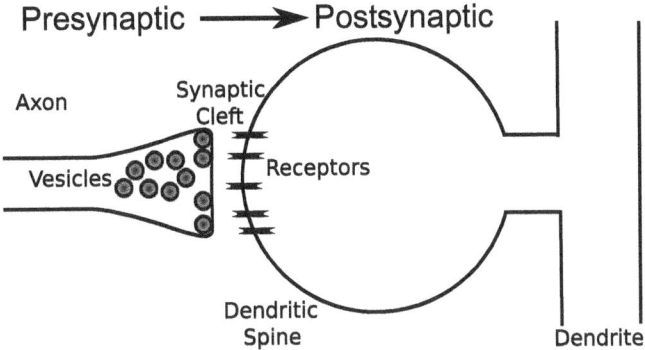

Figure 5.1
Sketch of a simplified chemical synapse. Vesicles containing neurotransmitter in the presynaptic axon terminal (left) can be docked, such that they abut the presynaptic cell's membrane and are ready to release neurotransmitter into the synaptic cleft following an action potential. Once they have released neurotransmitter they must be replaced by other nearby vesicles and recycled to make new vesicles. Neurotransmitter diffuses rapidly (in a few nanoseconds) across the synaptic cleft and binds to receptors in the postsynaptic neuron. The receptors shown here are on a dendritic spine, which is a small compartment with a narrow neck attaching it to the dendritic shaft, the main component of the dendrite. (Excitatory inputs to pyramidal cells, the most abundant excitatory cells in the cortex, are predominantly via synapses on spines.) The narrow neck permits flow of electrical current but can restrict diffusion such that spines are partially chemically isolated from each other while they are not electrically isolated.

influx initiates activation of proteins, which eventually cause *vesicles* (small, membrane-enclosed spherical containers) containing neurotransmitter to release their payload into the synaptic cleft.[1] The neurotransmitter diffuses across the small gap between cells and binds to receptors of the postsynaptic cell. When the receptors are bound by neurotransmitter they cause channels to open, either directly through a conformational change if they are ionotropic, or indirectly through a signaling pathway if they are metabotropic. The open channels cause electric current to flow, either into or out of the cell depending on the ion that flows, just as is the case with other ion channels in the cell membrane.

> **Receptor**   A protein that can be bound by a signaling molecule (such as a neurotransmitter) and, when bound, that either activates or changes its conformation in order to produce further change.

The series of steps described in the preceding paragraph (which omits several intermediate steps that provide fodder for many research careers) can seem like an overly tortuous pathway for one neuron to affect another. Amazingly, all of these biochemical processes and the intervening step of chemical diffusion can be completed within a millisecond, in part because they take place in a tiny volume of space. Indeed, when we simulate the interaction between two neurons it is common to treat the time from the action potential of one neuron to the conductance change in another as negligible. If any delays are added in simulations, typically these are due to propagation of the action potential from the soma to the end of the axon, or (either passive or active) propagation of the postsynaptic voltage change from a distal dendrite to the soma. Such delays may be important when considering spike-time dependent plasticity (section 8.3) in which the change in strength of a synapse can depend sensitively on the relative timings of a presynaptic and a postsynaptic spike.

The series of biochemical processes involved in synaptic transmission gives rise to some important computational consequences. The first is a short-term history dependence of the synaptic efficacy, which we will consider in section 5.3. The second is the introduction of randomness or noise, because an action potential does not guarantee the release of a vesicle of neurotransmitter, which depends on several proteins binding to the vesicle or activating vesicle-release machinery in concert. These stochastic biochemical reactions—whose ultimate source of randomness is the bouncing around of molecules due to thermal energy—are likely the major source of random variability in membrane potentials in vivo and hence, ultimately, a source of variability in animal behavior.

If there are many vesicles near the synaptic cleft, the variability in neurotransmission is reduced. For example, in the end plate junction—the connection between motor neurons and muscle fiber—many hundreds of vesicles can be released. This is important, because while we may not want perfect predictability in some of our thought processes, when we need to jump out of the way of a car, our leg muscles must be guaranteed to contract!

Synapses in the retina and in many invertebrates with thousands of vesicles allow for continuous and *graded release* of neurotransmitter, so that small changes in the membrane potential of a presynaptic cell can impact the current flow into a postsynaptic cell in the absence of any action potential.

Motor neuron   A neuron that helps control an animal's movement.

Graded release   Neurotransmitter release at a rate that can increase or decrease according to the presynaptic membrane potential.

The most important feature of a chemical synapse is whether synaptic transmission causes an increase or a decrease of the postsynaptic membrane potential. Synapses that cause a depolarization are *excitatory*, while synapses that cause a hyperpolarization or reduce any depolarization are *inhibitory*. Whether a synapse is excitatory or inhibitory depends on the neurotransmitter being released and the receptors in the postsynaptic cell. The effects of a neurotransmitter can be qualitatively different across species and even across the course of development between the same two cells when reversal potentials (i.e., equilibrium potentials) change due to changes in ionic concentrations (see equation 2.1, the Nernst equation).

Excitatory synapse   A synapse causing depolarization of the postsynaptic cell.

Inhibitory synapse   A synapse causing hyperpolarization of the postsynaptic cell.

Dale's Principle   All synapses of a presynaptic cell release the same types of neurotransmitter, sometimes extended to denote the situation in particular circuits where neurons are either only excitatory or only inhibitory in their effect on other cells.

Dale's principle states that the neurotransmitters released by a neuron at one synapse are the same as the neurotransmitters released by that neuron at another synapse. That is, while most neurons have synapse-specific receptors, receiving different types of input from different types of presynaptic cells, they typically have the same type of output at all axon terminals. In this form, Dale's principle has very few exceptions, but it is commonly rephrased to state, incorrectly in general, that a presynaptic cell provides only excitatory or only inhibitory input to other cells. In some circuits the latter statement is true, in which case it is convenient to classify neurons as excitatory or inhibitory according to the neurotransmitter that they release most abundantly.

The reason such classification is an oversimplification is twofold. First, many neurotransmitters have mixed effects—i.e., whether they excite a cell or inhibit it can depend on the postsynaptic receptor, the excitatory D1 and inhibitory D2 receptors for dopamine being the best-known example.[2,3] Second, neurons can release different types of, often modulatory, neurotransmitters together in a process called cotransmission, so the postsynaptic effects depend on the relative abundance of receptors to the different neurotransmitters.

It is reasonable to label a cell as either excitatory or inhibitory, when it packages just one type of neurotransmitter at its axon terminals and if, when that neurotransmitter can bind to more than one type of postsynaptic receptor, the different receptors cause qualitatively the same postsynaptic effect.

Glutamate   The dominant excitatory neurotransmitter in mammals.

AMPA receptor   A receptor, which when bound by glutamate produces a rapid excitatory response in the postsynaptic cell.

NMDA receptor   A receptor, which when bound by glutamate, and when the postsynaptic membrane potential is high, produces a slower excitatory response in the postsynaptic cell.

γ–Aminobutyric acid (GABA)   The dominant inhibitory neurotransmitter in mammals.

For example, the most common neurotransmitter released by excitatory cells in mammalian cortex is glutamate. Glutamate binds to two common receptors that are expressed together in postsynaptic cells, N-methyl-D-aspartate (NMDA) receptors and α–amino-3-hydroxy-5-methyl-4-isoxazolepropionic acid (AMPA) receptors. The two receptors each cause excitatory channels to open, allowing flow of a mixture of positive ions (cations) with a net flux into the cell, but otherwise they have distinct properties. AMPA receptors respond rapidly to glutamate binding, opening almost instantaneously and closing over a timescale of a couple of milliseconds. NMDA receptors have slower dynamics and are able to remain open for many tens of milliseconds. They also admit calcium ions, which can be important for biochemical signaling and they contain a magnesium block that requires postsynaptic depolarization to be released. Thus, their conductance depends not only on when the presynaptic cell spiked but also on the membrane potential of the postsynaptic cell. Still, while glutamate mediates excitatory synaptic transmission in mammalian brains, it is worth noting that it has an inhibitory effect in many insect brains where it causes chloride channels to open.

The most common inhibitory neurotransmitter in mammalian brains is γ–aminobutyric acid (GABA). The receptors for GABA cause chloride channels to open. Chloride ions are in higher concentration outside the cell, so they flow into the cell unless the cell is hyperpolarized. While the reversal potential of chloride channels is always negative, early in development it can be far enough above the resting membrane potential for the chloride current to be outward and have an excitatory effect on neurons. Later in development the reversal potential drops toward the resting potential or below—certainly far from threshold—such that channel opening produces an inhibitory effect on the postsynaptic cell.

In this book, we will follow the simplified rule of cells being either excitatory or inhibitory, because such a rule describes the dominant interactions in the circuits that we focus on. However, the reader should be aware of the existence of many exceptions, more of which are likely to be discovered each year.

5.2   Modeling Synaptic Transmission through Chemical Synapses

The effect of synaptic transmission can be simulated via the synaptic conductance in the post-synaptic cell. Just as we saw that sodium conductance and potassium conductance were treated separately in single-cell models, so too must each type of synaptic conductance be treated with a separate term in the differential equation. Different receptors of neurotransmitter cause channels to open and close with different dynamics and these channels may admit different ions with different reversal potentials. It is important to note that it is the reversal potential of the ion channels that determines whether a synapse is excitatory or inhibitory—even if those channels allow flow of negative ions, their conductance is always positive when they open.

The general form of the equation for current admitted to a cell due to a particular synaptic conductance is (compare to equation 2.2):

$$I_{syn}(t) = G_{syn}(t)[V_{syn} - V(t)] \tag{5.1}$$

where V_{syn} is the reversal potential, $G_{syn}(t)$ is the synaptic conductance which depends primarily on the presynaptic neuron, and $V(t)$ is the membrane potential of the postsynaptic neuron. As always, the parentheses (t) in the term $G_{syn}(t)$, as in the terms $I_{syn}(t)$ and $V(t)$, are optional and are used to indicate a quantity that varies with time.

5.2.1   Spike-Induced Transmission

The simplest time-dependence for $G_{syn}(t)$ is a step increase at the time of a presynaptic spike, followed by an exponential decay back to zero. This can be modeled by:

$$\frac{dG_{syn}(t)}{dt} = \frac{-G_{syn}(t)}{\tau_{syn}} \tag{5.2}$$

and

$$G_{syn}(t) \mapsto G(t) + \Delta G \text{ if } t \in \{t_{spike}\}, \tag{5.3}$$

where $\{t_{spike}\}$ is the set of presynaptic spike times so the term $t \in \{t_{spike}\}$ (literally "t is a member of that set") simply means add ΔG whenever t is the time of a presynaptic spike. Equation 5.2 produces exponential decay to zero with a time constant, τ_{syn}, that is specific to the type of synapse. We will be simulating the above model of synaptic transmission in tutorial 5.1.

The preceding equations for synaptic transmission are particularly conducive to large-scale simulations because once the conductance is incremented following a presynaptic spike, no further memory of that spike time is needed—the ensuing change in postsynaptic conductance only depends on its value at that time, not the earlier time of the spike. Moreover, the equation for rate of change of conductance (equation 5.2) is linear, so the effects of all synapses of a given type within a compartment can be summed together and a single equation (exactly like equation

5.2) solved for their net effect—i.e., we do not need to keep track of the conductance of each individual synapse.

One simplification in the above model is the immediate step increase in conductance. This could be mitigated by incorporation of a delay from presynaptic spike to the time of the step increase, without reducing the computational ease of combining multiple synapse-specific conductance terms into one total. However, a more realistic method, particularly for NMDA receptor-mediated channels with their slow time constant, is addition of a rise time to the model of the channel conductance.

An example of a method with such a rise time, shown in figure 5.2, is a simulation of the dynamics of the conductance following a presynaptic spike as:

$$\Delta G_{syn}(t) = \frac{\Delta G}{K} e^{-(t-t_{spike})/\tau_{decay}} \left[1 - e^{-(t-t_{spike})/\tau_{rise}}\right], \tag{5.4}$$

where τ_{rise} is the rise time and τ_{decay} is the decay time such that $\tau_{decay} > \tau_{rise}$. The factor K is included to ensure the peak height of the conductance change is ΔG. A bit of math (the interested reader can show this for herself) yields

$$K = \left(\frac{\tau_{decay}}{\tau_{rise} + \tau_{decay}}\right)\left(\frac{\tau_{rise}}{\tau_{rise} + \tau_{decay}}\right)^{\tau_{rise}/\tau_{decay}}.$$

In simulations with this method, each spike produces a change in $G_{syn}(t)$ that must be calculated either once for all later times at the time of the spike, or on each succeeding time point using the time of that earlier spike. Once the preceding formula has been used to calculate the contribution, $\Delta G_{syn}(t)$, from each spike, those contributions are summed together to produce $G_{syn}(t)$. Clearly the realistic method is more involved than the simple method, but it will be

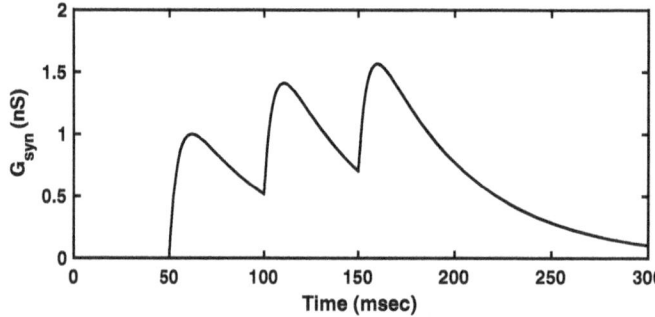

Figure 5.2
Synaptic conductance accumulation and decay. Conductance $G_{syn}(t)$, following equation 5.4 in response to a series of three spikes at intervals of 50 ms. Parameters are $\Delta G = 1$ nS, $\tau_{rise} = 5$ ms, and $\tau_{decay} = 50$ ms. This figure is produced by the online code synaptic_opening.m.

useful to see how to incorporate it into codes in tutorial 5.1. Except in situations where the precise timing of spikes is very important (e.g., in sound-source localization or in cases where network synchronization is possible) the simpler method should be fine—but one should always check if a simulated circuit's behavior depends on such simplifications.

5.2.2 Graded Release

Neurotransmitter release is always voltage-dependent, in the sense that it depends on calcium influx through voltage-gated calcium channels. However, in most neurons that produce discrete, fast action potentials, the membrane potential at the axon terminal only ever rises above the threshold for calcium influx at the discrete times of spikes. Until now we have just considered the modeling of such all-or-none events.

However, many neurons, for example in invertebrates and in the retina, can release neurotransmitter in a graded, voltage-dependent manner in the absence of a spike.

A simple model of channel opening due to graded release of neurotransmitter consists of a synaptic conductance variable with a voltage-dependent steady state and voltage-dependent time constant for its approach to the steady state:[4,5]

$$\frac{dG_{syn}(t)}{dt} = \frac{G_\infty(V_{pre}) - G_{syn}(t)}{\tau_{syn}(V_{pre})}. \tag{5.5}$$

In some ways, this provides current to a cell in a similar manner to a persistent conductance (such as the potassium conductance of equation 4.9 if the synapse is inhibitory, or the calcium conductance of equation 4.20 if the synapse is excitatory). However, since we are modeling a synaptic conductance here, its steady state during release is given as a positive monotonic function of the presynaptic (not postsynaptic) membrane potential. For example,

$$G_\infty(V_{pre}) = \frac{G^{(max)}}{1 + \exp\left[\frac{-(V_{pre} - V_{th})}{V_{range}}\right]} \tag{5.6}$$

produces a synaptic conductance that can be maintained over a range from zero to its maximum, $G^{(max)}$, while the presynaptic voltage, V_{pre}, varies over a range of order V_{range} around V_{th} (figure 5.3A).

The time constant is modeled so that the conductance increases quickly with a rise in V_{pre}, but falls more slowly when V_{pre} decreases (figure 5.3B, C):

$$\tau_{syn}(V_{pre}) = \tau_{rise} + (\tau_{decay} - \tau_{rise})\left[1 - \frac{G(V_{pre})}{G^{(max)}}\right] = \tau_{rise} + \frac{(\tau_{decay} - \tau_{rise})}{1 + \exp\left[\frac{(V_{pre} - V_{th})}{V_{range}}\right]}. \tag{5.7}$$

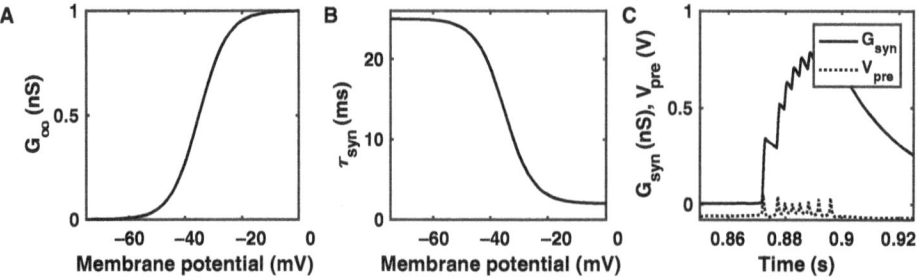

Figure 5.3
Model of synaptic transmission by graded release. (A) Steady-state conductance is a continuous function of presynaptic membrane potential. (B) Time constant is slow for unbinding of neurotransmitter at low presynaptic membrane potential and fast for binding at high presynaptic membrane potential. (C) Response of the model synaptic conductance (solid) to a single burst of presynaptic spikes (dotted, produced by the Pinsky-Rinzel model, figure 4.16). Parameters for this figure are $G^{(max)} = 1$ nS, $V_{th} = -35$ mV, $V_{range} = 5$ mV, $\tau_{decay} = 25$ ms, and $\tau_{rise} = 2$ ms. This figure is produced by the online code graded_release.m and the file PR_VS.mat.

Equation 5.7 models the rapid binding of neurotransmitter once released to produce rapid opening of channels, which can remain open for some time without further neurotransmitter release.

5.3   Dynamical Synapses

We use the term "dynamical synapse" here to refer to the rapid change in the effective strength of a synaptic connection that arises following each spike and recovers on short timescales—typically over hundreds of milliseconds, but occasionally over tens of seconds. Such variation is commonly referred to as short-term plasticity,[6-9] but in this book, we reserve the term "plasticity" for changes that persist (chapter 8).

Dynamical synapse   A synapse whose effective strength varies on a short timescale due to the history of incoming action potentials and vesicle release.

Short-term synaptic depression   A temporary reduction in synaptic strength.

Short-term synaptic facilitation   A temporary increase in synaptic strength.

5.3.1   Short-Term Synaptic Depression

When a certain number of vesicles release their neurotransmitter from the axon terminal, the number of release-ready vesicles must immediately decrease by that number. Such a reduction in the number of release-ready vesicles causes a reduction in the synaptic transmission produced by an action potential arriving immediately after vesicle release. Therefore, if one action potential were to immediately follow another, all other things being equal, the second action potential would produce a weaker increase in postsynaptic conductance. Such immediate reduction in synaptic efficacy following an action potential is called synaptic depression.

The reduced efficacy typically recovers on a timescale of 100–200 milliseconds, the timescale over which other full vesicles dock in the membrane to become release-ready. Synaptic depression is strong in those synapses with a high baseline probability of release such that a first action potential may cause a substantial reduction in the number of docked vesicles.

Saturation   The maximum level of a process at which no further increase is possible.

Short-term depression can cause synaptic transmission to depend more strongly on changes in the firing rate of a presynaptic cell than on its stable level of activity. In particular, if a presynaptic neuron is very active, synaptic depression leads to a saturation of synaptic transmission—the rate of neurotransmitter release becomes limited by the rate at which vesicles full of neurotransmitter can dock at the axon terminal rather than the rate of incoming spikes. Therefore, when the activity in the presynaptic cell switches from a low rate to a maintained high rate, the postsynaptic conductance might increase strongly with the initial rate change, but then decay back toward the value it had before the change.[10] We will simulate such effects in tutorial 5.1.

Mathematically, synaptic depression produces negative feedback, since the greater the firing rate the lower the synaptic efficacy. Negative feedback is a method for control of circuit activity, since by limiting the amount of synaptic transmission, the possibility of runaway feedback excitation is reduced. Synaptic depression also causes adaptation to ongoing inputs. Adaptation, which we have seen in the spike-rates of neurons (figures 2.7–2.8), is important in sensory systems, allowing us to perceive changes in inputs without being swamped by background levels.

5.3.2   Short-Term Synaptic Facilitation

Synaptic facilitation provides a counterpoint to synaptic depression, since it increases synaptic efficacy immediately after a presynaptic spike. Synaptic facilitation corresponds to a temporary increase in the release probability of those vesicles that remain docked in the membrane of the axon terminal following a spike. Release-ready vesicles may be bound by some but not all of the proteins necessary for release of neurotransmitter following an action potential. If a second action potential follows soon on the heels of the first, those proteins may still be binding the vesicle, making it much more likely for all the necessary steps to be completed for neurotransmitter release. Residual calcium in the presynaptic terminal also enhances release probability.[9] The timescale over which facilitation persists depends on the rates of dissociation of those proteins and return of calcium to baseline levels. In most observations, the longest time constant is about a half-second, but timescales of several seconds have been observed and used in models.[9,11]

Synaptic facilitation is a positive feedback process—the more rapidly spikes arrive at the axon terminal the more likely any remaining vesicles will release neurotransmitter, so the greater the synaptic efficacy. The positive feedback can allow the postsynaptic cell to remain insensitive to any low-rate spontaneous activity of a presynaptic cell ("noise") while producing very strong responses to higher firing rates or a burst of presynaptic activity ("the signal").

Synaptic facilitation can dominate over synaptic depression when initial release probability is low. For such a synapse, the fraction of vesicles lost after an initial spike is low, so synaptic depression is not strong—most vesicles still remain for the next spike. Moreover, if the initial release probability is low, it has a lot of "room" to increase toward one for a subsequent spike. Conversely, if the release probability is nearly one for an initial spike, it cannot increase further for a subsequent spike and facilitation is not possible, yet depression would be strong.

5.3.3 Modeling Dynamical Synapses

The expected number of vesicles releasing neurotransmitter following an action potential is the number of docked, release-ready vesicles multiplied by their individual probability of release. The first number is affected by synaptic depression, the second by synaptic facilitation, so we can simply multiply the two effects together to obtain the synaptic efficacy at any time.[7,8]

We let D be the variable for synaptic depression ($0 \leq D \leq 1$) denoting the fraction of docked release-ready vesicles compared to the number it attains after quiescence. We let F be the variable for synaptic facilitation ($1 \leq F \leq F_{max}$) such that the release probability at any point in time is $p_0 F$, where p_0 is the initial release probability. Note that $F_{max} = 1/p_0$ since the maximum release probability is one. By combining these terms, we find that the change in synaptic conductance following a spike is given by:

$$\Delta G = G_{max} p_0 FD. \tag{5.8}$$

Both the depression and facilitation variables return to the value of one in between spikes, so follow equations of the same form,

$$\frac{dD}{dt} = \frac{1-D}{\tau_D}, \tag{5.9a}$$

and

$$\frac{dF}{dt} = \frac{1-F}{\tau_F} \tag{5.9b}$$

which produce decaying exponentials between spikes. A qualitative difference arises between equation 5.9a and equation 5.9b, because D increases to 1 from lower values, while F decreases to one from higher values. Following each spike the values of F and D are updated via:

$$D \mapsto D - p_0 FD \quad \text{and} \quad F \mapsto F + f_{fac}(F_{max} - F). \tag{5.10}$$

The first update reduces D by the fraction of vesicles emptied during the spike. Notice that we are taking the mean here, a quantity that may not be valid for a particular synapse—for example, if a synapse has five vesicles with a release probability of $p_0 F = 0.5$, this equation suggests that 2.5 vesicles are released. Such a result is fine if we assume we are averaging over many synapses. However, when individual synapses are simulated independently then, in a much more computationally intensive procedure, the binomial distribution must be sampled for each synapse to produce probabilistically, the integer number of vesicles released. For such a simulation, a variable indicating the state of each synapse must be stored. Instead, here we just need a single variable to represent the mean state of synapses that can release neurotransmitter from the presynaptic cell.

The second update increases the facilitation variable toward its maximum by a factor f_{fac} that denotes the degree of facilitation ($0 \leq f_{fac} \leq 1$). If f_{fac} is zero, there is no facilitation, while if f_{fac} is one then a single presynaptic spike causes the release probability to jump to one immediately thereafter. In practice, those synapses with the most facilitation never have an increase in release probability of more than 4 following a spike. This limits the valid values of f_{fac} to be less than $3p_0/(1-p_0)$ for $p_0 \leq 1/4$ to ensure $F \leq 4$, and $f_{fac} \leq 1$ for $p_0 \geq 1/4$ so that $F \leq F_{max}$ and release probability is never greater than one.

5.4   Tutorial 5.1: Synaptic Responses to Changes in Inputs

Neuroscience goal: Understand the contrasting effects of synaptic depression and facilitation on the steady-state and transient responses of synapses to changes in presynaptic rates.

Computational goals: Gain more practice with a time-varying Poisson process; combine multiple variables with stepwise changes and continuous variation in coupled ODEs.

In this computer tutorial, you will treat the spike train of a presynaptic neuron as a Poisson process whose emission rate will change stepwise at discrete times. You will examine the synaptic response to these inputs first in a synapse without short-term changes in efficacy, second in a synapse with synaptic depression, and finally in a synapse with synaptic facilitation. If time permits you will simulate the response of a postsynaptic LIF cell to these inputs. This tutorial is based on figure 5.19 of the textbook by Dayan and Abbott.[12]

a.   Produce a time vector, with timesteps of size $\delta t = 0.1$ ms, from 0 to 4 s.

b.   Produce a vector of the same size as the time vector with presynaptic firing rates of 20 Hz, 100 Hz, 10 Hz, and 50 Hz for each one-second portion of the time period.

c. Produce an array of ones and zeros to represent a Poisson spike train (section 3.3.3), such that the probability of a spike in any time bin of size δt is equal to $\delta t \times r(t)$ where $r(t)$ is the firing rate at that time-bin generated in b) (you can ignore the possibility of multiple spikes in a time-bin).

d. Produce a synaptic conductance vector, $G_{syn}(t)$, which increments by $\Delta G = 1$ nS when each spike arrives and decays back to zero between spikes with a time constant of 100 ms. Plot $G_{syn}(t)$ versus time.

e. Assume an initial release probability of $p_0 = 0.5$ and produce a synaptic depression vector, $D(t)$, that is initialized to one, decreases by an amount $p_0 D(t)$ following each spike at time, t, and recovers via $(dD(t))/dt = (1 - D(t))/\tau_D$ with time constant of $\tau_D = 0.25$ s.

f. Produce a second synaptic conductance vector with decay time constant of 100 ms that increments by $\Delta G = \Delta G_{max} p_0 D(t_-)$ following each spike at time, t, where $\Delta G_{max} = 5$ nS, $p_0 = 0.5$, and where $D(t)$ is obtained from (e). The symbol t_- is to indicate the time immediately preceding the spike, so that you use the value of $D(t)$ calculated just before its decrease due to that spike. Plot the conductance versus time.

g. Assume an initial release probability of $p_0 = 0.2$ and produce a synaptic facilitation vector, $F(t)$, that is initialized to one, increases by an amount $f_{fac}(F_{max} - F(t_-))$ following each spike at time, t, and recovers via $(dF(t))/dt = (1 - F(t))/\tau_F$ with time constant of $\tau_F = 0.25$ s, facilitation factor, $f_{fac} = 0.25$, and maximum facilitation of $F_{max} = 1/p_0$.

h. Produce a new synaptic depression vector initialized to one, which decreases by an amount $p_0 F(t_-) D(t_-)$ following each spike at time, t, and which recovers as in e), with $p_0 = 0.2$ and $F(t)$ generated in (g).

i. Produce a third synaptic conductance vector with decay time constant of 100 ms that increments by $\Delta G = \Delta G_{max} p_0 F(t_-) D(t_-)$ following each spike at time, t, where $\Delta G_{max} = 4\ nS$, $p_0 = 0.2$, $F(t)$ is obtained from (g), and $D(t)$ is obtained from h). Plot the conductance versus time.

Comment

100 ms is a long synaptic time constant, but it is necessary in this simplified model to see accumulation of inputs from a single spike train. To observe such accumulation, we have assumed that the postsynaptic receptors are never completely saturated by neurotransmitter. However, this is unlikely when neurotransmitter is released at a high rate and the time constant for dissociation and closing of channels is as high as 100 ms. To include the effects of such saturation the update term should be replaced, for example in (i), by $\Delta G = (G_{max} - G_{syn}) p_0 F(t_-) D(t_-)$, with $G_{max} = 4$ nS.

Alternatively, we can assume that a time-varying stimulus impacts a neuron via multiple synaptic inputs. In option A we will simulate fifty such inputs via synapses with a 2 ms time constant—a short enough synaptic decay time that saturation can be neglected.

Challenge Problems
Option A

j. Initialize a new total conductance vector of the same size as the time vector. Using a `for` loop, repeat (c)–(d) fifty times (once each for each of the fifty inputs), each time using a new Poisson spike train. Make the alteration of a 2 ms time constant for decay of synaptic conductance. Within the loop accumulate the total conductance at each point in time by summing across the values generated for that time point by each of the fifty input vectors. Plot the summed vector and compare it to the one plotted in (d).

k. Repeat (j), but use synaptic depression—with a separate value for each input—while using all other variables as in (e) and (f). Compare the summed conductance vector to the one plotted in (f).

Option B

l. Simulate the postsynaptic cell as a leaky integrate-and-fire neuron with parameters $E_L = -65$ mV, $E_{syn} = 0$ mV, $G_L = 2$ nS, $C_m = 20$ pF. The model neuron will receive inputs $G_{syn}(t)$ given by each result produced above in (d), (f), and (i). In this example, E_{syn} is the reversal potential of an excitatory synapse. (If the synapse were inhibitory, we would use a value for E_{syn} in the region of E_L.) Use the equation

$$C_m \frac{dV}{dt} = G_L(E_L - V) + G_{syn}(t)(E_{syn} - V)$$

to simulate the membrane potential, V, which produces a spike whenever it crosses a threshold given by $V_{th} = -50$ mV, after which it is immediately reset to $V_{reset} = -80$ mV. Plot V versus t and count the number of spikes produced in the four intervals of distinct input firing rate, commenting on how the numbers differ across the three types of synapse.

5.5   The Connectivity Matrix

A remarkable property of neurons that distinguishes them from other cells is their physical structure. Dendrites and axons of a neuron can extend far from the cell body, ramify to generate hundreds of branches with thousands of possible connection points, and target specific regions and subregions or layers of the brain, producing contacts with a highly selective subset of other cells. In most models of neural circuits—including those produced in this course—all of the spatial intricacy and important cell biology needed to produce such extensive architectures is subsumed into a connectivity matrix, which simply tells us whether any two neurons are connected and ignores their actual position within the physical space of the brain.

> **Connectivity matrix**   A square matrix indicating the absence or presence, the direction, and
> sometimes the strength of all connections within a group of neurons.

One concession to the reality of physical space in some models is the inclusion of a delay from spike generation in the soma to the time of synaptic transmission, or from the time of EPSC in a dendrite to current flow in the soma. Multicompartmental models can and do include more realistic consequences of the physical dimension for each cell. As more data is accrued, not only of which cells are connected, but also where those connections occur relative to the cell body, such features can be incorporated into circuit models.

In this course, any model of a neural circuit is simulated with point neurons, so all that is needed is a single number to indicate the presence and strength of any connection, between each pair of neurons. The connectivity of an entire circuit can then be represented as a matrix of size $N \times N$, where N is the number of neurons. See section 1.4.2 for an introduction to the use of matrices if necessary.

Connectivity matrices can be defined at multiple scales. For a circuit containing a small number of neurons, each entry in the connectivity matrix represents a directed connection from one neuron to another—a single entry could represent multiple connections between the same presynaptic and postsynaptic cells. In models of larger circuits and networks, each entry can represent the effect of one functional unit on another, where a functional unit comprises a group of cells with similar responses to stimuli—for example, similar receptive fields in sensory areas. The average activity of such groups acting as functional units is considered via firing rate models in chapter 6. At the largest scale of whole human brain models, each element in the connectivity matrix corresponds to the connection or correlation between discrete volumes of brain tissue. Examples of such large-scale connectivity matrices include the functional connectivity based on correlated activity between regions acquired by functional magnetic resonance imaging (fMRI), and structural connectivity based on magnetic resonance imaging of fluid flow combined with tensor analysis (the whole process known as diffusive tensor imaging, DTI). In this section, we focus on connections at the level of individual neurons.

If we write W_{ij} as the strength of connection from neuron i to neuron j then each row of the matrix corresponds to a single presynaptic cell and each column corresponds to a single postsynaptic cell. Since synaptic connections are directional, a synapse from neuron i to neuron j can be present without a synapse from j to i. This means the matrix W_{ij} is not symmetric and rows cannot be switched with columns—so one must be careful about which (of rows or columns) correspond to presynaptic cells and which correspond to postsynaptic cells.

The effect of a connection is determined not by a single number—its strength—but also by other synaptic properties. In particular, given differences in the reversal potential, one needs separate connectivity matrices for inhibitory connections when simulating synaptic input as a change in conductance. However, in simpler models, synaptic input can be simulated as a current and it is possible to use a single matrix with negative entries for inhibitory connections. The negative entry corresponds to the amount of negative current that can be produced by inhibitory synaptic input. Dale's principle could then be observed by ensuring each row of the matrix only contains positive entries for an excitatory presynaptic cell and only negative entries for an inhibitory presynaptic cell.

5.5.1 General Types of Connectivity Matrices

The diagonal entries of a connectivity matrix correspond to self-connections or autapses, which are uncommon in practice (an exception being for fast-spiking inhibitory neurons).[13] Thus, diagonal entries are usually zero for connectivity matrices representing individual neurons. This is in contrast to connections between units or groups of similar cells, in which case connections within a group can be the strongest, leading to diagonal entries of the connectivity matrix being the largest.

It can be daunting to observe structure within a connectivity matrix of more than a few neurons. When there is underlying structure it can be made more apparent by appropriate ordering of the neurons. If two rows of a connectivity matrix are switched with each other and the corresponding two columns are switched, then the connections represented by the matrix are unchanged—only the labels of two of the neurons have been switched with each other (figure 5.4). In tutorial 5.2 we will see how structure can be made clearer within the connectivity matrix by appropriate ordering of the rows and columns.

Some particular terms for features of a connectivity matrix are:

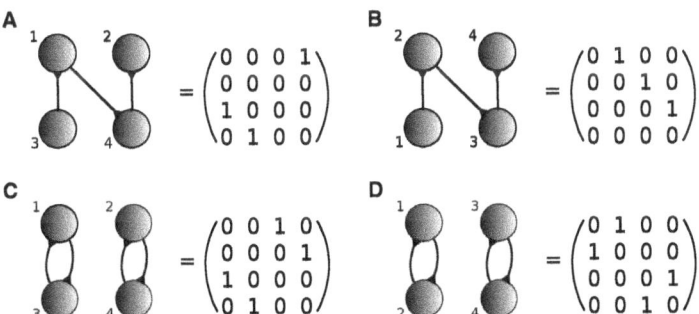

Figure 5.4
Connections between cells can be represented by a matrix. In this example, entries of 1 or 0 in the matrix indicate respectively the presence or absence of a connection between two cells. The row number of the matrix corresponds to the presynaptic cell and the column number the postsynaptic cell. (A) A feedforward circuit can be hidden within the connectivity matrix until (B) appropriate rearrangement of neurons reveals an upper-triangular connectivity matrix. (C) Disconnected groups of neurons produce a block diagonal connectivity matrix (D) after appropriate rearrangement. In the transition from C to D, cells 2 and 3 are swapped, so in the connectivity matrix the second and third columns are swapped *and* the second and third rows are swapped.

Sparse: Most possible connections are absent, so most entries of the matrix are zero.

Globally feedforward: Neurons can be ordered with no connection from a later neuron to an earlier one in the order. The matrix can be transformed to be upper-triangular (figure 5.5A).

Locally feedforward: A recurrent network can be locally feedforward if neurons connect to neurons without a reciprocal connection, but it may receive inputs from neurons that are a few connections downstream, so the architecture is one of a loop with a particular direction of information flow or activation around the loop (figure 5.5B).

Recurrent: Neurons receiving input from a cell can influence that cell in return, either by a direct connection or via connections to other cells (figure 5.5C).

Disconnected: Some sets of neurons neither receive input nor provide input to other sets. In such a case the connectivity matrix can be rearranged into a block diagonal form (figure 5.5D).

Clustered: The neurons can be arranged into groups with significantly stronger connections or greater proportion of connections within a group than between groups.

Local: Neurons can be labeled with indices such that neurons with similar indices are more likely to be connected than neurons with very different indices. If the indices are spatial coordinates then local connectivity is local in space, so connections are spatially structured (figure 5.5C).

Small world: Any neuron can reach any other neuron via a small number of connections and neurons that are connected with each other connect with a highly overlapping group of other neurons. Clustered networks can be small world networks.

5.5.2 Cortical Connections: Sparseness and Structure

Within cortical circuits the connections are sparse—even cells whose dendrites and axons intertwine connect with only 10–20 percent likelihood. The probability that any two cells are connected increases with the number of other cells with which they have connections in common. Another way of putting this is that if two cells are connected, the two sets of other cells to

A Globally feedforward

$$\begin{pmatrix} 0 & 1 & 1 & 0 & 0 & 1 \\ 0 & 0 & 1 & 0 & 1 & 0 \\ 0 & 0 & 0 & 1 & 1 & 1 \\ 0 & 0 & 0 & 0 & 1 & 0 \\ 0 & 0 & 0 & 0 & 0 & 1 \\ 0 & 0 & 0 & 0 & 0 & 0 \end{pmatrix}$$

B Locally feedforward

$$\begin{pmatrix} 0 & 1 & 1 & 0 & 0 & 0 \\ 0 & 0 & 1 & 1 & 0 & 0 \\ 0 & 0 & 0 & 1 & 1 & 0 \\ 0 & 0 & 0 & 0 & 1 & 1 \\ 1 & 0 & 0 & 0 & 0 & 1 \\ 1 & 1 & 0 & 0 & 0 & 0 \end{pmatrix}$$

C Recurrent and local

$$\begin{pmatrix} 0 & 1 & 0 & 0 & 0 & 1 \\ 1 & 0 & 1 & 0 & 0 & 0 \\ 0 & 1 & 0 & 1 & 0 & 0 \\ 0 & 0 & 1 & 0 & 1 & 0 \\ 0 & 0 & 0 & 1 & 0 & 1 \\ 1 & 0 & 0 & 0 & 1 & 0 \end{pmatrix}$$

D Disconnected

$$\begin{pmatrix} 0 & 1 & 1 & 1 & 0 & 0 \\ 1 & 0 & 1 & 1 & 0 & 0 \\ 1 & 1 & 0 & 1 & 0 & 0 \\ 1 & 1 & 1 & 0 & 0 & 0 \\ 0 & 0 & 0 & 0 & 0 & 1 \\ 0 & 0 & 0 & 0 & 1 & 0 \end{pmatrix}$$

Figure 5.5
Examples of connectivity matrices with different network architecture. (A) The globally feedforward network is upper-triangular (no nonzero elements on the diagonal or below it). (B) This locally feedforward network is globally recurrent, because a chain of connections commencing from the first cell eventually returns to the first cell. It is locally feedforward because neurons can be arranged in a circle with connections only directed clockwise. (C) In this recurrent and local network the neurons can be arranged in a circle with connections only to nearest neighbors. (D) In this disconnected network, none of the first four neurons are connected with any of the last two neurons.

which they each connect have more overlapping members than is expected by chance. Furthermore, if any two cells respond similarly to stimuli, they are more likely to be connected than is expected by chance. Such connectivity features can arise via correlational learning (Hebbian learning), which we will discuss in chapter 8, and can be summarized as "connections preferentially strengthen and remain intact between pairs of neurons that are active at the same time."

5.5.3 Motifs

When the presence or absence of connections between all pairs of neurons in a group is known, then various motifs can be enumerated and compared to chance or to predictions of models. Motifs represent all the distinct ways in which a small group of cells can be interconnected and are best considered visually. The simplest motifs are for two cells, which can be disconnected, or unidirectionally connected, or bidirectionally connected, producing three possible motifs. For a group of three cells of the same type, there are in fact sixteen distinct possible motifs (shown in figure 5.6).

Motif   The representation of a possible pattern of connections within a group of cells, with connections depicted as lines or arrows (edges) and the cells as points (nodes).

When considering the abundance of different motifs, one should consider first what would be expected a priori from random connections, and then what would be produced by different

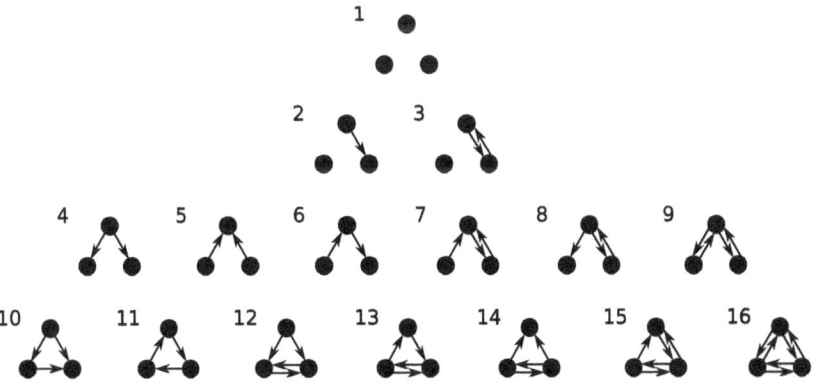

Figure 5.6
Connectivity motifs for groups of three neurons. Sixteen connection patterns are possible, differing in the numbers of unidirectional or bidirectional connections or in the pattern of cells so connected. Any other pattern can be rearranged into one of the depicted forms by swapping the positions of one or two pairs of cells (which does not affect the motif, since the positions are arbitrary). Empirical data from rat visual and somatosensory cortices[14,15] show that the triply connected motifs on the bottom row are overrepresented compared to chance.

types of connectivity matrix. Also, when measuring these connectivity patterns in real tissue, it is important first to account for spatial location of cells—if two cells are far apart then the likelihood they are connected is lower than if they are nearby, simply because axons and dendrites have limited ranges over which they are highly branched with a high density of potential contacts. The decrease in connection probability with spatial separation leads to an abundance of the highly connected motifs (e.g., motifs 12–16 in figure 5.6) compared to chance if the circuit being characterized spreads over a wide range of space. That is because if two cells are connected they are more likely to be near each other and both cells are more likely to connect to a third nearby cell than to other cells on average. Such factors mean that motifs are often measured only within local circuits where all of the cells have the potential to be highly interconnected.

However, one benefit of the computational approach advocated in this book, is that the effects of confounding features observed in experiments can be simulated and accounted for. In this case, spatial structure can be controlled for by comparison with simulated connectivity matrices that incorporate the observed separation-dependence of connection probability. Thus, one could assess whether particular motifs observed in brain tissue are overabundant in excess of what is expected from the spatial distribution of cells—and observed in simulated circuits. Moreover, the simulated circuits allow one to extract statistical significance, simply by enumerating the fraction of, say, 1,000 different random instantiations of a circuit that contain a particular motif with the observed abundance or greater.

Tutorial 5.2 provides experience in generating different types of connectivity matrix and analyzing a matrix for structure that might be hidden. Observing structure in a connectivity matrix is not always easy until the neurons have been ordered appropriately. Even a network containing two entirely disconnected circuits might not appear as such from a glance at the connectivity matrix if the number of cells is high (figure 5.7). Therefore, a major goal of tutorial 5.2 is to introduce you to routines that switch around the labels of connections to allow easy extraction then visualization of any underlying structure.

A

	a	b	c	d	e	f	g	h	i
a	1	0	0	0	1	0	0	0	1
b	0	0	1	1	0	0	1	0	0
c	0	0	0	1	1	0	0	0	0
d	0	1	1	0	0	0	1	0	0
e	1	0	0	0	0	1	0	0	1
f	0	0	0	0	1	1	0	1	0
g	0	0	0	1	0	0	1	0	0
h	1	0	0	0	1	0	0	1	0
i	0	0	0	0	1	1	0	0	1

B

	a	e	f	h	i	b	c	d	g
a	1	1	0	0	1	0	0	0	0
e	1	0	1	0	1	0	0	0	0
f	0	0	1	1	0	0	0	0	0
h	1	0	1	1	0	0	0	0	0
i	0	1	1	0	1	0	0	0	0
b	0	0	0	0	0	0	1	1	1
c	0	0	0	0	0	0	0	1	1
d	0	0	0	0	0	1	1	0	1
g	0	0	0	0	0	0	0	1	1

Figure 5.7
Hidden structure revealed by rearranging labels. The connectivity matrix in A contains two disconnected recurrent circuits that can be revealed by reordering the cells as shown in B. The resulting matrix is block-diagonal, meaning that entries are zero apart from specific squares along the diagonal, in this case a 5 × 5 square, then a 4 × 4 square. The two matrices correspond to identical circuit structure, but the particular ordering on the right makes it easier for us to visualize such underlying structure.

5.6   Tutorial 5.2. Detecting Circuit Structure and Nonrandom Features within a Connectivity Matrix

Neuroscience goal: Acquire techniques that are useful in the analysis of connectivity data.

Computational goals: Gain experience at manipulating, sorting, and shuffling rows or columns of matrices; practice using the `while` loop.

In this tutorial, you will produce a connectivity matrix with structure that is not clear by inspection of the matrix. You will write an algorithm to extract the "hidden" structure and to analyze it for motifs that are present at levels significantly above chance.

Part A

1. Create a vector of length N_{cells}, where $N_{cells} = 60$, such that each element of the vector is a random integer from 1 to 4. This vector indicates the group to which each cell belongs. (Hint, either use the functions `ceil` and `rand` or the single function `randi` to achieve this.)

2. Create a connection matrix, C, of size $N_{cells} \times N_{cells}$, where each element, C_{ij}, representing the presence or not of a connection from neuron i to neuron j, is 1 or 0. Let $C_{ij} = 1$ with a probability of $p_0 = 0.1$ if neurons i and j belong to different groups and with a probability of $p_1 = 0.5$ if neurons i and j belong to the same group. *Hint:* A line of MATLAB code such as A = rand(100)<0.2; will produce a 100×100 matrix of ones and zeros, with each element equal to 1 with probability 0.2.

3. Plot the connection matrix in 2D (e.g., using the function `imagesc`).

Part B

The goal is to rearrange the neurons into their respective groups so that the structure of the connection matrix can be made visible.

4. To begin, you will create two correlation matrices, one set of correlations between all the outgoing connections of all cells, the other set of correlations between all the incoming connections of all cells. The expectation is that if two cells belong to the same group their connections will be more strongly correlated than between two cells of different groups.

    a. Calculate the correlation matrix for outgoing connections by evaluating the correlations between all pairs of row vectors of C. The correlation matrix so produced will be of size $N_{cells} \times N_{cells}$ (with a diagonal of 1, because each row is perfectly correlated with itself).

    b. Repeat (b) for all pairs of column vectors of C to evaluate the correlation matrix between all incoming connections of cells.

    c. Sum the two correlation matrices to produce a matrix called `total_corr` and plot the histogram of values. (If it is bimodal, that is a good sign of clustering.)

5. We wish to set a threshold for the correlation between two cells such that if the correlation is above the threshold we will ascribe the same group identity to those two cells. The number

of resulting groups will depend on the threshold—if it is set too low then only one group will be found, whereas if it is set too high many groups might be found. Select a threshold such that 10 percent of all correlations are above the threshold (one method is to use `sort(total_corr(:))`—the `:` produces a vector of the matrix elements). *Alternative simplification*: Just choose a threshold of 0.5 and adjust it up or down to obtain the correct number of groups after running the code if necessary.

6. To begin assigning cells to groups, create a vector of zeros that will contain the group identity of each cell (this will be the discovered identities, not the known ones generated in part 1). Set a new variable to zero—this variable will act as a group counter. Create a `for` loop so that a variable, i, runs from 1 to N_{cells}.

7. Within the loop check the currently assigned group identity of cell number i, with an `if` statement—if the group identity is zero (as it will be initially, when i=1) then increase the group count by one to commence extraction of a new group. Do parts 8–10 within the `if` statement (i.e., only if group identity of the cell reached was zero, meaning it has not yet been allocated to a group). If the currently assigned group identity is above zero, it means this cell has been found within a completed cell group, so just increase i by 1 and test whether the next cell requires a new group.

The goal of these next three parts is to ascertain which other cells are in the same group as cell number i. This requires finding which cells have an above-threshold correlation with cell i, which other cells have an above-threshold correlation with these, and so on until no more new cells are found in the group. Have a go at doing this yourself if you are confident, without following the instructions in parts 8–10.

8. Initialize two vectors, one containing cells in the group whose high-correlation partners remain to be found (a "remaining" list), the other of which will contain the cells in the group whose high-correlation partners are already found (a "used" list). The first vector is initialized with the single element i, while the second vector is initialized as the empty vector `[]`.

9. Create a loop using a `while` statement, which will continue until the list of remaining cells is empty. Once the list of remaining cells is empty, the code should go to the next step of the `for` loop and increment i by one.

   a. Within the `while` loop extract the neuron that is the first element of the list of remaining cells.

   b. Then find all cells with correlation greater than the threshold with this neuron.

   c. Give the current group identity to all of those cells (in so doing you will give the group identity to the neuron being used as its self-correlation is 1, so is certainly above-threshold).

   d. Add the current neuron to the "used" list then add the group of cells just found to the "remaining" list.

e. Remove all double counts on the "remaining" list, so each cell on it appears only once. Then remove from the "remaining" list any cell that is on the "used" list. This is to ensure that we avoid an endless loop where we keep looking for high-correlation partners of the same set of cells over and over. A danger of `while` loops is that they might never end!

10. `end` the `for` loop begun in step 6, so that the code returns to step 7 with the cell number `i` increased by 1.

11. After the `for` loop in `i` has completed, use the `sort` command to sort the list of group identities—be sure to store the indices such that alongside the sorted list of group identities is the list of cells to which the group identities belong. Produce and plot an alternative connection matrix with the entries reordered according to the sorted list of group identities to observe the underlying group structure.

Challenge: Part C

Our goal is to extract motifs and compare their abundance with what one expects by chance given the average connection probability.

12. Find the average connection probability, $\bar{p}$, between pairs of cells in the original connection matrix.

13. Find the number of cell pairs with bidirectional connections and compare the number with what is expected by chance, given your answer in 12.

14. Use your answers to questions 12 and 13 to calculate the probability of a cell pair having two connections, p_2, a single connection, p_1, or no connection, p_0. Hint: total number of connections must be $N_{pairs}\bar{p} = N_{pairs}p_1 + 2N_{pairs}p_2$, where $N_{pairs} = N_{cells}(N_{cells}-1)/2$ is the total number of cell-pairs.

15. Look at the motifs in figure 5.6 and enumerate the number of each type found in the connectivity matrix. For example: initialize a vector of length 16 with zeros; using three nested loops running from 1 to N_{cells} for variables i, j (with j>i), and k (with k>j), test each triplet of connections, and add 1 to the count of the corresponding motif number.

16. To compare with chance, plot the ratio of counts enumerated in Q.15 with the expected number, E_i, for motif i, obtained from the following formula:

$$E_i = M(i) N_{triplets} p_0^{n_0(i)} p_1^{n_1(i)} p_2^{n_2(i)}$$

where $n_0(i)$, $n_1(i)$, $n_2(i)$ are the numbers of unconnected, unidirectionally connected and bidirectionally connected cells in the motif, $N_{triplets} = N_{cells}(N_{cells}-1)(N_{cells}-2)/6$ is the number of distinct cell-triplets, and $M(i)$ is the number of ways of producing the motif by shuffling connections between the cells, as given in table 5.1. Note that the sum of multiplicities in table 5.1 is $2^6 = 64$, the total number of patterns given the six possible connections, each of which is either absent or present, between three cells.

Table 5.1
For each three-neuron motif (figure 5.6), the number of distinct sets of connections between the three cells that produce the same motif is given as the multiplicity

Motif label	1	2	3	4	5	6	7	8	9	10	11	12	13	14	15	16
Multiplicity	1	6	3	3	3	6	6	6	3	6	2	3	6	3	6	1

5.7   Oscillations and Multistability in Small Circuits

The response of a neural circuit depends not only on the external inputs it receives but also, and perhaps more so, on the internal connections within the circuit.

Once neurons are connected with each other, they can exhibit patterns of activity that are not present in the individual disconnected neurons. We have seen that many single neurons oscillate, either with a single sodium spike per cycle, or with a burst of sodium spikes riding on a slower calcium spike. Neurons that do not oscillate alone can oscillate when connected to other neurons, in particular if there is inhibitory feedback.

Alternatively, a system with more than one stable state can result from the connections between cells, the simplest example being a "flip-flop," a bistable switch in which one of two neurons, or groups of neurons, suppresses the other. The behavior of such small circuits can depend subtly on attributes of the synaptic connections, such as their strengths, time constants, and whether short-term depression or facilitation dominate, as well as on the intrinsic properties of each cell.

Multistability   The existence of more than one stable pattern of activity or stable level of activity in a system in response to a particular set of inputs.

Quasistable   A system that reaches a state that is almost stable, but, perhaps due to a slowly varying property, eventually changes away from that state.

The simplest form of multistability is bistability, meaning that under constant stimulus conditions two states of neural activity are possible. A widely appreciated phenomenon in the fields of vision and psychology is perceptual bistability, whereby a single image can appear as one of two possible objects or scenes. Some examples of images giving rise to bistable percepts are shown in figure 5.8. If an observer fixes her eyes on the image then occasional transitions from one percept to the other are common. It should be noted that mathematically, if the percept does switch at regular intervals then the system is an oscillator and not bistable (a state is not stable if

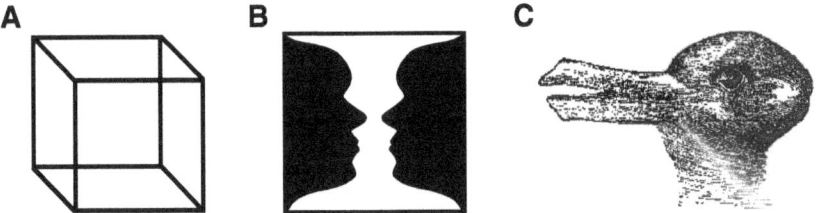

Figure 5.8
Images producing bistable percepts. (A) The Necker cube can be seen with its front face as bottom-right or top-left. The 3D percept produced by such a 2D line drawing is experience-dependent—a fifty-year-old who gained sight after being blind from birth did not perceive a cube. We will consider how experience-dependent synaptic strengthening can produce multistable circuits in chapter 8. (B) Rubin's vase can appear as a single white vase in the center or two black faces looking at each other. Danish psychologist Edgar Rubin created the illusion in 1915.[16] (C) The image by Joseph Jastrow in 1899 can appear either as a leftward-facing duck or a rightward-facing rabbit.[17]

it does not persist). In tutorial 5.3 we will see the close relationship between bistable systems and oscillators. Oscillators that are based on regular switching between states that are almost stable (quasistable) are called relaxation oscillators (section 7.6).

5.8   Central Pattern Generators

A central pattern generator is a neural circuit that can produce, autonomously, a pattern of activity, such as an oscillation, which can drive muscles in a reliable manner.[19] While external input can switch the central pattern generator on or off, or switch it between modes, and neuromodulation can adjust features of the pattern, the neural circuit of a central pattern generator does not rely on any external input to produce the activity pattern. The oscillator in part B of tutorial 5.3 (figure 5.9A) is a simple example of a central pattern generator, related to that found in the heartbeat control system of the medicinal leech.[18,20] When animals have stereotypical coordinated muscle contractions accompanying a behavior—such as breathing, peristalsis, locomotion, or chewing—central pattern generators are typically involved.

The most important goal of a central pattern generator is to ensure the muscles contract in the correct order. This requires that the neurons in the circuit are active in a regular, predictable order. The connectivity pattern between neurons can determine the order of activation of cells. When it is essential that different muscles do not contract at the same time, then cross-inhibition between the neurons controlling those muscles is a solution.[21] Thus, in the two examples we consider here, the rhythms are produced by inhibitory cross-connections.

Two other characteristics of a central pattern generator that should be controlled are the frequency of oscillation (or duration of activity if nonoscillating) and the duty cycle of each neuron, that is the relative time it spends active versus quiescent. In many cases in nature, the frequency of oscillation can vary in response to external conditions or internal control, without any change in the order of firing and the duty cycles of the neurons involved. It is far from trivial for us to

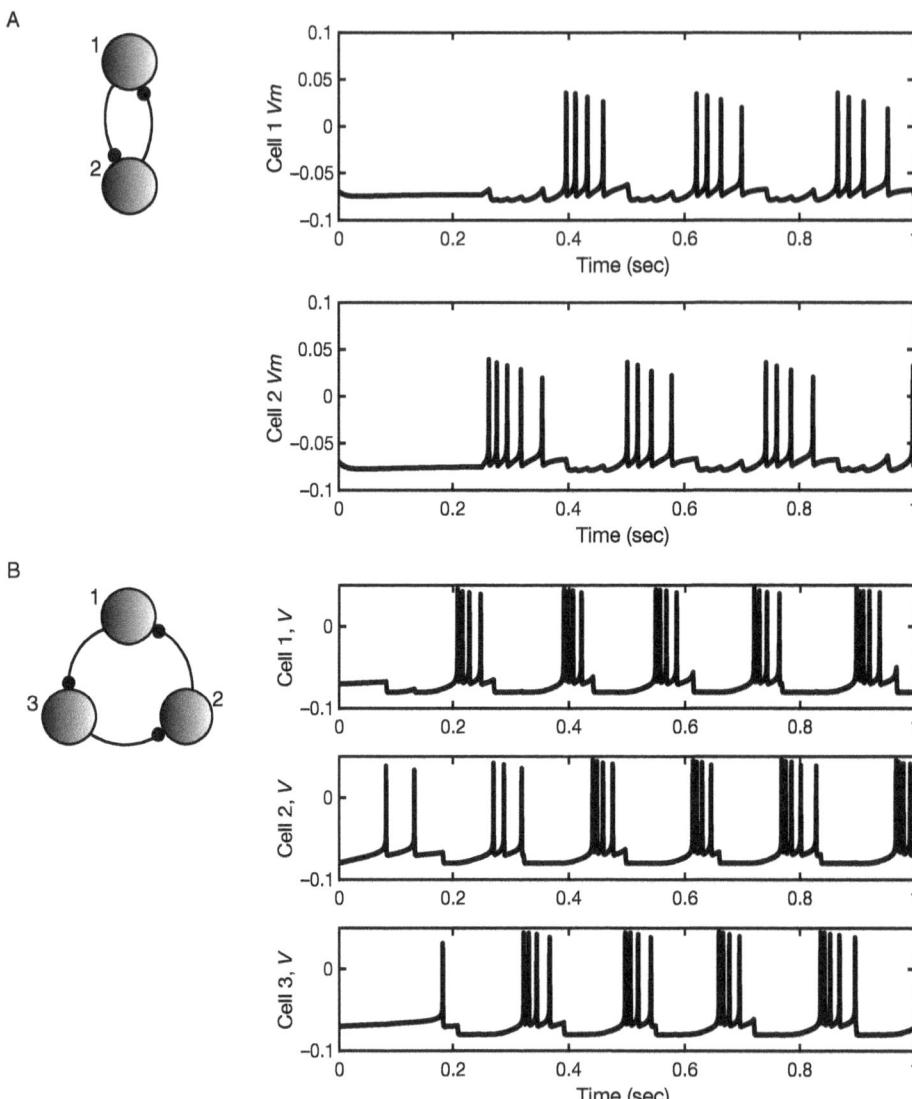

Figure 5.9
(A) The half-center oscillator. Activity of two neurons connected by cross-inhibition can oscillate in antiphase.[18] Spike trains are produced by the online code `coupled_PIR.m`. (B) The triphasic rhythm. Activity of three neurons oscillates with a phase relationship determined by the connectivity pattern. Order of activation is 1–2–3–1– ... with inhibition from the active cell suppressing the cell that precedes it in the cycle. Spike trains produced by the online code `three_coupled_PIR.m`.

produce models of central pattern generators with such invariant properties,[22] but biological evolution often achieves solutions that we might not think of.

5.8.1   The Half-Center Oscillator

The half-center oscillator (figure 5.9A) is achieved with two neurons, each of which inhibits the other. Each neuron's activity oscillates in antiphase with the other, so the neurons can produce muscle contractions that alternate between two antagonistic muscles. Such alternating contractions between the two sides of an animal's body comprise the basic unit for locomotion by slithering or wiggling, most common among long legless animals (lampreys, eels, snakes, worms etc.).[23] It appears that evolution has ensured the neural circuits are wired, not only to produce the individual half-center oscillators that comprise the basic units, but also to ensure the correct phase relationship between successive units that is necessary to produce effective motion.[24]

The frequency of a half-center oscillator depends on the time from when one of the neurons commences its activity until the other one takes over. The switch in the dominating neuron can occur either because the active neuron cannot sustain high activity for very long or because, following suppression, the inactive neuron becomes more excitable over time. The first mechanism for switching is called *release*, and the second is called *escape*.[18]

Release is so-called, because as the activity of the active neuron decreases, so the inhibitory input to the inactive neuron decays away. Such weakening and eventual loss of inhibition from the active cell is the release. The period of the oscillation is dominated by the timescale of inactivation of the active cell.

Switching by escape depends on a regenerative process, such as deinactivation of an excitatory current, within the inhibited neuron. The time for the inhibited neuron to overcome the inhibition and become active again depends strongly on the strength of the inhibitory connection, but also the strength and timescale of the regenerative process.

5.8.2   The Triphasic Rhythm

The simplest models of a triphasic rhythm require three neurons, which activate in a reliable cyclic order (figure 5.9B). As with the half-centered oscillator, the rhythm can be produced with inhibitory connections alone, if each cell can fire spontaneously or via postinhibitory rebound. In this case, the inhibition of one neuron suppresses the activity of the neuron preceding it in the cycle. The time for a neuron to rebound from such inhibition is a key determinant of the oscillation frequency.

The best-studied triphasic rhythm is the pyloric rhythm produced in the stomatogastric ganglion of the lobster or crab. The rhythm controls stomach contractions and dilations. Although the numbers and types of cells and their connectivity pattern are conserved across individuals of any species, the synaptic connection strengths and values of intrinsic cell conductances vary over wide ranges across individuals. An intriguing question is how, in spite of the huge individual variation, the important characteristics of the rhythm are maintained—in particular, because in models, if any one parameter is varied to the degree observed in nature, then the rhythm produced by the circuit changes dramatically or disappears.

5.8.3 Phase Response Curves

A useful technique for understanding the behavior of coupled oscillators is analysis of their phase response curves.[25] A phase response curve (figure 5.10) indicates how an oscillator responds to a perturbation, such as a brief excitatory synaptic input. The phase response is quantified by the amount of advancement or delay of ensuing oscillations caused by the perturbation. The response can depend on the point in a cycle at which the perturbation occurred.

> **Perturbation** A small, externally produced, change in a system.

For example, a brief pulse of excitatory input might advance the next burst of action potentials by hastening the positive feedback of voltage-dependent activation of calcium or sodium channels. Yet the same pulse of input might delay the next burst if it were to occur earlier, when it would slow the deinactivation of those channels, or slow the deactivation of potassium

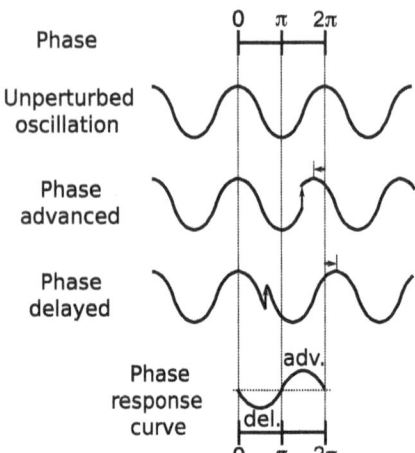

Figure 5.10
The phase response curve (PRC). A phase response curve can be produced for any system undergoing periodic activity—i.e., an oscillator (top)—such as a regularly spiking or bursting neuron. If the oscillator is perturbed by an input, such as a brief pulse of excitation (vertical arrow), it can be advanced in phase (a positive phase response, upper curve) or delayed in phase (a negative phase response, lower curve). The phase response curve (bottom) is produced by measuring the change in phase (horizontal arrows) as a function of the point in the cycle at which the perturbation arises. Regularly bursting and type-II neural oscillators typically have biphasic phase response curves similar to the one sketched here.

channels. So, one might expect that shortly before a spike or burst, excitatory input advances the next cycle of an oscillation, whereas shortly after a spike or burst, the same excitatory input delays the next cycle. Such behavior is shown in figure 5.10 and can be seen in simulated neurons in figure 5.11.

The phase response curve can be used to find whether an oscillator will entrain to a periodic input and to determine the phase relationship of any eventual entrainment. If the oscillator has a longer period than the input, then, in order to entrain, it must receive sufficient phase advance in each cycle to reduce its period to that of the input. Conversely, if the oscillator's period is shorter than the input's then it must receive sufficient phase delay on each cycle in order to entrain.

Given the cyclic nature of phase response curves, any required level of advancement or delay can be reached at an even number (typically two) values of phase offset—for each offset at which the curve crosses a particular phase shift from below, there is another phase offset at which the curve recrosses that particular phase shift from above. It turns out that only sections of the curve with negative gradient—so that the oscillator's phase advancement decreases, or its phase delay increases, with increasing phase delay of the input—can give rise to stable phase offsets (shaded regions of figure 5.11).

To understand the stability requirement, consider a periodic input at a faster rate than an oscillator's natural frequency. Suppose the input on one cycle arrives at a phase offset that corresponds to a point on the curve with positive gradient and insufficient phase advancement for the oscillator to match the periodic input. Because the advancement of the oscillator is insufficient—that is, the oscillator has not been sped up enough to match the input—the next input will arrive earlier in the oscillator's cycle. The earlier input produces even less phase advancement for a later cycle, so the system moves away from the point of potential entrainment.

On the other hand, if insufficient phase advancement occurs at a phase offset where the phase response curve has negative gradient, then on the following cycle, the earlier arrival of the pulse produces greater phase advancement, so the oscillator approaches the frequency of the inputs until it is entrained. If the oscillator receives too much advancement on one cycle, the input arrives at a slightly later phase on the following cycle, producing less advancement and slowing the oscillator relative to the inputs until it is entrained again.

When two or more oscillators are coupled, a similar analysis can be carried out to find the final phase relationship and eventual frequency of the coupled oscillators. The impact of each oscillator on each other must be considered. For any two oscillators (A and B) the sum of phase offsets (A to B plus B to A) is, by necessity, 2π, or a multiple thereof.

When the two oscillators have identical phase response curves, the phase response of an offset angle, $\theta_{AB} = \theta$, must be identical to that at an offset angle of $\theta_{BA} = 2\pi - \theta$, in order for the two oscillators advance or delay by the same amount per cycle and remain entrained.

The requirement for stability in the previous section still holds, so the curve should have negative gradient for such a phase offset to be stable. Except in rare cases, the only possible stable solutions are $\theta_{AB} = \theta_{BA} = 0$ (in-phase or synchronous) or $\theta_{AB} = \theta_{BA} = \pi$ (out-of-phase or antiphase). In the example of figure 5.10, the phase offset of 0 corresponds to a point of negative

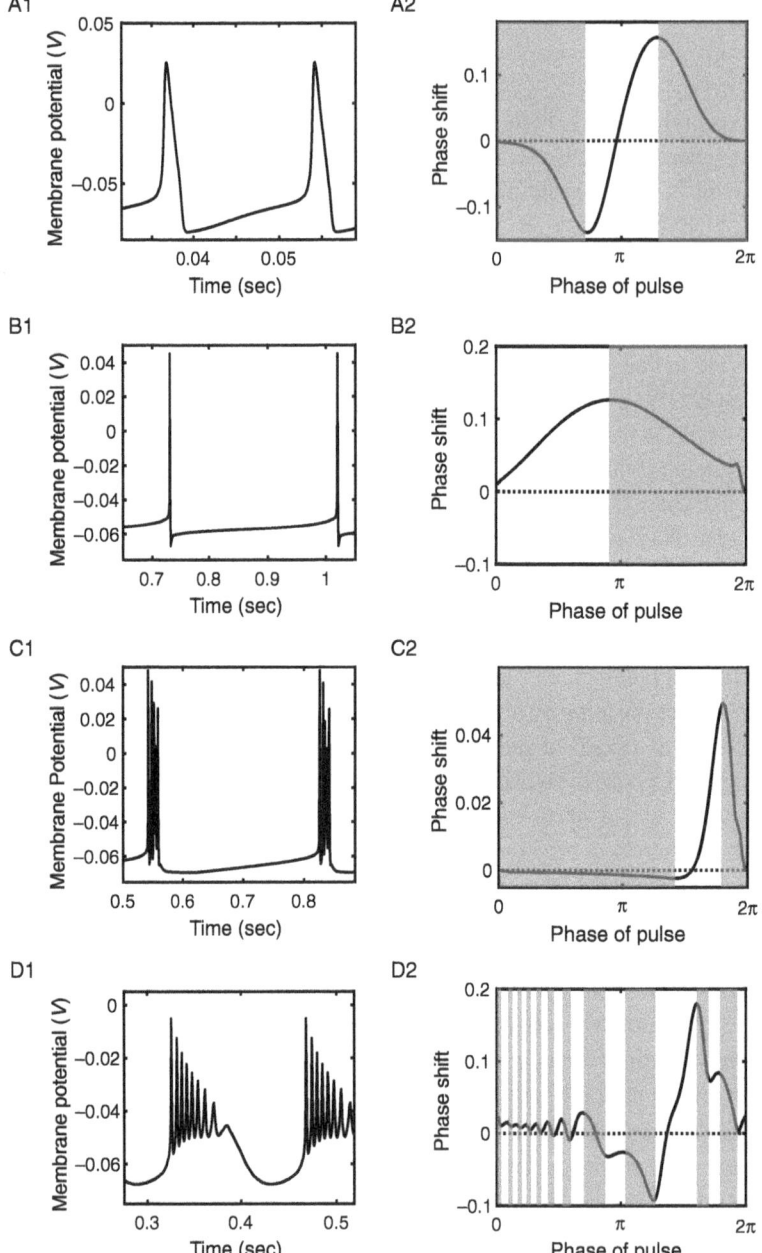

Figure 5.11
Phase response curves of simulated neurons. Model neurons in a regularly oscillating state (left) receive a 10 pS pulse of excitatory current for 5 ms at different points in their oscillating cycle, with a phase of 0 defined to be at the upswing of a spike or the first spike of a burst. Shaded regions on the right indicate where stable entrainment to a periodic sequence of such pulses is possible. (A) Hodgkin-Huxley model, a type-II neuron with biphasic PRC. (B) Connor-Stevens model, a type-I neuron with only phase-advance. (C) Pinsky-Rinzel model with current applied to the dendrite. While biphasic, the delay portion of the PRC is of longer duration and twentyfold smaller amplitude than the advance portion. (D) Thalamic rebound model with altered parameters to render it regularly oscillating produces a more intricate PRC, with the timing of the pulse with respect to individual spikes within a burst impacting the phase shift, on top of a general delay then advance when input arrives between bursts. These panels are produced by the online codes A: `Phase_Response_HH.m`, B: `Phase_Response_CS.m`, C: `Phase_Response_IHPR.m`, D: `Phase_Response_PIR.m`.

slope on the phase response curve, so is a stable phase offset for excitatory coupling between these oscillators, whereas the phase offset of π is unstable. However, if the coupling were inhibitory, the phase response curves would flip sign and the "in-phase" offset of 0 would be unstable, while the "antiphase" offset of π would become the stable one. This is one account of the antiphase behavior of many inhibition-coupled oscillators.

The resulting frequency of the coupled oscillators is obtained from the phase response curve by reading out the phase response, $\Delta\theta$, at the stable phase offset. The phase response is calculated using the same units as the phase offset, so an advancement of 10 percent of a cycle corresponds to a phase response of $\Delta\theta = 0.1 \times 2\pi$. Since such an advance would occur on each cycle, the coupled frequency, f_c, would be 10 percent higher than the uncoupled frequency, f_u. In general, then, the coupled frequency is given by:

$$f_c = f_u \left(1 + \frac{\Delta\theta}{2\pi} \right). \tag{5.11}$$

5.9 Tutorial 5.3: Bistability and Oscillations from Two LIF Neurons

Neuroscience goals: Learn how cross-inhibition can produce either oscillations or bistability in a single circuit, depending on synaptic properties (i.e., circuit structure does not determine behavior or function); observe how noise can produce transitions between activity states.

Computational goals: Keep track of two modeled neurons and be careful to use the output of one as the input of the other; record and plot the distribution of durations of states.

In this tutorial, you will couple together two leaky integrate-and-fire neurons with inhibitory synapses. We will see that if the inhibition is strong enough only one neuron can fire—as it fires it suppresses any activity in the other neuron. If we add noise, random switches in the active neuron arise. We will then see that adding short-term synaptic depression generates oscillations whose regularity is affected by the strength of noise. Similar models have been used to describe the alternations between percepts of visual or auditory stimuli that are ambiguous, a phenomenon known as "perceptual rivalry" or "perceptual alternation" (figure 5.8).

Set up the simulation to solve the following leaky integrate and fire equations for the membrane potentials, V_1 and V_2, of the two cells, with synaptic gating variables, s_1 and s_2, and depression variables, D_1 and D_2, respectively:

$$C\frac{dV_1}{dt} = \frac{E - V_1}{R} + G_{21}s_2\left(E_{21}^{rev} - V_1\right) + I_1^{App} + \sigma \cdot \eta(t), \quad \frac{ds_1}{dt} = \frac{-s_1}{\tau_{syn}}, \quad \frac{dD_1}{dt} = \frac{1 - D_1}{\tau_D},$$

$$C\frac{dV_2}{dt} = \frac{E - V_2}{R} + G_{12}s_1\left(E_{12}^{rev} - V_2\right) + I_2^{App} + \sigma \cdot \eta(t), \quad \frac{ds_2}{dt} = \frac{-s_2}{\tau_{syn}}, \quad \frac{dD_2}{dt} = \frac{1 - D_2}{\tau_D}.$$

In the preceding equations, G_{12} and E_{12}^{rev} denote the maximal conductance and the reversal potential respectively, of the synapse from cell 1 to cell 2, while G_{21} and E_{21}^{rev} denote those properties of the other synapse (from cell 2 to cell 1). $\eta(t)$ are white noise terms of unit standard deviation (obtained by using the function `randn` in MATLAB) and σ is the amplitude of noise (see section 1.6.6 for noise implementation in the simulation). I_1^{App} and I_2^{App} are the applied currents to each cell, that will include a constant baseline term and a transient term to cause a switch.

Spikes are simulated as follows:

If $V_1 > V_{Th}$ then: $V_1 = V_R$, $s_1 \mapsto s_1 + p_R D_1 (1-s_1)$ and $D_1 \mapsto D_1 (1- p_R)$.

Similarly, if $V_2 > V_{Th}$ then: $V_2 = V_R$, $s_2 \mapsto s_2 + p_R D_2 (1-s_2)$ and $D_2 \mapsto D_2 (1- p_R)$.

Part A: Bistability with No Synaptic Depression

(i) Set the membrane capacitance, $C = 1$ nF, the membrane resistance, $R = 10$ MΩ, the leak/reversal potential, $E = -70$ mV, the threshold, $V_{Th} = -54$ mV and the reset potential $V_R = -80$ mV for each of the identical cells. Make the synapses identical and inhibitory, such that $E_{12}^{rev} = E_{21}^{rev} = -70$ mV, $G_{12} = G_{21} = 1$ μS, and $\tau_{syn} = 10$ ms. Set the baseline applied current to be 2 nA. Set $p_R = 1$ and remove the effects of synaptic depression by ensuring that $D_1 = D_2 = 1$ throughout this simulation.

(ii) Simulate the two, coupled neurons for a total time of 6 s with an extra 3 nA of applied current to one cell for the first 100 ms of the simulation, then a pulse of 3 nA to the other cell for 100 ms at the midpoint of the simulation. Initially begin with no noise ($\sigma = 0$). Plot the membrane potential against time and the synaptic gating variables against time. Comment on your results.

(iii) Now add noise by setting $\sigma = 50$ pA.s$^{-1/2}$ (5×10^{-11} A.s$^{-1/2}$)—that is, in each timestep add a different random number to the current into each cell, from `randn` multiplied by $5 \times 10^{-11} / \sqrt{dt}$. Rerun the simulation with only the baseline current (no additional transient currents). Plot and comment on your results.

(iv) Run the simulation for long enough to see over 1,000 switches if you can and plot a histogram of the duration in each state—you will need to set a variable that records the state you are in and changes when the other cell spikes, at which point you record the time of state-switching. Taking differences in the switching times (use `diff`) will produce a vector whose odd entries correspond to durations of one state and whose even entries correspond to durations of the other.

Part B: Oscillations with Synaptic Depression

(v) Repeat (i) and (ii), but include the full effects of synaptic depression by setting $p_R = 0.2$ and updating the depression variables according to the equations above, with the time constant for recovery set as $\tau_D = 0.2$ s. Plot and comment on your results.

(vi) Add a small amount of noise, setting $\sigma = 5$ pAs$^{-1/2}$. Repeat (iv) and comment on any differences in the distribution of state durations between parts (iv) and (vi).

Questions for Chapter 5

1. If the number of calcium channels at the axon terminals of a neuron were increased substantially, would the synapses near those terminals become more depressing, more facilitating, or neither? Explain.

2. What happens to the total conductance of a neuron's membrane and its effective time constant for change in the membrane potential when:
   a. It receives a lot of excitatory synaptic input?
   b. It receives a lot of inhibitory synaptic input?

3. A paired-pulse ratio is a measure of the synaptic transmission produced by a second action potential relative to that produced by an immediately preceding first action potential. If a synapse has twelve release-ready vesicles, a baseline release probability of 1/3 and a facilitation factor of 1/3, what is its paired-pulse ratio? Does this correspond to synaptic facilitation or depression?

4. In a simplified model of the CA3 region of the hippocampus of a rat, 250,000 excitatory pyramidal neurons connect with each other randomly with each possible connection present at a 1 percent probability. If one neuron excites any other neuron which also returns the excitation, we consider that neuron to be in a two-step feedback loop.
   a. What is the mean number of two-step feedback loops per neuron?
   b. What is the probability that a neuron has no such two-step feedback loops and how many such neurons do you expect altogether in such a model of CA3?

5.10. Appendix: Synaptic Input Produced by a Poisson Process

In this section, we will obtain mathematical expressions for how the synaptic input to neurons depends on presynaptic firing rate when biophysical processes such as saturation of receptors, synaptic depression, and synaptic facilitation are taken into account. Each process responds to a series of presynaptic events that are either the spikes or the release of vesicles. The mean effect of each of these processes alone can be calculated in response to either a regular series of events or, if the events are uncorrelated, as a Poisson process. Here we will derive the formulae for the response to a Poisson process. The results can be used in firing rate models (chapter 6), which require the mean synaptic inputs as a function of presynaptic firing rates.

5.10.1 Synaptic Saturation

To account for synaptic saturation, the synaptic response to each action potential is given by $s \mapsto s + \alpha p_r (1 - s)$, or equivalently by $s_+ = s_- + \alpha p_r (1 - s_-)$, where α is the fraction of unbound receptors bound following maximal vesicle release and p_r is the release probability for each vesicle following a spike. s is the fraction of synapses bound by neurotransmitter, which unbinds with a time constant, τ_s, such that $\tau_s (ds/dt) = -s$ between the times of spikes. s_+ denotes the value of s immediately after a spike and s_- denotes its value immediately before the spike. In this formulation α, which depends on the amount of neurotransmitter released per spike, is assumed to be constant and independent of spike history (i.e., depression and facilitation are not included here).

When presynaptic spikes arrive so as to release vesicles as a Poisson process, the value of s is always fluctuating and the values of s_+ and s_- are different for every release event. However, by using the fact that the changes in s are linear functions of s, we can write equations for the mean values when the distribution of s has reached a steady state. In this case, using $\langle s \rangle$ to denote the mean of s, we can relate the mean value before and after a spike via:

$$\langle s_+ \rangle = \langle s_- \rangle + \alpha p_r (1 - \langle s_- \rangle). \tag{5.12}$$

We also know that the dynamics of s between spikes is exponential decay such that

$$s_-^{(i+1)} = s_+^{(i)} \exp\left(-\frac{T_i}{\tau_s}\right) \tag{5.13}$$

where $s_+^{(i)}$ is the value of s immediately after the i-th spike and T_i is the i-th interspike interval. We can now calculate the mean reduction in s between spikes by using the important Poisson property that the interspike interval, T_i, is independent of the prior value, $s_+^{(i)}$. Such independence means the average of the product of terms is equal to the product of their individual averages:

$$\langle s_- \rangle = \left\langle s_-^{(i+1)} \right\rangle = \left\langle s_+^{(i)} \right\rangle \exp\left(-\frac{T_i}{\tau_s}\right)_i = \langle s_+ \rangle \left\langle \exp\left(-\frac{T_i}{\tau_s}\right)\right\rangle \tag{5.14}$$

The tricky term in equation 5.14, $\left\langle \exp\left(-\dfrac{T_i}{\tau_s}\right)\right\rangle$, requires an average over the interspike intervals, T_i. We use the Poisson formula for the probability distribution of interspike intervals, $P(T_i) = r \exp(-rT_i)$, where r is the rate of the Poisson process to obtain:

$$\left\langle \exp\left(-\frac{T_i}{\tau_s}\right)\right\rangle = \int_0^\infty \exp\left(-\frac{T}{\tau_s}\right) P(T) dT = \int_0^\infty \exp\left(-\frac{T}{\tau_s}\right) r \exp(-rT) dT$$
$$= r \int_0^\infty \exp\left(-T \frac{1 + r\tau_s}{\tau_s}\right) dT = \frac{r\tau_s}{1 + r\tau_s}. \tag{5.15}$$

We can then insert this result into equation 5.14 to find

$$\langle s_- \rangle = \langle s_+ \rangle \frac{r\tau_s}{1 + r\tau_s} \tag{5.16}$$

and use equation 5.16 to substitute for $\langle s_- \rangle$ in equation 5.12 to obtain

$$\langle s_+ \rangle = \langle s_+ \rangle \frac{r\tau_s}{1+r\tau_s} + \alpha p_r \left(1 - \langle s_+ \rangle \frac{r\tau_s}{1+r\tau_s} \right). \tag{5.17}$$

Some rearrangement leads to

$$\langle s_+ \rangle = \frac{\alpha p_r (1+r\tau_s)}{1+\alpha p_r r\tau_s} \tag{5.18}$$

such that

$$\langle s_- \rangle = \frac{\alpha p_r r\tau_s}{1+\alpha p_r r\tau_s}. \tag{5.19}$$

Interestingly, the overall mean value of s, $\langle s \rangle$ to be calculated later, is also $\alpha p_r r\tau_s / (1+\alpha p_r r\tau_s)$, the mean of the values of s immediately before an event, $\langle s_- \rangle$. This may seem counterintuitive because across each interspike interval the average of s is always higher than s_- to which it monotonically decays just before the next event. However, the occurrences of a lower than average value of s_- follow longer intervals that contribute more to the mean value of s than do typical intervals. Since, for a Poisson process, an event can occur equally likely at any point in time, the value of a function immediately before an event is sampled equally across time, so the mean of values just before the event is equal to the mean across time.

For the aficionados, this result can be verified by a calculation that uses the average over an interspike interval, T, of an exponential decay:

$$\left\langle \exp\left(-\frac{t}{\tau_s}\right) \right\rangle_T = \frac{1}{T} \int_0^T \exp\left(-\frac{t}{\tau_s}\right) dt = \frac{\tau_s}{T}\left[1 - \exp\left(-\frac{T}{\tau_s}\right)\right] \tag{5.20}$$

to obtain the weighted mean of this over all possible interspike intervals:

$$\langle s \rangle = \langle s_+ \rangle \frac{\int_0^\infty \left\langle \exp\left(-\frac{t}{\tau_s}\right) \right\rangle_T TP(T)dT}{\int_0^\infty TP(T)dT} = \langle s_+ \rangle \frac{r\tau_s \int_0^\infty \left[1 - \exp\left(-\frac{T}{\tau_s}\right)\right]\exp(-rT)dT}{r\int_0^\infty T\exp(-rT)dT} \tag{5.21}$$

$$= \langle s_+ \rangle r^2 \tau_s \left[\frac{1}{r} - \frac{\tau_s}{1+r\tau_s}\right] = \langle s_+ \rangle \frac{r\tau_s}{1+r\tau_s} = \frac{\alpha p_r r\tau_s}{1+\alpha p_r r\tau_s} = \langle s_- \rangle$$

In summary, the mean synaptic activation increases linearly with the presynaptic firing rate, r, when the firing rate is low ($r \ll 1/\alpha p_r \tau_s$), but synaptic activation saturates, approaching its maximum value of 1 at high firing rates ($r \gg 1/\alpha p_r \tau_s$).

5.10.2   Synaptic Depression

Synaptic depression reduces the effective strength of a synapse temporarily when firing rates are high, because as we saw in the last section, the supply of release-ready vesicles gets depleted. We can quantify the mean effect of such depletion if presynaptic spikes arrive as a Poisson process of rate r and if the release probability, p_r, is held constant. The calculation is similar to the one for synaptic saturation (section 5.10.1).

We can replace the factor, αp_r, the fraction of receptors bound following a presynaptic spike, by the factor $\alpha_0 D p_r$, where $0 < D \leq 1$ is the depression variable—i.e., $\alpha = \alpha_0 D$. As we saw (equation 5.9a), the dynamics of D between spikes follows $\tau_D (dD/dt) = 1 - D$ and following each spike (equation 5.10), $D_+ = D_- (1 - p_r)$, where D_+ is the value of D immediately after and D_- the value of D immediately before the spike.

As in section 5.10.1, we can consider the mean of values before and after spikes, writing

$$\langle D_+ \rangle = \langle D_- \rangle (1 - p_r) \tag{5.22}$$

and using the solution for the value of D at a time t since the previous spike as

$$D(t) = 1 - (1 - D_+) \exp(-t / \tau_D) \tag{5.23}$$

to produce

$$\langle D_- \rangle = 1 - (1 - \langle D_+ \rangle) \langle \exp(-T_i / \tau_D) \rangle, \tag{5.24}$$

where the final term is averaged over all interspike intervals, T_i.

We use the result of equation 5.15 for a Poisson process of rate r,

$$\langle \exp(-T / \tau_D) \rangle = \frac{r \tau_D}{1 + r \tau_D} \tag{5.25}$$

to rewrite equation 5.24 as

$$\langle D_- \rangle = \frac{1}{1 + r \tau_D} + \langle D_+ \rangle \frac{r \tau_D}{1 + r \tau_D}. \tag{5.26}$$

Using equation 5.22 to substitute for $\langle D_+ \rangle$ in equation 5.26 and rearranging leads to

$$\langle D_- \rangle = \frac{1}{1 + p_r r \tau_D}. \tag{5.27}$$

It is this value, representing the fraction of the complete stock of release-ready vesicles present when an action potential arrives, that determines the amount of neurotransmitter released and the synaptic efficacy. At low rates, $r \ll 1 / p_r \tau_D$, $\langle D_- \rangle$ is approximately equal to one, but at high rates,

$r \gg 1/p_r\tau_D$, $\langle D_-\rangle$ decays inversely with rate. Therefore, at high rates the total synaptic transmission of $\alpha_0 p_r \langle D_-\rangle r\tau_s = \alpha_0 p_r r\tau_s/(1+p_r r\tau_D)$ (ignoring synaptic saturation) approaches, but never exceeds a maximum value of $\alpha_0 \tau_s / \tau_D$.

5.10.3 Synaptic Facilitation

The term for synaptic facilitation, F, scales the vesicle-release probability, $p_r = Fp_0$, where p_0 is the baseline value, so it also provides a multiplicative contribution to the fraction of receptors bound per spike. The dynamics of F between spikes is identical to the dynamics of D—namely an exponential decay toward 1—but with a time constant of τ_F and with the understanding that $F(t) \geq 1$ for all t. Therefore, the calculation for the average over an exponential decay to 1, relating the mean value of F immediately before a spike to its mean value immediately following a spike leads to a similar result to the one for depression (equation 5.26):

$$\langle F_-\rangle = \frac{1}{1+r\tau_F} + \langle F_+\rangle \frac{r\tau_F}{1+r\tau_F}. \tag{5.28}$$

The change in F following a spike (equation 5.10), $F_+ = F_- + f_{fac}(F_{max} - F)$ allows us to write

$$\langle F_+\rangle = \langle F_-\rangle(1-f_{fac}) + f_{fac}F_{max}. \tag{5.29}$$

Using equation 5.29 to substitute for $\langle F_+\rangle$ in equation 5.28 and rearranging leads to

$$\langle F_-\rangle = \frac{1+f_{fac}F_{max}r\tau_F}{1+f_F r\tau_F} = 1 + \frac{(F_{max}-1)f_{fac}r\tau_F}{1+f_{fac}r\tau_F}. \tag{5.30}$$

At low rates ($r \ll 1/f_{fac}\tau_F$) the facilitation factor, $\langle F_-\rangle$, increases from 1 linearly with firing rate and at high firing rates rates ($r \gg 1/f_{fac}\tau_F$) $\langle F_-\rangle$ approaches its maximum possible value, F_{max}.

5.10.4 Notes on Combining Mechanisms

The derivations for each of the three mechanisms above rely on the incoming series of events being Poisson processes. However, once mechanisms are combined the Poisson assumption can no longer hold. For example, if presynaptic spikes arrive as a Poisson process but there is synaptic facilitation then vesicles are not released as a Poisson process. In fact, facilitation tends to increase the coefficient of variation of release events, increasing the relative probability of very short intervals at the same time as increasing the relative probability of very long intervals between vesicle-release times compared to a Poisson time series. This is because release is more likely after a short interval than on average, enhancing the number of short intervals, while release is less likely after a long interval, which makes long intervals even longer. Conversely, synaptic depression tends to even out the intervals between times of vesicle release as fewer vesicles are available to be released shortly after a period of high release rate, but more can be released when release-rate is low.

Given the inappropriateness of the Poisson formulae for the complete synaptic dynamics, alternative approaches are either: (1) to improve the calculations by including the correlations induced at each stage; or (2) simply to simulate a single synapse and empirically fit the mean response (and perhaps also its fluctuations) as a function of presynaptic rate. Such fitting can be carried out for any fixed set of parameters (time constants, facilitation factor, base release probability) but has the disadvantage that for any change in underlying parameters a new set of simulations are required to fit the output before the firing rate model can be updated.

6

Firing-Rate Models and Network Dynamics

In order for us to understand the functioning of a brain composed of billions or tens of billions of neurons, we must first understand simpler systems. In the earlier chapters, we simplified by looking at the processing of just a single neuron or of two to three neurons connected together. Another method of simplification is to ignore the idiosyncrasies of individual neurons and to aggregate the behavior of many neurons with similar properties. Such methods enable us to gain insight so that we can generate hypotheses for the mechanisms by which a variety of cognitive processes arise in neural circuits—insight that would be impossible when surveying voltage traces or spike trains from hundreds or thousands of neurons in a simulation or experiment.

Therefore, in this chapter we explore models of network dynamics at a level of detail that does not require simulation of the membrane potential of every neuron. If the information processed by a network of neurons is represented by the number of spikes produced by different subsets of neurons in small time windows, then a simplified description of the network is possible. Such a description is the firing-rate model.

6.1 Firing-Rate Models

The key variables of a firing-rate model are the firing rates, which correspond to the average number of spikes per unit time of a subset of similarly responsive cells. This is in contrast to spiking models in which the key variables are the membrane potentials of individual cells. The subset of similarly responsive cells in a firing-rate model can be called a unit, such that the unit's firing rate variable represents the mean firing rate of its constituent neurons.

Unit In a firing-rate model, a group of neurons whose mean activity is simulated.

In firing-rate models, the impacts of all the spikes from one unit on the mean firing rate of another unit combine to produce an effective connection strength. Self-connections (i.e., from a unit to itself) are possible, indeed common, in firing-rate models, because neurons within a unit, being similarly responsive and often spatially proximal to each other, are often connected with each other. Therefore, the spikes from neurons within a unit alter the firing rate of other neurons within the unit, so the unit's mean firing rate is impacted by a contribution that depends on its own firing rate. Even if no individual neuron were connected with itself, the units of a firing-rate model would do so. Such recurrent feedback is an important feature of many models of neural circuits that we will consider in more detail in this chapter.

Recurrent feedback   Input to a unit that depends on the unit's own activity.

Firing-rate models are valid if the effects of spikes from different neurons in the unit can be combined together linearly and if the spike times from different neurons in the unit are distributed uniformly across the typical time window. The time window of importance corresponds to the timescale over which the network can change its firing rate, which can be on the order of the neural membrane time constant for increases in rate, or on the order of the synaptic time constant for decreases in rate.

Firing-rate models have two main advantages over membrane-potential models. First, and most important, being simpler, they allow us to better understand the behavior of many model circuits. Second, having fewer variables and with slower dynamics, they allow for simulations that are orders of magnitude faster than the corresponding spiking network model. The second advantage enhances the first, for rapid simulations allow us to explore large regions of parameter space, in some cases exhaustively so that we can exclude certain hypotheses. In many cases a firing-rate model can be treated by mathematical analysis without simulations, so that the direct effect of a parameter—such as connection strength on mean firing rate or network stability—can be seen in the resulting formula. Thus, firing-rate models provide an important step in allowing us to *explain* network behavior, when in more complicated simulations we might just be able to *describe* the behavior.

Of course, firing-rate models have corresponding disadvantages. In reality spike times do matter, and the response of a downstream neuron to a barrage of spikes does depend on the correlations between those spikes on a short timescale, not just the total number within time windows on the order of 10 ms or more, as assumed in firing-rate models.

Moreover, the combining of many neurons into a unit is only valid if either (1) the neurons have identical firing rate curves or (2) if neurons with different firing rate curves have a fixed ratio of connection strengths to such different neurons in other units. If both of these two requirements are broken, then the effective connection strength from one unit to another depends on which subsets of cells within a unit have been excited, so a single mean firing rate is a poor description of the unit. Therefore, distinct classes of neurons (which typically have different connectivity patterns), as well as neurons with different tuning curves, should be treated as different units.

Finally, behavior that does depend on the membrane potential—for example, the conductance of the NMDA receptor channel (section 5.1.2)—is omitted from the firing-rate model, though in principle it can be included if mean membrane potential can be directly extracted from mean firing rate (see figure 2.9C).

Mean-field theory   An analysis that ignores correlations between variables so that the effect of fluctuations can be "smoothed out" by averaging and the system's behavior described according to the mean effect of all variables that comprise a unit.

Any firing-rate model is a mean-field theory in the sense that only the mean firing rate is represented—where the mean is the rate averaged across all neurons in a unit. That is, the influence of one unit on another is assumed to depend only on the activity averaged across all neurons in the unit. The assumption can be valid if spike times are asynchronous in the limit of large numbers of neurons per unit.

One effect of the finite number of neurons per unit that is absent in a standard firing-rate model formalism, is the impact of individual spikes from the unit. As we have seen in chapter 5, each spike from a presynaptic cell produces a rapid change in synaptic conductance of any post-synaptic cells, which results in a small uptick in the mean input conductance of any connected unit. These discrete events at random times can be incorporated into a firing-rate model with the addition of input noise to each unit. Since the mean firing rate of a neuron depends on the level of its input noise (see tutorial 2.1), in more sophisticated firing-rate models such noise should be incorporated in a consistent manner. However, within this course, we will just add a noise term in the differential equations and treat its magnitude as a parameter that can be adjusted to the level needed to produce realistic network behavior.

6.2   Simulating a Firing-Rate Model

Firing-rate models are based on the input-output function of a neuron (figure 6.1), which describes its firing rate as a function of either its total input current (as in figures 2.8 and 2.9B) or as a function of its excitatory and inhibitory incoming synaptic conductance. The second formulation is perhaps better since the spikes of one cell directly impact the synaptic conductance rather than synaptic current of another cell. In many cases a generic input-output function is used, such as a "sigmoid," which increases from zero at very low or very negative input and saturates at a maximum for large positive input. Alternatively, a power law can be used, or an empirical fit can be made to the response of either a real neuron or a spiking model neuron.

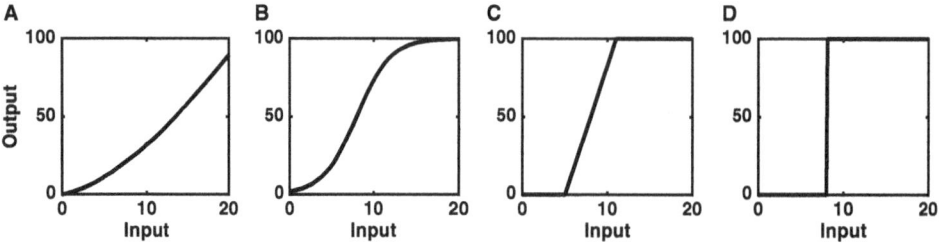

Figure 6.1
Examples of input-output functions for firing-rate models. (A) Power-law function, $y = A[I - I_0]_+^\alpha$ with amplitude $A = 1.2$, exponent $\alpha = 1.5$ and threshold $I_0 = 0$. The square brackets $[I - I_0]_+$ equate to 0 if $I < I_0$ and to $I - I_0$ otherwise. (B) An example of a sigmoid, the logistic function, $y = \dfrac{r_{max}}{1 + \exp\left[-\left(\dfrac{I - I_{0.5}}{\sigma}\right)\right]}$ with maximum output, $r_{max} = 100$, input for half-max, $I_{0.5} = 8$, and inverse steepness, $\sigma = 2$. (C) Threshold linear response function, with $y = 0$ for $I < I_0$, $y = (r_{max}(I - I_0))/\Delta I$ for $I_0 \le I \le I_0 + \Delta I$, and $y = r_{max}$ for $I > I_0 + \Delta I$. Saturation is at $r_{max} = 100$, responsive range has $\Delta I = 6$, and the threshold is $I_0 = 5$. (D) Binary response function, with $y = 0$ for $I < I_0$ and $y = r_{max}$ for $I > I_0$. Saturation is at $r_{max} = 100$ and the threshold is $I_0 = 8$. This figure is produced by the available online code `firing_rate_curves.m`.

> **Sigmoid**   Any function shaped as a forward-leaning "S" monotonically increasing from a minimum value to a maximum value over a certain range, such as the logistic function, $1/[1 + \exp(-x)]$, which is often used in firing-rate models.

As well as the unit's input-output function, which is its firing rate response, it is also important to know how the input to other units depends on the firing rate of a particular unit. In firing-rate models, such input is the product of a fixed connection strength and a synaptic gating function. The interunit connection strength depends on the numbers and mean strengths of excitatory and inhibitory connections between the neurons that constitute each unit. The synaptic gating function depends on the mean firing rate of cells in the unit supplying input and on the synaptic dynamics (e.g., equation 5.21).

Simulation of a firing-rate model entails solving a set of coupled ordinary differential equations just as in prior simulations. The underlying procedure follows two steps. First, the inputs to the units determine their firing rates. Second, the firing rates of units determine the new inputs of the units to which they are connected. Iteration by repeated cycles of this two-step process reveals the network's behavior in a simulation. If the behavior is a steady state of constant firing rates or one of regular oscillation, the properties of these states may also be revealed by mathematical analysis of the underlying coupled equations.

The general form of the dynamics of a firing-rate model is:

$$\tau_{r_i} \frac{dr_i}{dt} = -r_i + f_i\left(\{W_{ji}s_j\}\right) \tag{6.1}$$

coupled with

$$\tau_{s_i} \frac{ds_i}{dt} = -s_i + F(r_i), \tag{6.2}$$

where W_{ji} indicates the strength of connection from unit j to unit i, s_i denotes the fraction of downstream synaptic channels open as a function of incoming firing rate, and r_i is the mean firing rate of cells in each unit labeled i.

Writing the firing rate function as $f_i\left(\{W_{ji}s_j\}\right)$, means the firing rate of unit i depends on each other unit, j, in a manner determined by the connection strength, W_{ji}, from unit j to unit i, multiplied by the synaptic gating variable, s_j, of unit j. The general form of the function allows for different types of unit to have different types of effect on each other—for example input via some synaptic connections could add to the firing rate, whereas input via others could cause a multiplicative or divisive change in firing rate.

However, most often, and in this course, the inputs are just summed so that

$$f_i\left(\{W_{ji}s_j\}\right) = f_i\left(\sum_j W_{ji}s_j\right) = f_i(S_i). \tag{6.3}$$

Equation 6.3 means that the firing rate depends on the total synaptic input to a unit, $S_i = \sum_j W_{ji}s_j$, which is calculated by adding together the effects of all connected units. Inhibitory connections between units can then just be treated with negative values for the corresponding connection strength, W_{ji}.

The synaptic gating function, $F(r_i)$ in equation 6.2, is usually taken as a linearly increasing function of the presynaptic firing rate. This is an approximation, since synaptic gating, being the fraction of synaptic channels open, should saturate (i.e., never surpass a maximum of 1) even if the firing rate were to increase to very high levels. However, in practice, if the firing rate itself saturates at a maximum value, the synaptic input implicitly has a corresponding maximum value too. In this case, we can write $F(r_i) = r_i / r_i^{(max)}$, where $0 \leq F(r_i) \leq 1$, which is a linear dependence with an inherent maximum value.

Finally, if one of the time constants is assumed to be much shorter than the other—for example, if synaptic input changes on the timescale of 1 ms following a change in presynaptic firing rate, but the rate takes 10 ms to respond to changes in synaptic input—then one can approximate the shorter time constant as zero and require the right-hand side of the corresponding equation to evaluate to zero. For example, if $\tau_{s_i} \ll \tau_{r_i}$, then from equation 6.2: $-s_i + F(r_i) \approx 0$. This allows us to avoid any simulation of equation 6.2 simply by setting $s_i = F(r_i)$ and using this value in

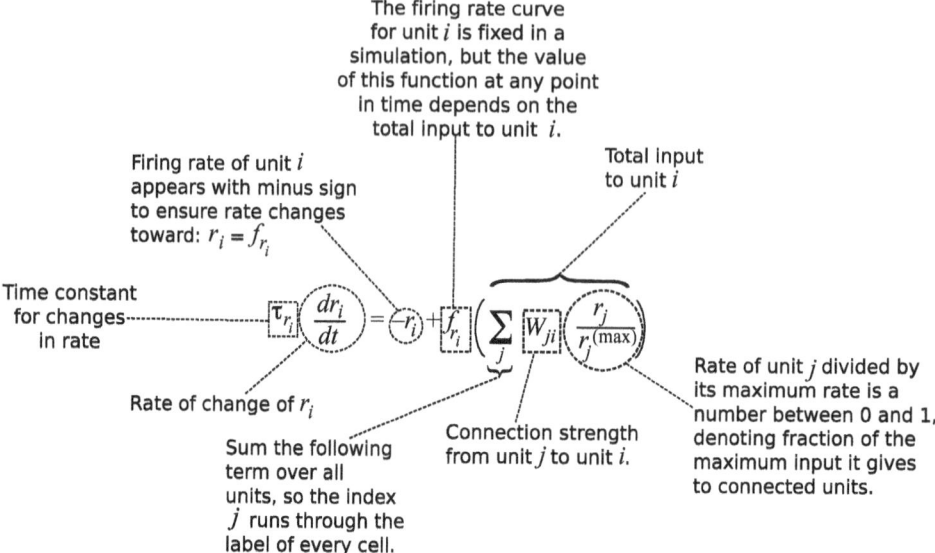

Figure 6.2
Annotation of equation 6.4. The reduced firing-rate model describes the rate of change of firing rate of each unit in terms of the sum over all of its inputs from other units.

equation 6.1. If we proceed in this manner and incorporate the linear approximation for $F(r_i)$, then $s_i = r_i / r_i^{(max)}$ and the set of differential equations to solve in equation 6.1 becomes (see figure 6.2):

$$\tau_{r_i} \frac{dr_i}{dt} = -r_i + f_i \left(\sum_j W_{ji} \frac{r_j}{r_j^{(max)}} \right).$$

(6.4)

In equation 6.4 the number of variables that must be simulated is equal to the number of units.

6.2.1 Meaning of a Unit and Dale's Principle

Many individual neurons, especially in cortical circuits, have either a predominantly excitatory or predominantly inhibitory effect on all other neurons to which they provide input. Whether the effect is excitatory or inhibitory depends on the dominant type of neurotransmitter they release, which is conserved across a neuron's axonal terminals in nearly all cases according to Dale's principle (section 5.1.2). The principle can be extended to state that neurons are either excitatory or inhibitory, a classification which is valid for many circuits. When valid, such separation into two classes requires the connectivity matrix between neurons to contain rows of either all non-negative entries (for excitatory neurons) or all nonpositive entries (for inhibitory neurons).

However, such a restriction on the connectivity matrix need never apply to firing-rate models, because each entry in the matrix can represent the net impact of a unit made up of both excitatory and inhibitory neurons on another such unit. For example, if unit A connects to unit B through a preponderance of excitatory connections to excitatory cells then that interunit connection, A-to-B, would be excitatory. Yet unit A could also connect to unit C through a preponderance of inhibitory connections to excitatory cells, in which case the interunit connection, A-to-C, would be inhibitory (such an inhibitory interunit effect can also arise from excitatory connections to inhibitory cells). Thus, if units comprise a mixture of excitatory and inhibitory cells, then there is no restriction on the signs of connections between units.

6.3   Recurrent Feedback and Bistability

In this section, we will see how excitatory recurrent feedback can generate bistability. In this case, bistability means that neurons can maintain spiking activity at two distinct firing rates when receiving the same level of input (which is typically zero). The lower firing rate state is either quiescence, the absence of spikes, or activity in the range of 0 to 10 Hz corresponding to the spontaneous activity of neurons in the absence of a stimulus. The higher firing-rate state—typically in the tens of hertz in models, but lower in empirical data—can be initiated by an excitatory stimulus and persists once the stimulus is removed.

Such a bistable unit acts as a simple memory circuit, since its level of activity is history-dependent. Its firing rate is high if its most recent input was excitatory (to switch on activity), and its firing rate is low if its most recent input was inhibitory (to switch off its activity). Amazingly, neurons with similar stimulus-specific sustained responses have been recorded in monkeys during short-term memory tasks.[1] Moreover, an antagonist of NMDA receptors (which disrupts their normal function) prevents such sustained responses.[2] These data suggest that small circuits with excitatory feedback are the basis of short-term memory maintenance in the brain.

6.3.1   Bistability from Positive Feedback

Bistability requires a process that, following a small shift of a system away from equilibrium, acts to accelerate the system even further from equilibrium until a tipping point is reached and a return to the prior equilibrium is not possible. For a new equilibrium to be reached—as it must for the system to be bistable rather than unstable—a second process must kick in to limit the overall change. The first, accelerating process is positive feedback, as it increases the rate of change in the same direction as the change. The second, limiting process is negative feedback, since further change decreases the rate of change.

In neural circuits with excitatory feedback (figure 6.3), we can consider how much extra synaptic input is produced by an initial small change in firing rate. If that extra synaptic input is sufficient to produce a bigger change in firing rate than the initial small change, then even more feedback ensues. Greater firing rates followed by even greater feedback would then follow in

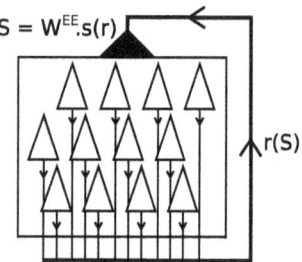

$S = W^{EE}.s(r)$

$r(S)$

Figure 6.3
Firing-rate model unit with recurrent feedback. A unit of many neurons (triangles) produces spikes at a mean rate, r, which is a function of the total synaptic input, S. The total synaptic input, S, depends on the mean rate, r, via the term $s(r)$, which corresponds to the fraction of open synaptic channels when the presynaptic cells fire at the mean rate, r. The fraction of open channels, $s(r)$, which varies between 0 and 1, is multiplied by the maximum effective feedback strength, W^{EE}, to produce the total synaptic input, S. The superscript "EE" in W^{EE} denotes excitatory feedback to excitatory cells in this model.

a positive feedback loop. The initial firing rate at which such positive feedback dominates is a point of no return.

The amount of positive feedback arising within a circuit depends on the steepness of the firing-rate curve and the amount of extra synaptic input per spike. At low firing rates (such as zero), the firing-rate curve is almost flat, since a small amount of extra current does not generate more spikes when neurons are far below threshold. Therefore, a low firing rate state with minimal feedback can be a stable state in which the positive feedback is too small to cause runaway excitation. However, as the input and the firing rate increases, firing-rate curves become supralinear (of increasing slope). It is possible for them to reach sufficient steepness that the positive feedback dominates—each additional spike produces enough input current to generate more than one additional spike—and a point of no return is reached. The neurons in the unit then begin to fire ever more rapidly. The firing rates are prevented from rising inexorably by one or more limiting processes, whose impact we will consider in some detail.

> **Supralinear function**   A function whose gradient increases along the x-axis, also called a convex function.

The simplest of limiting processes is firing rate saturation—neurons and all reasonable models of neurons have a maximum firing rate, on the order of several tens of hertz for excitatory pyramidal cells and on the order of 100–200 Hz for many "fast-spiking" inhibitory interneurons. Once the limit is reached, in simple models the firing-rate curve is flat, with spiking activity remaining at its maximum rate. The flaw in such simple models is that empirically spiking activity can cease with excessive input, e.g., due to a loss of sodium deinactivation (section 4.2.3).

Figure 6.4 depicts the behavior of a single unit, i, with excitatory feedback $W_{ii} = W^{EE}$, which increases in strength across panels A to C. In the left-hand panels the firing-rate as a function of input is plotted as a solid curve with a sigmoidal shape. In the absence of input the rate is low, the gradient of the curve increases with input at low rates, but decreases at high rates until it saturates at the maximum rate (100 Hz here). The dashed lines in each of the left panels indicate the recurrent excitatory feedback. These lines are plotted using a flipped orientation of the axes, because the synaptic input plotted on the x-axis is due to recurrent feedback, so depends on the firing rate, which is plotted on the y-axis. When the firing rate is zero there is no synaptic input from recurrent feedback, so the dashed lines pass through the origin. If feedback is weak (panel A1) the synaptic input is low even when the firing rate is maximal. At higher feedback strength (panels B1 and C1) a given firing rate produces greater synaptic input, so the lines are less steep.

Fixed point   A set of values for all of the variables of a system, at which their time derivatives (their rates of change with time) are all zero, so the system does not change when at the fixed point.

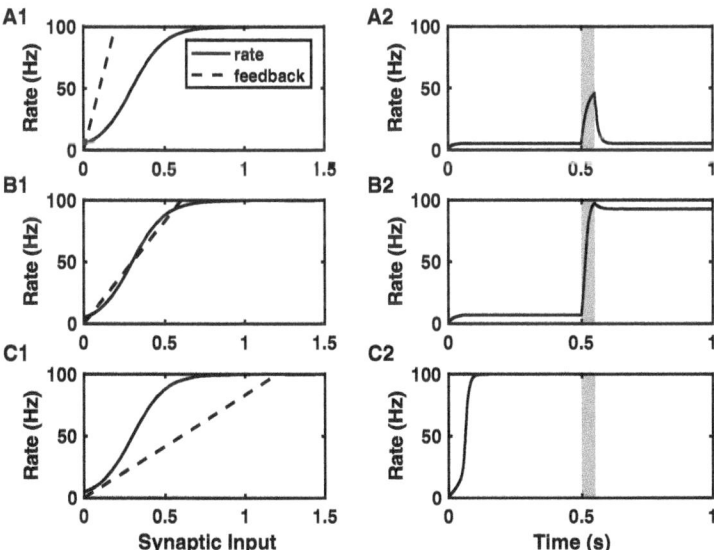

Figure 6.4
Recurrent feedback strength determines the stable states of a circuit. (A1, B1, C1) Firing rate response of the unit is a sigmoidal function (solid line). For the feedback function (dashed straight line) the firing rate on the y-axis determines the level of synaptic input (x-axis). The stronger the connection strength, the greater the synaptic input produced by a given firing rate, so the shallower the feedback curve. (A2, B2, C2) Dynamics of firing rate as a function of time for the models with rates initialized at zero and a pulse of input between 0.5 s and 0.55 s (gray bar). Note the switch in activity, representing memory of the stimulus in B2. (A) With weak feedback, $W^{EE} = 0.2$, only a low-rate state is stable. (B) With medium feedback, $W^{EE} = 0.6$, both a low-rate and a high-rate state are stable so a transient stimulus can switch the system between states. (C) With strong feedback, $W^{EE} = 1.2$, only a high-rate state is stable. This figure is produced by the available online code bistable1.m.

The points of intersection of the two curves (dashed line and solid curve) in panels A1, B1, and C1, of figure 6.4, indicate fixed points of the system, meaning the firing rate and the synaptic input do not change if they are set to those values (for more on fixed points see chapter 7). This can be seen in the corresponding right-hand panels (A2, B2, and C2) where the firing rates are stationary when they reach their values at those intersections in the left-hand panels.

In panel A2, the temporary stimulus between 0.5 s and 0.55 s causes a temporary increase in firing rate. However, once the stimulus is removed, the firing-rate returns to its prior stable value. We can understand this from the curves in figure 6.4, panel A1, where we see from the dashed line that at the higher firing rate the synaptic input is relatively low. We then see from the solid line of panel A1 that the low synaptic input generates a low firing rate, which leads to even lower synaptic input. Eventually, at the point of intersection, the small amount of synaptic input produced is exactly the amount needed to maintain the low firing rate, and the system does not change further. The single stable state, with low firing rate, is reached.

Stable fixed point   A fixed point toward which the system returns following any small changes in its variables.

Unstable fixed point: A fixed point away from which the system changes following the slightest change in one or more of its variables. An unstable fixed point can be the point of no return when a system changes from one stable fixed point to another.

In the bistable system of figure 6.4, panel B1, the two curves intersect at three points. Therefore, the system has three fixed points, that is, three distinct sets of firing rate and synaptic input at which their rates of change are zero. However, the system only has two stable states because the intermediate fixed point is an unstable one. An unstable fixed point is akin to a pencil standing vertically on its tip—while theoretically it could be aligned so that its center of mass is exactly above the tip, in which case it has no preferred direction in which to fall, in practice such perfect alignment is impossible and random air movements or vibrations would disturb any perfect alignment were it possible. Such unstable fixed points would not be observed in a simulation (as in panel B2) unless the simulation were started with the exact values of the fixed point and were noise-free.

That the intermediate fixed point is unstable can be understood from panel B1, by considering the system's response to small deviations above or below the fixed point. Following any deviation from a curve, the system will move back toward the two curves, with the firing-rate curve (solid) indicating the component of change along the y-axis, and the feedback curve (dashed) indicating the component of change along the x-axis. That is, a point on the figure representing the instantaneous firing rate and synaptic input would move vertically toward the solid curve and horizontally toward the dashed curve. Wherever the firing-rate curve is above (greater y-coordinate) and the feedback curve is to the right (greater x-coordinate), the net effect is an increase in both firing rate and feedback until the curves cross again. Wherever the firing rate curve is below (smaller y-coordinate) and the feedback curve is to the left (smaller x-coordinate), the net effect is a decrease in both firing rate and feedback until the curves cross again. These two effects combine to mean that the state of the system moves away from the intermediate fixed point following any small deviation.

The intermediate fixed point is, in fact, the point of no return mentioned earlier. If the system, having been in the low-firing rate state, is provided enough stimulus to pass the intermediate fixed point then it will head to the high-firing rate state, even after any stimulus is removed.

6.3.2   Limiting the Maximum Firing Rate Reached

In the previous section, a leveling off of the firing-rate curve at its maximum value provided the only limit on runaway activity. Indeed, as can be seen from figure 6.4, panel B1, if the linear feedback curve intersects the firing-rate curve at three points, the high-rate point is inevitably close to the unit's maximum rate (of 100 Hz in this example). However, in vivo, firing rates are observed to switch by smaller amounts (10–30 Hz), even though the individual neurons can reach rates on the order of 100 Hz given sufficient input. One possible cause of a reduction in the system's maximum persistent rate is a saturation of synaptic feedback.

One limit on synaptic feedback is the number of receptors available to be bound by neurotransmitter at the surface of the postsynaptic neuron. Once all of the receptors at feedback synapses are bound by ligand and their corresponding channels are opened, there is no further possible increase in synaptic input due to feedback. Therefore, once neurons are firing at sufficient rate to keep the postsynaptic receptors bound, no further feedback input is possible, so the positive feedback remains constant and the rate does not increase any further. If such feedback is predominantly mediated via receptors from which neurotransmitter is slow to dissociate, or whose associated ion channels are slow to close once opened, then the presynaptic rate at which feedback begins to saturate can be low. Thus, NMDA receptors, with a slow time constant of 50–75 ms (at in vivo temperatures) for unbinding of glutamate neurotransmitter, could play an important role in stabilizing persistent activity at low rates[3] (figure 6.5).

A second limit on synaptic feedback occurs at the axonal terminals of the presynaptic neuron. The limited number of release-ready vesicles that causes synaptic depression (section 5.3.1) also limits the amount of sustained synaptic input to postsynaptic cells. Once a presynaptic cell is producing action potentials at a rate quicker than vesicles can "dock" in the membrane or be replenished, any further increase in presynaptic firing rate does not increase the rate of release of vesicles with neurotransmitter, so it does not produce more synaptic input to postsynaptic cells (equation 5.27 and following). The firing rate at which synaptic depression begins to limit the synaptic feedback is proportional to the rate of replenishment of release-ready vesicles, and

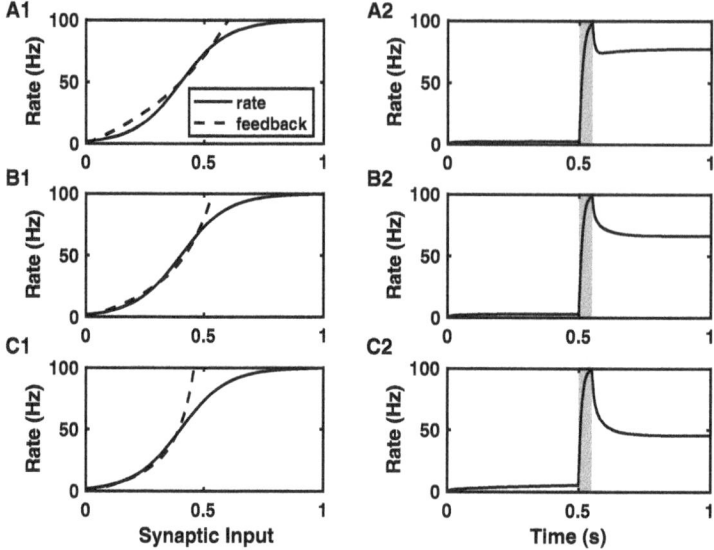

Figure 6.5
Persistent activity at lower rates via synaptic saturation. When synaptic feedback saturates (dashed curves become vertical at high rates) bistability is possible without the more active state's firing rate being so unrealistically high. Synaptic saturation occurs at lower rates (C) if the fraction of receptors bound per spike, α, is high rather than low. (A1-A2) $\alpha = 0.2$, $W^{EE} = 1.2$. (B1-B2) $\alpha = 0.5$, $W^{EE} = 0.75$. (C1-C2) $\alpha = 1$, $W^{EE} = 0.55$. For all curves, $p_r = 1$ and $\tau_s = 50$ ms. This figure is produced by the available online code `bistable2.m` and are in the same format with the same stimuli as figure 6.4.

inversely to the probability, for each release-ready vesicle, that an action potential causes its release.

In the following section, we will consider how such synaptic saturation can be incorporated into a firing-rate model, so that we can simulate the processes and assess their impact on the rate of sustained activity in tutorial 6.1.

6.3.3 Dynamics of Synaptic Response
In the appendix of chapter 5 (section 5.10.1), we showed that the synaptic time constant, τ_s, limits the mean synaptic response, s, due to a presynaptic neuron firing spikes as a Poisson process at rate r via

$$s = \frac{\alpha p_r r \tau_s}{1 + \alpha p_r r \tau_s},$$
(6.5)

where p_r is the release probability of each docked vesicle per spike and α is the maximum fraction of postsynaptic receptors bound by neurotransmitter when all presynaptic docked vesicles are released. In a firing-rate model that omits the presynaptic dynamics of depression and facilitation, the mean synaptic response at steady state, s, can be used as $F(r)$ (equation 6.2) when the synaptic gating variable is set instantaneously as $s = F(r)$. Alternatively, if the dynamics of s are simulated, the corresponding equation (whose steady state matches equation 6.5) is:

$$\frac{ds}{dt} = -\frac{s}{\tau_s} + \alpha p_r r(1-s) \tag{6.6}$$

where the first term on the right is the decay of synaptic conductance due to unbinding of neurotransmitter, and the second term is the amount of increase in s per spike multiplied by the mean rate of spike arrival.

The effective time constant for changes in s following changes in r becomes $\tau_{eff}(r) = \tau_s / (1 + \alpha p_r r \tau_s)$. In the absence of spikes ($r = 0$), the effective time constant approaches τ_s, the timescale for decay of synaptic input to zero. When the firing rate is high, the effective time constant is smaller, allowing s to increase more quickly in proportion to the rate of spike production.

In a circuit with excitatory feedback, we can replot the two steady state curves, $r(S)$ and $s(r)$, where $S = W^{EE}s$ and see that the rate of sustained activity is reduced if τ_s is relatively large (figure 6.5).

6.3.4 Dynamics of Synaptic Depression and Facilitation

Synaptic depression and facilitation can also be incorporated in firing-rate models. In the appendix of chapter 5 (sections 5.10.2–3) we showed how the steady state values for the depression variable, D, and the facilitation variable, F, depend on presynaptic firing rate for a Poisson process. These variables modify respectively the amplitude of synaptic response, $\alpha = \alpha_0 D$, and the individual vesicle release probability, $p_r = p_0 F$. Their dynamics can be modeled according to the equations (cf. equations 5.9–5.10)

$$\frac{dD}{dt} = \frac{1-D}{\tau_D} - p_r D r \tag{6.7}$$

and

$$\frac{dF}{dt} = \frac{1-F}{\tau_F} + f_F (F_{max} - F) r. \tag{6.8}$$

These dynamical equations have stable steady states at fixed rate, r, that match the mean response to spikes arriving as a Poisson process of rate r (cf. equation 5.27):

$$D(r) = \frac{1}{1 + p_r r \tau_D} \tag{6.9}$$

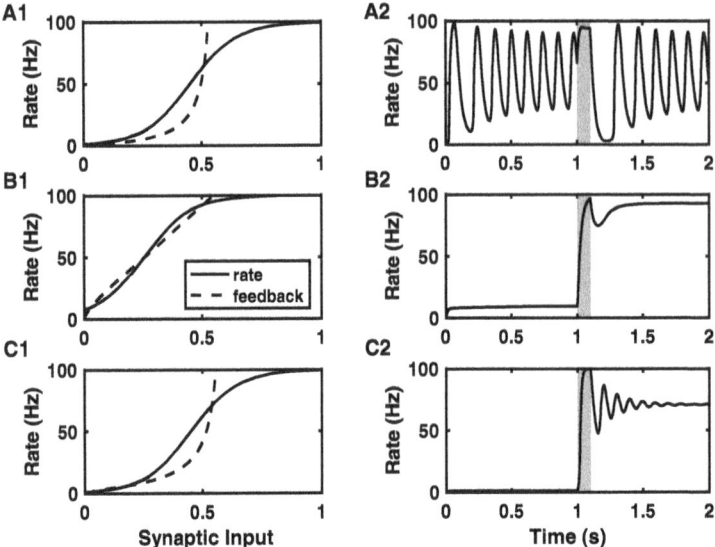

Figure 6.6
Activity states in a circuit with excitatory feedback and short-term synaptic dynamics. (A) Synaptic depression alone can destabilize the persistent activity state and lead to slow oscillations. (B) Synaptic facilitation enhances the stability of a bistable system, so the low-rate state can have activity above 10 Hz. (C) Facilitation and depression combine to generate a bistable system even when synaptic saturation is absent. Format of the figure is the same as that of figure 6.4 and with the same stimulus. This figure is produced by the available online code `bistable_fac_dep.m`.

and (cf. equation 5.30)

$$F(r) = 1 + \frac{(F_{max} - 1) f_{fac} r \tau_F}{1 + f_{fac} r \tau_F}. \tag{6.10}$$

Synaptic depression causes the mean steady-state synaptic response to saturate at high presynaptic firing rates in a manner mathematically equivalent to synaptic receptor saturation. However, the rate at which such saturation becomes important is on the order of $1/p_r\tau_D$, which is typically lower than $1/\alpha p_r\tau_s$, the corresponding rate for saturation of the postsynaptic response. While such saturating feedback can limit the firing rate of persistent activity states to the observed lower levels, the slow time constant of synaptic depression tends to destabilize the persistent state and can generate oscillations (figure 6.6A). We will investigate this process in tutorial 6.1.

Synaptic facilitation increases the effective synaptic strength when firing rate increases, so it can stabilize a bistable system. The additional stability arises because the low firing-rate state has effectively weaker synapses—making it harder for firing rate to increase—while the high firing-rate state has effectively stronger synapses—making it harder for firing rate to decrease. Because of this, synaptic facilitation allows the spontaneous, low-firing-rate state of a bistable system to have activity that is significantly above zero (figure 6.6B) without the excitatory feedback causing a runaway increase in firing rate.

6.3.5 Integration and Parametric Memory

In bistable systems, the feedback curve crossed the firing rate curve at three points, two of which indicated the firing rate and synaptic input of the two stable states (figure 6.4B). In another scenario, rather than crossing, the two curves could lie on top of each other for a range of firing rates. In this case, any points in the range that lies on both curves correspond to values of the firing rate and synaptic input that the system would remain at if set to have those values. Moreover, if the curves overlap or cross nowhere else, then the system's variables move to a point where the curves overlap. In this situation, the system possesses a line attractor (also called a continuous attractor or a marginal state).

> **Line attractor** A continuous range of fixed points that can be plotted as a curve or line; also called a continuous attractor, or marginal state. The system returns to the line, but not generally to the original point on the line following any small deviation.

The line attractor is said to possess marginal stability, because following perturbations of the system the firing rate and synaptic input return to a point on the line, suggesting stability, but they may not return to the original point on the line. Movement along the line is easy.

Such easy movement can be beneficial, since the system can respond to—even sum up, or integrate—a series of small inputs, each of which kick the system's state a little further along the line. Indeed, if a property of the circuit, such as its final firing rate, depends monotonically on the amount of external excitatory input the unit receives, then the circuit is acting as an integrator. That is, a line attractor can act as an integrator.[4]

> **Integrator** A circuit that integrates its inputs in the mathematical sense, summing the values of discrete inputs.

> **Parametric memory** Memory of a continuous quantity, or parameter of a stimulus.

An integrator provides memory of a continuous quantity—namely the integral over time of the stimulus—because when the stimulus is removed, the integral over time no longer changes.

Therefore, the memory of the integrated value at stimulus offset remains (figure 6.7). We shall investigate such integration in an excitatory feedback circuit in tutorial 6.1.

Self-excitation   Excitatory recurrent feedback to a group of cells or a unit.

Cross-inhibition   An inhibitory interaction between two cell groups or units that respond to different or opposite stimuli.

Integration has received considerable attention in computational neuroscience because integrators appear to be necessary in some small circuits—e.g., to maintain a fixed eye position in some species[5]—they play a large role in the field of decision making,[6,7] and they appear to underlie the ability of animals to retain short-term memory of continuous parameters.[8,9] However, a circuit based on self-excitation of a single unit (or cross-inhibition between two units) requires

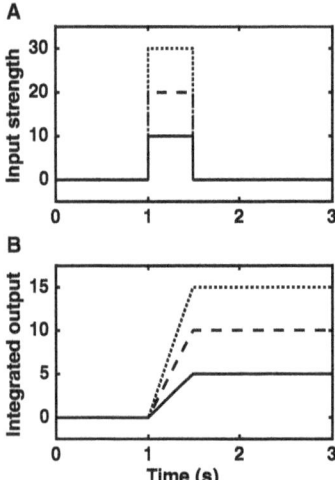

Figure 6.7
Connection between integration and parametric memory. (A) Transient stimuli of different amplitudes are provided as possible inputs to an integrator. (B) The rate of change of an integrator depends on its inputs and is zero in the absence of input. The integrator's output increases at a stimulus-dependent rate to reach a level that depends on the amplitude of the stimulus. The output remains as a constant when the stimulus is removed. Therefore—so long as input duration is fixed and the output is reset to zero before any input—the final output of the integrator provides a memory of the amplitude of its prior input. In practice the output could be simply the mean firing rate or, more generally, any function of the firing rates of a group of neurons. This figure is produced by the available online code `integrator_cartoon.m`.

fine-tuning in order for firing-rate and synaptic-feedback curves to align to an even greater extent than that shown in figure 6.5C1. Whether neural circuits possess the necessary mechanisms for the fine-tuning needed to produce a line attractor remains an open question.

6.4 Tutorial 6.1: Bistability and Oscillations in a Firing-Rate Model with Feedback

Neuroscience goals: Learn how synaptic dynamics impacts the behavior of a bistable circuit with excitatory feedback; see how an approximate line attractor can behave as an integrator.

Computational goal: Practice at simulating firing-rate models.

In this tutorial, we will explore the conditions for bistability and the effects of short-term synaptic plasticity in a firing-rate model with excitatory feedback. The three coupled variables for the single unit will be firing rate, r, depression variable, D, and synaptic gating variable, s. The equations that you will simulate are an extension of equation 6.1:

$$\tau_r \frac{dr}{dt} = -r + f(S)$$
$$\frac{dD}{dt} = \frac{1-D}{\tau_D} - p_r D r \qquad\qquad (6.11)$$
$$\frac{ds}{dt} = -\frac{s}{\tau_s} + \alpha_0 D p_r r \cdot (1-s)$$

with the additional condition $r \geq 0$ and where the total synaptic input, S, is given by $S = W^{EE}s + s_{in}$. The firing rate curve, $f(S)$, in the top equation corresponds to the steady state firing rate as a function of total synaptic input. It is given as a power law with saturation:

$$f(S) = r_0 + r_{max} \frac{S^x}{S^x + \sigma^x} \text{ if } S > 0, \text{ and } f(S) = r_0 \text{ if } S \leq 0.$$

The exponent is $x = 1.2$ and the rate with no input is $r_0 = 0.1$ Hz in the first simulation (but r_0 will vary thereafter). The maximum possible firing rate is $r_0 + r_{max}$, with $r_{max} = 100$ Hz. The input needed to reach the midpoint of the firing-rate range is given by $\sigma = 0.5$, and the time constant for changes in firing rate is $\tau_r = 10$ ms. Other parameters in the model are α_0, p_r, W^{EE}, and τ_D, and τ_s, which all vary from question to question.

For each question with a distinct parameter set, you will plot the steady state response functions in the absence of input: $f(W^{EE}s)$ as a function of s, and $s(r) = \alpha_0 D p_r r \tau_s / (1 + \alpha_0 D p_r r \tau_s)$ as a function of r. In the calculation of $s(r)$, you will need to use the steady state response of the depression variable, $D(r) = 1/(1 + p_r r \tau_D)$. After plotting the steady state values as two distinct curves (whose intersections are fixed points of the system) you will then simulate the system and its response to transient inputs, s_{in}.

1. Fix the depression variable so that it remains unchanged from 1and set the other parameters
    as follows: $\alpha_0 = 0.5$, $W^{EE} = 8$, $p_r = 1$, and $\tau_s = 2$ ms.

    a. Plot $f\left(W^{EE}s\right)$ for a range of values of s from 0 to $1/W^{EE}$. On the same figure, for a range
        of values of r from 0 to r_{max} plot $s(r)$, with r on the y-axis and s on the x-axis.

    b. Simulate the full set of three coupled ODEs (equation 6.11) for 20 s, with initial condi-
        tions $s = 0$ and $r = 0$. Add a temporary input of strength $s_{in} = 0.05$ for a duration of 50
        ms beginning at a time of 10 s. Plot the resulting firing rate versus time. Check that the
        stable rates correspond to the crossing points of the two curves in a).

2. Include synaptic depression with $\tau_D = 250$ ms and $p_r = 0.2$; keep $r_0 = 0.1$ Hz and $\alpha_0 = 0.5$;
    and increase W^{EE} to 60 (to compensate for the reduced release probability and for the weak-
    ening of synapses due to depression). Repeat parts 1a and 1b, except set the temporary input
    to be of strength $s_{in} = 0.002$ and of duration 2 s. Comment on any behaviors you see, in
    particular any difference from 1.

3. Increase p_r to 0.5, reduce W^{EE} to 35, and set r_0 to -0.1 Hz. Reset the temporary input to
    $s_{in} = 0.05$ for a duration of 50 ms. Keeping all other parameters the same as in part 2, repeat
    parts 1a and 1b and comment on your results.

    c. Set the initial value of the firing rate to 9 Hz and the initial values of all other variables
        to their steady state at that rate (the steady state can be calculated rather than simulated).
        Repeat the simulation of part 3b with the new initial conditions. Comment on your
        results.

4. Oscillations are enhanced with slow negative feedback. To see this, repeat 3a and 3b with
    the following parameters: $\tau_D = 125$ ms (half of its prior value); $\alpha_0 = 0.25$ (half of its prior
    value); and $p_r = 1$ (twice its prior value). You should find the steady state curves are identical
    to those in question 3. Try to explain why the active state is now stable (oscillations decay)
    whereas in question 3 it was unstable (oscillations grow).

Challenge

5. If the feedback curve overlaps with the firing-rate curve then the system can be almost stable
    at a lot of different firing rates. This allows for the firing rate to slowly accumulate when a
    stimulus is applied and for the final rate to be an approximately continuous readout of the
    time-integrated stimulus. You will omit synaptic depression, fixing $D = 1$ as in question 1,
    and set the following parameters: $\tau_s = 50$ ms, $W^{EE} = 1.6$, $\alpha_0 = 1$, $p_r = 0.1$, and $r_0 = 0.5$ Hz.
    Repeat 1a with these parameters.

    b. Repeat 1b with these parameters using an input of strength $s_{in} = 0.001$ and a duration of
        6 s.

    c. Repeat 5b except with a series of 30 inputs, each of amplitude $s_{in} = 0.005$ of duration
        100 ms, starting each input every 500 ms. Comment on the behavior of the circuit.

d. Suppose that following an unknown number of such stimuli, you are given the firing rate of the unit 500 ms after the onset of the last stimulus. Estimate how many distinct numbers of stimuli you could distinguish using the firing rate at this time.

6.5 Decision-Making Circuits

The importance of decision making for us as humans is widely acknowledged because our decisions have consequences that shape our lives. However, it can be hard to define decision making from a biological perspective, when we study a system from the outside and just consider how the functional response of a system depends on its inputs. Biologists can reduce the meaning of the term "decision making" to include random events pertaining to inanimate matter. For example, it is common to suggest that a bacterium "decides" which protein to express or to name the genetic event of a cell becoming one type or another a "cell fate decision." Given such examples, one could equally ask whether a storm "decides" which path to take.

The difficulty lies in that decisions do appear to include a random component—when an animal is given identical choices, the time taken to make a choice and the actual choice made vary from trial to trial. The deterministic components—such as the level of an ongoing stimulus, any prior belief based on the animal's experiences, the value of options based on perhaps genetic disposition as well as history—conspire together to shape the probability of different choices being made, while allowing (it appears) each individual choice to be undetermined.

In general usage, a decision suggests a point of no return—a commitment to a particular course of action—so has a lot in common with action selection. Often the selection is between discrete alternatives: Shall I eat one fruit or the other? Shall I accept a new job offer or stay in my current job? Shall I buy this house or keep looking? Other decisions are from a continuous range of possibilities: In which direction shall I point? How fast shall I drive? How long shall I wait?

Models of decision making address three distinct issues. First, theorists produce models so as to account for the experimental data. In this case, the data include behavioral measures, such as the proportion of different responses and the times taken to make those responses.[10] They also include electrophysiological measures, in particular the dynamics of neural activity during decisions.[11,12] A good model accounts for the distribution of these features over many trials and for how these distributions change as the stimulus changes.

Second, theorists analyze different models to ascertain which would produce more optimal behavior.[13,14] Addressing the second issue is valuable, because it can provide an explanation for why animals would make decisions a certain way. It also allows us to test the degree of optimality achieved by any animal in any task.

Third, theorists produce models constrained by the properties of neurons and neural circuits engaged in decision making.[11] While psychologists may be content with a verbal or purely mathematical model, which accurately describes a behavior, neuroscientists are not satisfied until they understand the mechanisms by which the brain generates the behavior. Theorists can then proffer explanations for any suboptimal behavior observed in terms, perhaps, of the underlying biological constraints.[15]

Perceptual decisions are ones that require identification of an ambiguous stimulus—the action is then completely determined by a previous learned rule that matches stimulus to action. Historically, one of the most common stimuli used has been a display of randomly flickering and moving dots on a screen with a net bias or drift in their motion toward one direction over another.[16] The animal—in this task, typically monkey or human—indicates the direction of motion it perceives. Often only two choices are available, in which case the design is known as a two-alternative forced-choice task.

The relative simplicity of the task, whose difficulty is easily titrated by adjusting the coherence of the dot motion, has produced a wealth of behavioral and electrophysiological data. The distribution of response times for correct choices and errors, as well as their relative probability, was first described by mathematical models such as the drift diffusion model, based on integration of noisy evidence to a threshold.[17] In these models the evidence refers to the difference between two stimuli, one representing rightward motion, the other representing leftward motion. The threshold corresponds to a point of no return at which the decision is made. Such a model is optimal in many conditions.

Integration of evidence   The optimal method of accumulating successive samples of a stimulus such that information from each sample is equally weighted.

In this section we shall consider a circuit (figure 6.8), based on the Wang model,[11] which accounts for much of the behavior and neural activity observed in perceptual decisions between two alternatives, and is constrained by the properties of neural circuits. The circuit can be extended trivially to model more than two alternatives when necessary. The Wang model showed how two groups of neurons corresponding to the two alternatives could both reproduce the behavior and the ramping activity of the "winning" group of neurons using a "winner-takes-all" circuit (see figure 6.9).

Winner-takes-all circuit   A circuit with multiple distinct cell groups or units within which only one unit can sustain high activity, because in so doing it suppresses activity of other units through cross-inhibition.

The winner-takes-all circuit (figure 6.8) is based on a combination of self-excitation and cross-inhibition. The self-excitation is sufficient that each unit alone would be bistable (as in figures

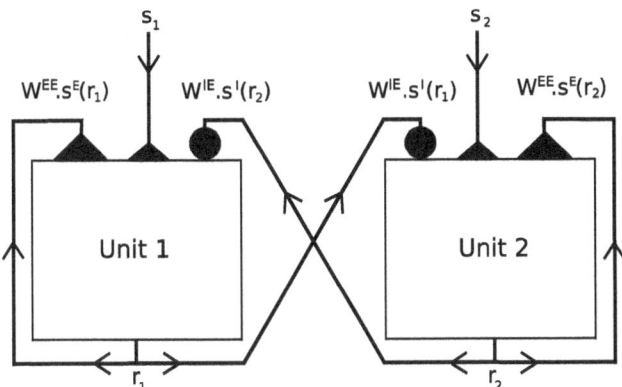

Figure 6.8
Decision-making circuit. Two units, that are each separately bistable via recurrent excitation of strength W^{EE}, are coupled by cross-inhibition of strength W^{IE}. If one unit becomes active it suppresses the activity of the other unit in a "winner-takes-all" manner, with the active unit representing the decision. If noisy inputs s_1 and s_2 are applied to units 1 and 2 respectively, the unit with greater mean input is more likely to become the active one.

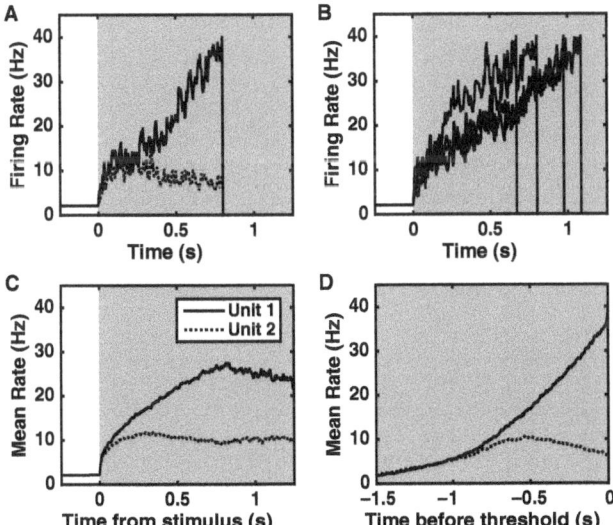

Figure 6.9
Dynamics of a decision-making circuit. (A) Upon stimulus delivery, the firing rate of unit 1 (solid) rises gradually over the course of almost a second until the threshold rate of 40 Hz is reached. The firing rate of unit 2 (dotted) rises initially, but is then suppressed by inhibition from unit 1. The gray shaded region indicates the time of input, during which unit 1 receives slightly greater excitatory input than unit 2. (B) The firing rate of unit 1 on four successive trials demonstrates considerable variability in the time to reach threshold. (C) The mean rate as a function of time from onset of the stimulus. (D) The mean rate as a function of time aligned to the threshold-crossing point—the curve of unit 1 is steeper, and the suppression of the rate of unit 2 is clearer, than in C (dotted). These figures are produced using the circuitry of figure 6.8, by the available online code decision_making_dynamics_figure.m.

6.4B and 6.5), able to remain in a state of low spontaneous activity or, once excited, to persist in a state of high activity. The cross-inhibition (i.e., reciprocal inhibition) between the two units is strong enough to ensure that if one of the units is in the active state the other cannot be highly active. Thus cross-inhibition ensures the processing of the circuit is competitive and can only produce a single winner, generating a binary outcome when two units are involved.

6.5.1 Decisions by Integration of Evidence

During perceptual decisions, neural activity in parietal cortex appears to ramp up gradually over a time period that can be on the order of a second.[12] Achieving such a slow time course of variation in a circuit that comprises neurons and synapses with time constants on the order of a few milliseconds requires fine-tuning (as with the integrator in tutorial 6.1, question 5). The degree of fine-tuning is reduced if feedback within the circuit is predominantly through NMDA receptors with a long time constant (50 ms is a reasonable value in vivo). The necessary degree of fine-tuning can be understood from a linear firing-rate model, such as the leaky-competing accumulator model (LCA).[18] The LCA is based on a circuit like that of figure 6.8, but with the following simplified dynamics for the two competing groups:

$$\tau \frac{dr_1}{dt} = -r_1 + W_s r_1 + W_x r_2 + i_1 \tag{6.12a}$$

$$\tau \frac{dr_2}{dt} = -r_2 + W_s r_2 + W_x r_1 + i_2, \tag{6.12b}$$

where W_s is a positive parameter representing self-excitation and W_x is a negative parameter representing cross-inhibition. If we define the difference between inputs as $i_d = i_1 - i_2$ and difference between rates as $r_d = r_1 - r_2$ then subtracting equation 6.12b from equation 6.12a leads to:

$$\tau \frac{dr_d}{dt} = -r_d + (W_s - W_x) r_d + i_d. \tag{6.13}$$

This can be rearranged to give

$$\frac{dr_d}{dt} = -\frac{r_d}{\tau_{eff}} + \frac{i_d}{\tau} \tag{6.14}$$

where the effective time constant, $\tau_{eff} = \tau/(1 - (W_s - W_x))$, determines the timescale that the firing rate decays to zero in the absence of input. The effective time constant is a circuit property, longer than the neural time constant, τ, because $W_s > W_x$ in these circuits.

Notice that when $W_s - W_x = 1$ the effective time constant is infinite and the system is a perfect integrator—i.e., r_d is proportional to the integral over time of i_d. However, in the absence of such perfect tuning, the system's time constant, τ_{eff}, is extended by a factor proportional to the closeness of $W_s - W_x$ to 1. For example, when $W_s - W_x = 0.95$ then the circuit's time constant, τ_{eff}, is 20 times greater than τ, the longest cellular time constant.[19] So, with fine-tuning at this 5 percent level (0.95 being 5 percent away from a perfect 1) the system's time constant would be 1 s if feedback were via NMDA receptors with a time constant of 50 ms. This would set 1 s as the timescale over which activity leaks away and over which any prior information about a stimulus encoded in the circuit's activity is forgotten.

6.5.2 Decision-Making Performance

Decision-making performance can be judged by different criteria, depending on the details of the task. If the stimulus duration is fixed (forced response), then the fraction of correct responses is the appropriate metric. If the subject chooses when to respond (free response) then performance can be judged via the mean time taken per decision when a required fraction of correct responses is produced. Alternatively, when a correct response yields more reward for an animal than an incorrect one, then the rate of reward is a suitable metric.

Forced response paradigm   A study in which the time of a required response is fixed.

Free response paradigm   A study in which the subject is free to make a response whenever ready.

The relative cost of an incorrect response versus a slower correct response determines whether faster decisions with more errors or slower decisions with fewer errors are optimal. When the goal is to maximize reward rate, an experimenter can adjust the intertrial intervals to manipulate a subject's speed or accuracy of responses. Slower, more accurate responses are favored if the delay between trials is lengthened.

When simulating a model neural circuit, a correct decision is indicated when the activity of the group of neurons receiving greater mean stimulus is above the activity of the other group of neurons at the decision time. In the free-response paradigm, the decision time is determined by the activity of either group of neurons reaching a response threshold.

For any given model circuit, the response threshold can be varied to determine the circuit's optimal performance. If the threshold is raised, then decisions are slower and more accurate. They are slower because it takes activity more time to reach threshold. They are more accurate, because noise accumulates as the square root of time, while the signal accumulates linearly with time in an integrator—so given sufficient time the signal can dominate over the noise.

> **Speed-accuracy tradeoff**   The common situation in which faster responses are less accurate, while slower responses are more accurate.

In a neural circuit a change in response threshold could be achieved by modulation of synaptic strengths at the outputs of the decision-making circuit such that a higher level of activity in the circuit is required to produce a response. As an alternative to altering the threshold, the inputs could be scaled to achieve the same speed-accuracy trade-off. Such an example can be seen in figure 6.10 by comparing the solid curves with the dashed curves. For an integrator, a scaling down of its inputs acts exactly like a scaling up of its threshold, assuming both the stimulus and noise terms to be integrated scale identically. The two types of scaling are only distinguishable by measuring activity in the neural circuit producing the integration.

Regardless of whether scaling is at the inputs or outputs of a circuit, the ability to modulate the strength of synaptic conductance allows an animal to adapt to circumstances and produce rapid inaccurate decisions or slower more accurate ones as desired. The neuromodulator norepinephrine has been proposed to serve this purpose.[20,21]

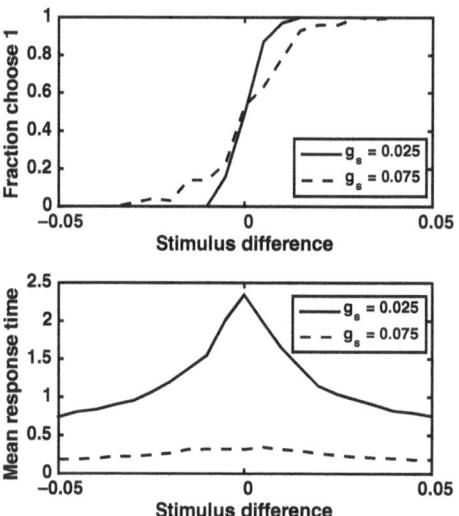

Figure 6.10
The speed-accuracy trade-off. (A) The fraction of trials in which unit 1 reaches threshold first increases in a sigmoidal manner from 0 to 1 as the stimulus difference, $s_1 - s_2$, increases. With reduced input conductance (solid line versus dashed line) a smaller stimulus difference is required for accurate responses. (B) The cost of improved accuracy is a longer time to reach threshold when input conductance is weaker (solid line versus dashed line). Responses are slowest when the stimuli are most equal. These figures are produced using the circuitry of figure 6.8, by the available online code decision_making_vary_g_stim.m.

6.5.3   Decisions as State Transitions

An alternative to the gradual ramping of neural activity arising from integration is the possibility that neural activity changes by abrupt transitions between different stable states. The abruptness might correspond to the suddenness with which percepts can change and ideas can "pop" into our heads. If the times of such transitions vary from trial to trial, such "jumping" behavior could underlie the observed gradual ramping once neural activity is averaged over many trials.

> **State-transition**   A rapid change in neural activity from one state, such as one stable fixed point, to another.

Transitions between stable states depend on noise fluctuations, so as the level of noise increases the time to make a transition decreases, corresponding to faster decisions. Random noise can enhance the performance of such a decision-making circuit, because the speeding of response time can more than compensate for the decrease in response accuracy when low levels of random noise are added. Such behavior of the circuit produces "stochastic resonance"—an optimal level of noise for producing responses in a given time—which is in contrast to the usual expectation that random noise detracts from information processing.

It is worth noting that perfect integration of evidence (equations 6.13 and 6.14) with $W_s - W_x = 1$ and $\tau_{eff} \to \infty$) is mathematically the optimal method for discerning the greater of two noisy inputs. In perfect integration, there is neither forgetting (or leak) of early information, nor an overreliance on the initial input, either of which can occur in other models. Rather, in perfect integration, input at all time points is weighted equally. However, when biological constraints are included so that the decision-making circuit contains its own intrinsic noise (so is never a perfect integrator) and the threshold cannot be raised indefinitely (because neural firing rates are bounded) it turns out that decisions by state transitions can be optimal.[15] In tutorial 6.2 we compare these two different types of decision making in a single circuit.

6.5.4   Biasing Decisions

If, ahead of any stimulus, recent history or other prior information indicates that one choice is more likely to be the correct one then we should bias our choice according to such prior information (see Bayes' theorem, equation 1.32). Similarly, if one of our choices is more rewarding than the other, we should bias our choice toward the more rewarding outcome (see equation 3.6). As we saw in section 3.5.2, the bias impacts the probability of a choice being correct multiplicatively, so if either the reward for one response, or its prior probability, is double that of the other, we should be twice as likely to make the more rewarding or more likely response, all other things being equal.

Bias   In decision making, a tendency to prefer one alternative over others before the stimulus is presented. Such bias can improve performance if appropriate for the task.

One can achieve such bias qualitatively in a model by applying additional constant input to the favored alternative. Such bias simply produces a horizontal shift in the choice probability curve (shown without bias in figure 6.10A) such that if the stimulus difference were zero, the probability of choosing the favored alternative would be greater than 0.5.

Alternatively, the circuit could be initialized in a state that is closer to one threshold than the other. For a perfect integrator, the ratio of the distances to threshold for the two competing alternatives is the ratio of choice probabilities produced when the stimuli are identical. For example, if the circuit were initialized to require half the change in rate to produce response A compared to that needed to produce response B, then response A would be produced on 2/3 of trials and response B on 1/3 of trials given a stimulus with equal support for A and B.

6.6   Tutorial 6.2: Dynamics of a Decision-Making Circuit in Two Modes of Operation

Neuroscience goals: See how noise can lead to significant trial-to-trial variability in behavior; compare the behavior of an integrator circuit with a state-transition circuit.

Computational goals: Simulate coupled firing-rate units with different types of noise; record event times and realign vectors with different event times and pad them for further analysis.

In this tutorial, we will produce a decision-making circuit of the type shown in figure 6.8 and study its response to inputs. The circuit contains two units, labeled $i=1,2$. When one of the two units reaches its decision-threshold we assume one of the two alternatives has been chosen.

We will be able to switch the mode of decision making by altering the connection strengths and thresholds of the units. In order to produce a perfect integrator, the firing-rate curve of the units will be a linear function of their inputs,

$$\tau \frac{dr_i}{dt} = -r_i + I_i - \Theta,$$

with the constraint $0 \leq r_i \leq r_{max}$. In the preceding equation, Θ is the firing-threshold (the minimum input needed to cause the unit to fire, not to be confused with the decision-threshold) and I_i is the input to unit i given by

$$I_i(t) = W_s \cdot r_i(t) + W_x \cdot r_j(t) + G \cdot s_i(t) + \sigma_{int} \cdot \eta_i(t),$$

where $r_j(t)$ is the rate of the other cell (so if $i=1$ then $j=2$ and if $i=2$ then $j=1$). Each unit will receive independent internal noise of strength σ_{int} as well as a scalable independent noise within the stimulus $s_i(t)$, of strength σ_s, as described in the following.

The speed-accuracy tradeoff is achieved by altering the parameter G, which scales the input synapses to the circuit (like an input conductance).

The stimuli, when present, are given by $s_1(t) = \bar{s} + \Delta s/2 + \sigma_s \cdot \eta'_1(t)$ and $s_2(t) = \bar{s} - \Delta s/2 + \sigma_s \cdot \eta'_2(t)$, with mean $\bar{s} = 1$, and variable difference Δs.

The terms $\eta_1(t)$, $\eta_2(t)$, $\eta'_1(t)$, and $\eta'_2(t)$ each represent unit variance noise to be generated independently on each trial, as described in question 2.

For the integration mode, set $W_s = 0.975$, $W_x = -0.025$, and $\Theta = -0.5$ for each unit.

For the jumping mode, set $W_s = 1.05$, $W_x = -0.05$, and $\Theta = 4$ for each unit.

For both circuits set a decision threshold of 50 Hz and the maximum rate, $r_{max} = 60$ Hz. In all simulations simulate until either the threshold is reached by one of the units or a maximum time of 10 s is reached. Use a time constant of $\tau = 10$ ms.

Answer all questions with the circuit in integration mode, then repeat them all with the circuit in jumping mode. Comment on any similarities or differences between the two sets of results.

1. In integration mode set $G = 1$ and in jumping mode set $G = 2.5$. Simulate a single trial without noise ($\sigma_{int} = 0$ and $\sigma_s = 0$) as follows:

   a. Initialize the firing rates of the two units to their identical steady state value in the absence of a stimulus or noise by solving for r in

   $$\tau dr/dt = -r + W_s \cdot r + W_x \cdot r - \Theta = 0$$

   and enforcing the bounds on the range of allowed values of r.

   b. Set $\Delta s = 0.1$ for a stimulus that commences at 0.5 s and remains on. Simulate the differential equations with a timestep of 0.5 ms, then plot the firing rate of both units as a function of time, and separately plot the difference in their firing rates.

2. Add independent noise to the entire duration of each stimulus, by repeatedly choosing two random numbers (one for each stimulus) from the unit Gaussian distribution (e.g., by using randn() in MATLAB) and keeping these two noise values fixed for each 2 ms of the stimulus duration. Multiply each noise value by $\sigma_s = 0.2$ and divide by the square root of 0.002, before adding to the stimulus. Add intrinsic noise in a similar manner, again keeping it fixed every 2 ms, and using the magnitude $\sigma_{int} = 0.25$. Ensure the intrinsic noise is present across all of the time simulated (not just when the stimulus is on). Simulate 200 trials, for each trial recording the firing rate as a function of time.

   a. Plot the mean firing rate aligned to stimulus onset for both units, being careful to include in the across-trial averaging at each time point only those trials yet to produce threshold-crossing.

   b. Plot the mean firing rate during the decision-making period, aligning across trials to the time of decision-threshold crossing. Again, be careful to either include at each time point only those trials being simulated at that time point, or to set the rates to the initial firing-rate value at all time points before the simulation started. This is necessary, since the firing-rate arrays for each trial will be offset from each other by large amounts according to the range of threshold-crossing times.

    c.   Record which unit first crosses the decision threshold and the time of threshold-crossing. Plot a histogram of the times to threshold on trials for which unit 1 (which received greater input) was the first to reach threshold.

3.   Now repeat 2c, looping through values of $\Delta s = 0, 0.25, 0.5, 0.75, 1$. Record the mean time for unit 1 to reach threshold (on trials when it does do so) as well as the fraction of trials in which unit 1 is the winner for each value of Δs. Plot these two quantities separately against stimulus difference, Δs.

4.   Now repeat 1–3, with $G = 0.5$ using the parameters for integration mode then $G = 2$ using the parameters for jumping mode.

Note 1: Notice that in integration mode, when the input is reduced, integration is only possible over a small range of rates before the weaker unit is silenced. This limits a circuit's ability to perform perfect integration over a longer period simply by weakening input synapses. Such deviation from perfect integration explains why it is possible to find the jumping mode to be both faster and more accurate in question 4, something that should not be possible when compared to perfect integration (except perhaps by "luck" from the randomness inherent in these simulations).

Note 2: The methods used here simulate each trial individually to build up statistics of the times to threshold and probability for each unit to reach threshold first. It is possible, using the methods of partial differential equations to simulate the probability of each unit having a given rate at a given time, and thus generate the probability that a unit crosses threshold at a given time. Such methods generate exact probabilities, akin to calculating the probability of two coin tosses being heads as $\frac{1}{2} \times \frac{1}{2} = \frac{1}{4}$, rather than simulating 1,000 pairs of coin tosses to obtain approximately 250 pairs of double-heads. These mathematical tools can be acquired in courses in advanced calculus or stochastic processes.[22]

6.7   Oscillations from Excitatory and Inhibitory Feedback

If sensitive electrodes are placed on our heads, they are likely to reveal oscillatory electric fields. Such external fields arise when oscillating neural activity is synchronized among many neurons aligned in a similar direction. We have seen in chapter 4 how the spiking of individual neurons can be an oscillatory process, or how specific intrinsic channels can lead to periodic bursts of spikes. We have also seen in chapter 5 how the interplay between two neurons can produce oscillations that would not arise in either cell alone. All of these oscillations depended on specific types of intrinsic channels within the neurons, so even in examples where the coupling between cells was essential, the oscillations could not be attributed entirely to the network effect. In this section, we will consider oscillations that are entirely due to network properties and thus can be simulated in firing-rate models in which each unit has no internal dynamics.

The frequency, power, and coherence of oscillations vary with our mental state and are task dependent. A relatively high-frequency oscillation is the gamma rhythm, in the range 30–80 Hz, which is associated with increased attention and memory processing.[23] For example, when a subject attends to a visual stimulus, the power of gamma oscillation in the responsive area of visual cortex increases,[24] whereas when the subject attends to an auditory signal, the gamma power in auditory cortex increases.[25]

Gamma rhythm Oscillatory neural activity in the range of 30–80 Hz, which can be localized to specific brain areas, indicative of attention and cognitive processing.

One subset of models of the generation of gamma oscillations relies on the coupling between excitatory neurons (pyramidal cells) and inhibitory neurons (fast-spiking interneurons).[26] The operation of these PING models (standing for Pyramidal-InterNeuron Gamma) can be understood using a firing-rate model (figure 6.11), because the oscillation depends entirely on network feedback, not on any intrinsic oscillation properties of the cells. In a PING model, self-excitation among excitatory pyramidal neurons destabilizes any low level of spontaneous activity, generating a burst of excitatory activity. The burst of excitation causes the interneurons to fire even more actively. The high interneuron activity produces delayed feedback inhibition to suppress the burst of excitation among the pyramidal cells. Once the excitation ends in this manner, the interneurons are no longer excited and their activity also drops, allowing a new burst to arise once the inhibitory input to pyramidal cells decays (figure 6.12).

In alternative models, oscillations of a similar frequency can arise among inhibitory neurons alone.[27,28] Such network oscillations require inhibitory neurons to have high spontaneous activity in the absence of input from other neurons in the oscillating circuit, because any periodic inhibitory input from other neurons can only produce periodic dips in a neuron's firing rate. The high spontaneous activity can be due to intrinsic properties of the cells themselves (high conductance of sodium and/or calcium channels relative to the conductance of potassium channels) or due to tonic (meaning constant, nonoscillating) input from excitatory cells from outside their circuit. In such a purely inhibitory circuit, the neurons can synchronize to produce pulses of inhibition that hyperpolarize all cells transiently. The time to recover from inhibition is determined predominantly by the synaptic time constant. If this is similar enough for all cells, then many of them will, upon recovery, produce action potentials at a similar time. These synchronized action potentials generate a new pulse of inhibition that prevents further spiking for a short amount of time (figure 6.13).

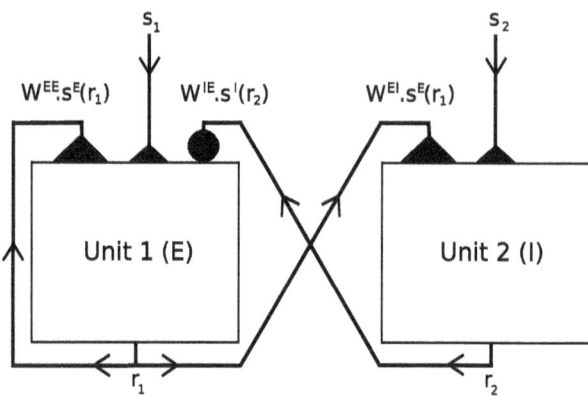

Figure 6.11
Two-unit oscillator. An excitatory unit (E) and an inhibitory unit (I) are coupled to produce an oscillator. The E-unit has strong self-excitation but also receives disynaptic inhibitory feedback, because it excites the I-unit, which in return inhibits the E-unit. The inputs, s_1 and s_2, determine the excitability of the units, and can be altered to switch on or off oscillations and control their frequency.

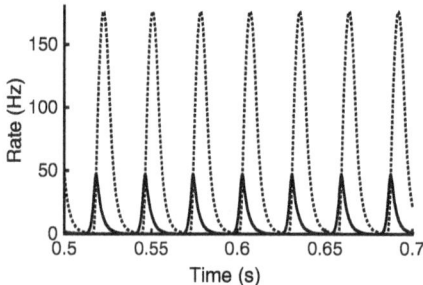

Figure 6.12
Gamma frequency oscillations via a PING mechanism. The firing rate of excitatory cells (solid) rises because of recurrent self-excitation. The inhibitory cells (dotted) are strongly excited, so rise with little delay. The inhibitory firing is so strong it proceeds to shut down the excitatory cells, after which inhibitory firing decays. The behavior shown here is only possible if units comprise many cells, because the mean firing rate of excitatory cells shown here is 13 Hz, below the oscillation frequency of 40 Hz. This ratio means that any individual cell fires spikes on only about 1/3 of the gamma cycles. This figure is produced by the available online code `oscillator_single_stim.m`.

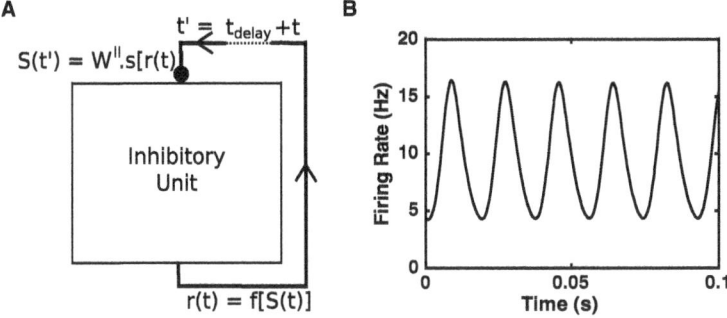

Figure 6.13
High-frequency oscillations in an all-inhibitory network with delayed feedback. (A) A single unit comprising inhibitory cells with feedback can generate oscillations if a delay is included between the firing rate and the inhibitory synaptic feedback. Such a delay arises from a summation of the time for action potential propagation, the time for dendritic input current to reach the soma, and a rise time for maximal synaptic channel opening to be reached following vesicle release. These separate components can be included individually in spiking models. (B) Firing rate oscillations from such an inhibitory circuit at a rate primarily determined by the synaptic feedback delay, t_{delay}. Results are produced by the available online code ineuron_oscillator_single.m with $t_{delay} = 3$ ms.

It is important in the firing-rate model of interneuron gamma to incorporate a synaptic delay—corresponding to a delay in time from when the soma of a neuron produces a spike to the time that the inhibitory current impacts the soma of a downstream cell. Without such a delay, as the firing rate of an inhibitory unit rises, it suppresses its own firing rate, preventing further rise. With a delay, the rate has time to rise before the inhibition kicks in. In a spiking model, the time for the membrane potential to recover from an inhibitory pulse would add to any delay in a synaptic input. In a firing-rate model, the total delay can be mimicked if the synaptic feedback is set to depend on the firing rate of the unit a short time, t_{delay}, earlier (figure 6.13A).

Power spectrum   A plot of oscillatory power versus frequency, to indicate the dominant frequencies in any time-dependent signal with some oscillating components.

Fourier transform   A process by which a time-dependent function can be separated into the contributing amplitudes of oscillating components of different frequencies, which together combine to produce the original function.

6.8   Tutorial 6.3: Frequency of an Excitatory-Inhibitory Coupled Unit Oscillator and PING

Neuroscience goal: Discover how the frequency and amplitude of oscillations in a circuit can depend on external inputs.

Computational goals: Addressing limiting cases of "0/0" analytically; introduction to the Fourier transform.

In this tutorial, we will simulate the circuit of an excitatory firing-rate unit (1) coupled to an inhibitory firing-rate unit (2), as in figure 6.11, with parameters that lead to oscillations. We will measure the amplitude and frequency of the oscillations and in so doing we will see how a Fourier transform operates.

The firing rate function for each unit will be based on the empirical fit by Chance and Abbott to the response of leaky integrate-and-fire neurons:[29]

$$f(V_{ss}) = \frac{V_{ss} - V_{th}}{\tau(V_{th} - V_{reset})\{1 - \exp[-(V_{ss} - V_{th})/\sigma_V]\}}.$$

In this model, V_{ss} is the steady state membrane potential for the LIF model in the absence of spiking (equation 2.9):

$$V_{ss} = \frac{G_L E_L + G_I E_I + G_E E_E}{G_L + G_I + G_E},$$

where G_L, G_I, and G_E are respectively the leak, inhibitory, and excitatory conductances (which can vary with time) and E_L, E_I, and E_E are their corresponding fixed reversal potentials. V_{th} is the threshold of the membrane potential, V_{reset} is the reset potential, τ is the membrane time constant, and the parameter σ_V (the noise in the membrane potential) determines the steepness and sharpness of curvature of the firing-rate curve.

As written, the function $f(V_{ss})$ is undefined if $V_{ss} = V_{th}$ (both numerator and denominator are zero). However, one can show that in the limit of very small δ (that is if $\delta \ll \sigma_V$), if $V_{ss} = V_{th} + \delta$ or $V_{ss} = V_{th} - \delta$, then $f(V_{ss}) \sim \sigma_V / [\tau(V_{ss} - V_{reset})]$. Therefore, the function $f(V_{ss})$ should be defined to have this limiting value when $\delta = 0$ (and $V_{ss} = V_{th}$). The computer cannot figure out such a limit though, so when evaluating the function, we must include a conditional statement in the code that produces the correct limiting value instead of dividing by zero whenever $V_{ss} = V_{th}$.

Use the following parameters: $V_{th} = -50$ mV, $V_{reset} = -80$ mV, $\sigma_V = 1$ mV, $\tau = 3$ ms, $E_L = -70$ mV, $E_I = -65$ mV, $E_E = 0$ mV and $G_L = 50$ pS. The excitatory and inhibitory conductances, G_E and G_I, will be time-dependent and differ between the two cells:

$$G_E^{(1)}(t) = W^{EE} s_E^{(1)}(t) + G_{in}^{(1)} \ G_I^{(1)}(t) = W^{IE} s_I^{(2)}(t); \ G_E^{(2)}(t) = W^{EI} s_E^{(1)}(t) + G_{in}^{(2)} \ G_I^{(2)}(t) = 0.$$

The connection strengths are set as $W^{EE} = 25$ nS, $W^{EI} = 4$ nS, and $W^{IE} = 800$ nS. The excitatory input conductances, $G_{in}^{(1)}$ and $G_{in}^{(2)}$ will vary across trials, but initially set $G_{in}^{(1)} = 1$ nS and $G_{in}^{(2)} = 0$ nS.

1. Set up the simulation of two units with parameters given in the preceding paragraphs, such that for unit i, the firing rate varies according to:

$$\tau \frac{dr_i}{dt} = -r_i + f(V_{ss,i})$$

where $V_{ss,i}$ is the instantaneous steady state that depends on the values of conductances for unit i as described above. The synaptic variables depend on presynaptic rate according to:

$$\frac{ds_E^{(1)}}{dt} = -\frac{s_E^{(1)}}{\tau_E} + \alpha r_1 \left(1 - s_E^{(1)}\right)$$

and

$$\frac{ds_I^{(2)}}{dt} = -\frac{s_I^{(2)}}{\tau_I} + \alpha r_2 \left(1 - s_I^{(2)}\right),$$

where the synaptic time constants are $\tau_E = 2$ ms and $\tau_I = 5$ ms. For all synapses set $\alpha = 0.2$. Simulate the equations for 2.5 s using a timestep of $\Delta t = 0.1$ms, assuming initial values of zero for all variables, and plot the firing rates of each unit on the same graph. If the firing rates oscillate, what is the oscillation frequency?

2. In this section, you will attempt the first of two methods for automatically calculating the frequency of oscillation.

   a. Copy the firing rate of the first unit into a new vector that omits an initial transient period of 0.5 s (or longer than 0.5 s if the oscillations take longer to stabilize).

   b. Determine the maximum and minimum values of the firing rate during the oscillation and use those values to select two thresholds, one a little below the maximum rate to indicate the rate is approaching a peak and the other a little above the minimum rate to indicate the rate is approaching a trough.

   c. Define a variable, a threshold-crossing indicator, which you initialize with a value of 0, unless the first value of firing rate is above the higher threshold, in which case you initialize the indicator with a value of 1 (because the system is already above-threshold).

   d. Loop through the firing rate vector. Within the loop set a conditional such that if the threshold-crossing indicator is zero and the rate is greater than the higher threshold, then the time point is recorded and the indicator is set to one. If the rate is lower than the lower threshold, reset the threshold-crossing indicator to zero.

e.   Calculate the oscillation period using the times of the first and last crossings of the higher threshold and the number of oscillations in between these crossings.

3.   In this section, you will use a computationally simpler method that is equivalent to taking the Fourier transform.

a.   Truncate the firing rate vector (as in 2a) to obtain the periodic portion of the trial.

b.   Produce a vector of frequency values from 0 Hz to 100 Hz with steps of 0.5 Hz, and set up a loop through all values of the vector.

c.   Within the loop, for each value of frequency, f, create two vectors, each of the same length as the truncated firing rate vector, one equal to $\sin(2\pi ft)$ and the other equal to $\cos(2\pi ft)$ where the time points, t, span the time interval of the truncated firing rate vector in steps of Δt.

d.   Find the overlaps between the sine vector and the truncated firing rate vector by multiplying each corresponding element of the two vectors and taking the mean of the result. Record this overlap in a vector of coefficients,

$$A(f) = \overline{\sin(2\pi ft) \cdot r_1(t)},$$

one value for each frequency (the overbar in the above equation indicates that the mean value is taken, in this case over the portion of the trial where $r_1(t)$ is periodic). Repeat to find the overlap of the firing-rate vector with the cosine vector, to produce a set of coefficients,

$$B(f) = \overline{\cos(2\pi ft) \cdot r_1(t)}.$$

e.   You now have two vectors; one with coefficients that each represent how similar the oscillating firing rate is to a sine wave of a given frequency, the other with coefficients that each represent the similarity to a cosine wave. It turns out that the sum of the squares of these two coefficients provides a measure of the oscillating power that is independent of the phase offset—the oscillation can be a sine or a cosine or anywhere in between and the sum of the squares is not altered. Therefore, produce a power spectrum as a vector such that each element, $P(f)$, is the sum of the squares of the corresponding sine coefficient and cosine coefficient:

$$P(f) = A^2(f) + B^2(f).$$

f.   Plot $P(f)$ as a function of the frequency vector and extract the value of f other than zero for which $P(f)$ is greatest. Compare this frequency to the oscillation frequency you calculated in question 2. Why does $f = 0$ produce the greatest value for $P(f)$ (so must be discarded) even when the rate oscillates? Do you see multiple peaks? If so, why?

4.   Let the applied stimulus to excitatory cells range from $G_{in}^{(1)} = 0$ nS to $G_{in}^{(1)} = 10$ nS (while keeping the input to inhibitory cells at $G_{in}^{(2)} = 0$ nS). Plot relevant figures (such as figure 6.14) to determine how the oscillation frequency, oscillation amplitude for both excitatory and

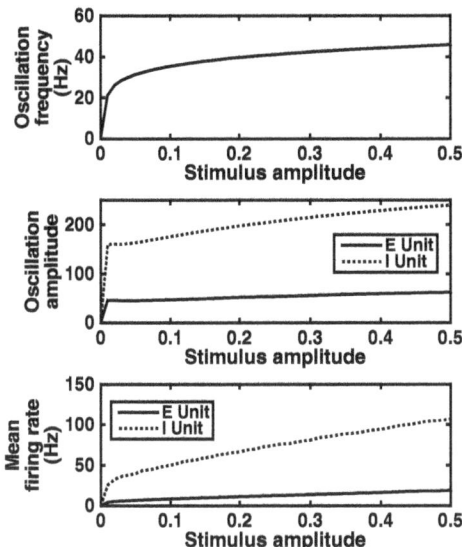

Figure 6.14
Background input affects the frequency and amplitude of an oscillating circuit. (A) The frequency of oscillation increases within the gamma range, from below 30 Hz to 50 Hz, as excitatory input to the excitatory unit is increased. (B) The amplitude of oscillation, in particular of the inhibitory cells, rises as more inhibition is needed to overcome the activity of the E-unit when the latter receives greater input. (C) The mean firing rates of both units increase as the oscillation amplitude increases, though mean firing rate of the E-unit remains below the oscillation frequency, an indication that individual cells fire spikes on intermittent gamma cycles. This figure is produced by the available online code `EI_oscillator.m`.

inhibitory cells, and the mean firing rates of both excitatory and inhibitory cells depend on $G_{in}^{(1)}$. Explain any trends or relationships you notice.

5. Fix $G_{in}^{(1)} = 2$ nS and let the applied stimulus to inhibitory cells range from $G_{in}^{(2)} = 0$ nS to $G_{in}^{(2)} = 25$ pS. Plot the same figures as in question 4 and explain any differences in the results of questions 4 and 5.

6.9 Orientation Selectivity and Contrast Invariance

A notable feature of many neurons in primary visual cortex (V1) is their strong activity in response to an edge or a bar of a particular orientation within their visual receptive field (figure 1.4B). As the orientation of a bar appears at an angle rotated from the preferred orientation (the one producing the greatest response in a neuron), a smooth decline in the neuron's firing rate is observed as a function of stimulus angle. When a bar is presented with the null orientation, which is perpendicular to the preferred orientation, the neuron's response can be lower than the spontaneous activity it produces in the absence of any stimulus.

Orientation selectivity   The preferred firing of many neurons in V1 to visual inputs that are edges or gratings oriented at a particular angle.

Primary visual cortex, V1   Also called occipital cortex, or striate cortex, the region of cortex at the back of the mammalian brain, which receives most direct visual input from the retina via the thalamus, with perhaps the most studied neural circuitry of mammals.

Lateral geniculate nucleus (LGN)   A region of the thalamus receiving inputs from the optic nerve, which sends excitatory outputs to V1.

Such orientation-tuned responses are observed in layer 4 neurons, which receive excitatory but not inhibitory input from the lateral geniculate nucleus (LGN) of the thalamus. The firing rates of neurons in the LGN increase with stimulus contrast—early stages of visual processing from the pupil to the retina compensate for changes in overall brightness or luminance of a visual stimulus such that light-dark boundaries generate downstream responses.

One puzzling question has been how the responses of orientation-selective neurons in V1 can be explained in terms of their inputs from the LGN. In particular, can V1 responses be explained in terms of the summation of different inputs from the LGN, or are intracortical connections within V1 essential for shaping the V1 responses?[30,31]

6.9.1   Ring Models

In answering this question, one feature that has required explanation is the observation of contrast invariance. The tuning curve of a neuron in V1 does not change its shape—it just increases its amplitude multiplicatively—as the contrasts of stimuli increase. Since projections from the thalamus to the cortex are only excitatory, the observation of contrast invariance rules out any model in which the V1 neurons only receive input from thalamic neurons. In such models, as contrast increases the excitatory thalamic input to the V1 neurons increases, even when the stimulus is at nonpreferred orientations. Therefore, while a tuning curve with a peak at a preferred orientation could be produced for a neuron, somewhat like an iceberg rising out of the water, the tuning curve would broaden with increased contrast (figure 6.15).

Contrast invariance   Tuning curves that do not change their shape—such that the firing rates can be multiplicatively scaled—when the contrast of a set of visual stimuli is changed.

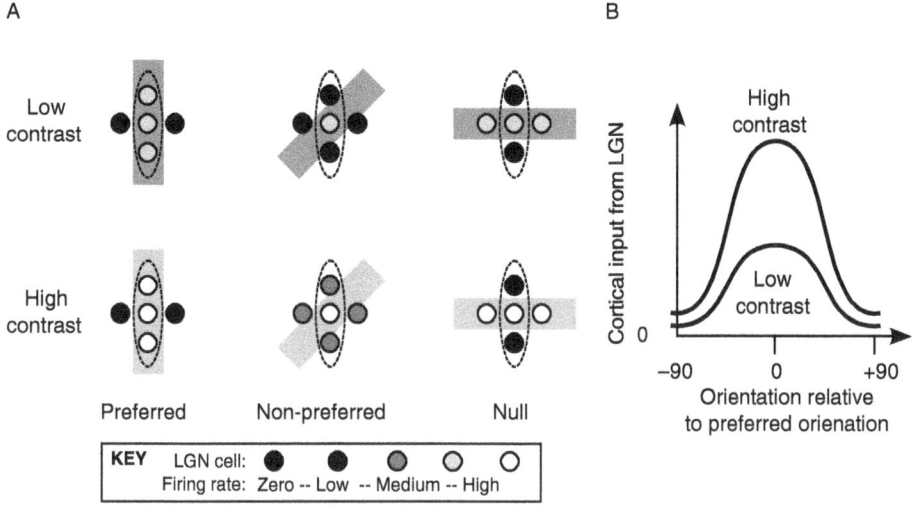

Figure 6.15
Feedforward excitatory input from thalamic cells increases with contrast across all orientations. (A) The circular receptive fields of five LGN neurons are indicated as circles. A bar of light (rectangle) at low contrast (top) or high contrast
(bottom) causes those neurons to become active (lighter circles indicate greater activity) when passing through the
center of the receptive fields of those LGN neurons. The vertically orientated ellipse indicates the receptive cell of a
cortical neuron that receives feedforward input from the three LGN neurons within the ellipse. Importantly, when the
bar of light appears at the null orientation at high contrast (bottom-right), the central, highly active LGN neuron could
provide as much input to the cortical neuron as the three aligned less active LGN neurons when the bar of light is at the
preferred orientation but at low contrast (top left). (B) The feedforward excitatory input from the LGN to a cortical neuron scales up across all orientations with increasing contrast. Given that the cortex receives only excitatory projections
from the LGN, panel B shows how the total input from LGN to cortex depends on orientation. If a cortical neuron's
activity were determined by this input alone, then its tuning curve would broaden as input increases, because the range
of orientations that produce input above any fixed threshold would increase. Such an "iceberg effect" is not observed in
neural recordings.

Such a model (figure 6.16A) is a bit of a straw man (i.e., expected to fail) because inhibitory
neurons also receive thalamic input. Therefore, in an alternative feedforward model, the tuned
excitatory V1 neurons could receive direct excitation from the thalamus, but also indirect inhibition from the thalamus. If the indirect inhibition were to target neurons with the opposite stimulus
preference and were to scale with contrast in the same manner as the direct excitation (figure
6.16B), then the increase in inhibition with contrast could counteract the increase in excitation
with contrast at nonpreferred orientations. We will investigate this possibility in tutorial 6.4.

The final type of model (figure 6.16C), which is now widely accepted, is one that includes
recurrent feedback in V1. Again, this should not be a surprise when considering the known
anatomy, because a typical V1 cell, even in layer 4, receives the vast majority of its inputs
(approximately 90%) from other cortical neurons rather than from the thalamus.[32]

A key feature of models based on recurrent feedback is that excitatory connections between
excitatory cells are more effective—being more prevalent, or stronger, or both—between neurons

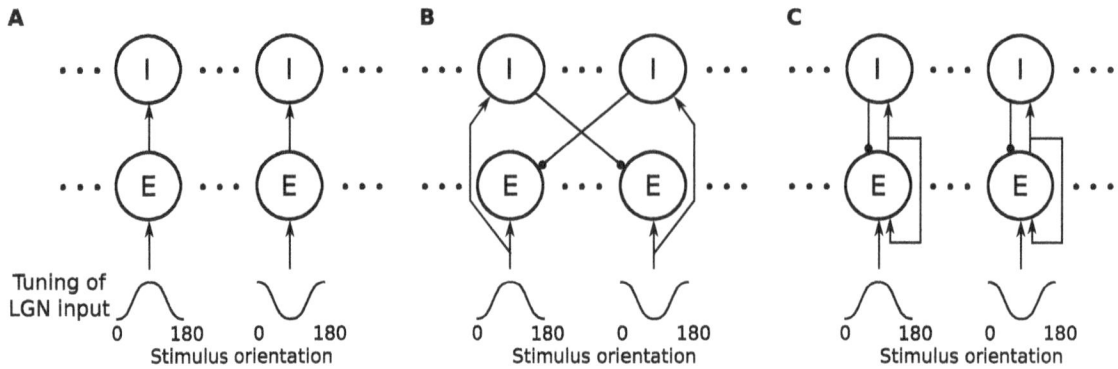

Figure 6.16
Circuit motifs for investigation of contrast-invariant orientation selectivity in V1. (A) Feedforward excitation to two excitatory cells (labeled E) with opposite orientation preferences. Tuning curves of excitatory cells are unaffected by cortical connections. (B) Feedforward excitation and inhibition to excitatory cells. Inhibitory input to excitatory units is from interneurons (labeled I) receiving oppositely tuned LGN input. (C) Feedforward and feedback excitation and inhibition. Feedback is via recurrent excitatory and inhibitory connections from similarly tuned cells within the cortex. (A–C) In all figures, just two pairs of cells are shown, whereas the complete set would be arranged according to orientation preference in a ring, with all orientations present. Only the strongest connections are shown, yet in all models the local connections also spread to neurons with similar, not just identical or opposite, orientation preferences, as shown in figure 6.17.

that have similar tuning curves (figure 6.17). Such local excitatory feedback increases the gain of responses to input, enhancing any supralinearity of the response, such that the effect of increased contrast is much greater at the preferred orientation (where the response is high) than at nonpreferred orientations.

Given that the connectivity between neurons in these models depends on the difference in their preferred orientations, neurons are labeled and arranged according to their preferred orientation. Such an arrangement produces a ring model, because as the bar-stimulus is rotated by π radians (i.e., by 180°), the stimulus returns to the original one. Therefore, neurons with preferred orientations of slightly greater than 0 are strongly connected to neurons with preferred orientations of slightly less than π.

Here it is worth noting that a connectivity structure can be ringlike without the neurons' being physically located in a ringlike structure. However, it turns out that for orientation-selective neurons in V1 a ringlike arrangement of the neurons' physical positions does arise, producing what are known as pinwheels (figure 6.18).[33,34] Such correspondence between the response properties of neurons and their physical location is called a topographic map, a common feature of primary sensory areas (see also figure 1.4B).

The second key component of contrast-invariant models of orientation selectivity is inhibitory feedback within the cortical network that spreads beyond the excitatory feedback. The inhibitory feedback must be sufficiently strong to prevent inactive units from becoming active whenever contrast increases and the firing rates of the most responsive units increase. In models of such

A   **B**

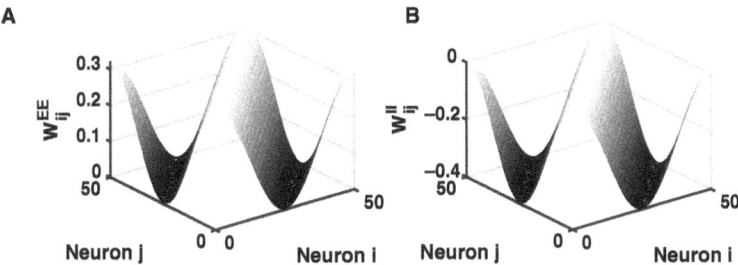

Figure 6.17
Structured connectivity in a ring model. (A) Structured excitatory-to-excitatory network. The strength of recurrent excitatory connection from neuron i to neuron j, is $W^{EE}[1+\cos(2\pi i/N-2\pi j/N)]$ in a ring model, producing cyclic structure. For an orientation selectivity model the preferred input of unit i is given by $\theta_i = \pi i/N$ (since rotation by π produces an identical stimulus). For a spatial location model (such as head direction, where rotation by 2π results in no change) the preferred input of unit i is given by $\theta_i = 2\pi i/N$. In the example shown (used to produce the results of figure 6.20 and found in the code `ring_attractor.m`) the mean connection strength, $\overline{W^{EE}} = 8/N$ where the number of units, $N = 50$. (B) Structured inhibitory-to-inhibitory network. In an alternative circuit containing only inhibitory neurons that are spontaneously active, cross-inhibition can produce a similar connectivity profile and similar network behavior. The strength of inhibitory connection from neuron i to neuron j, is $\overline{W^{II}}[1+\cos(\pi+2\pi i/N-2\pi j/N)]$, which is of maximal strength (but negative) to neurons of opposite tuning. In this example, the mean connection strength is $\overline{W^{II}} = -10/N$ and other parameters are defined as in A.

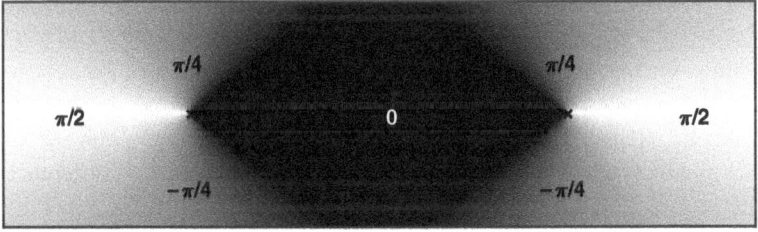

Figure 6.18
Pinwheel arrangement of neurons in V1. The figure represents an area of cortex containing numerous neurons shaded by their preferred orientations, with two pinwheel centers (marked by crosses) contained within the region. Neurons are arranged by orientation preference (indicated by labels on the figure) counterclockwise around the left pinwheel center and clockwise around the right one. Brightness of shading indicates the absolute value of a neuron's preferred orientation, which ranges from $-\pi/2$ to $\pi/2$. The pinwheel arrangement observed in many mammals[33,34] is efficient in ensuring the neurons most likely to be connected with each other in a ring model—those with similar orientation preferences—are nearby each other in the physical space of V1. This figure was made with the online code `pin_wheel.m`.

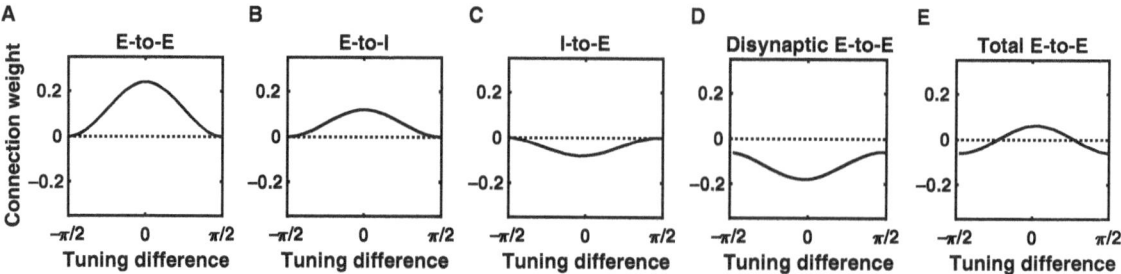

Figure 6.19
Local excitation and surround inhibition produced in the ring model. (A, B) Broad excitation to excitatory cells (A) and to inhibitory cells (B) decays with tuning difference. (C) Inhibition from inhibitory cells decays with tuning difference to the inhibited excitatory cell. (D) Combining the effects of E-to-I (B) and I-to-E (C) through convolution, produces a disynaptic inhibitory effect from excitatory cells to excitatory cells that does not drop to zero for oppositely tuned cells. Convolution takes into account and sums together every combination of two successive differences in tuning (labeled on the x-axes of B and C) to produce each single resulting difference in tuning on the x-axis of (D). (E) The sum of direct E-to-E (A) and indirect E-to-E (D) produces the total effective connectivity between excitatory cells, with net excitation between similarly tuned cells and net inhibition between oppositely tuned cells. This figure was produced by the online code `ring_attractor.m`.

"surround suppression" the extent of all connections can be identical (they are simply included as a cosine in our model). The inhibitory feedback spreads farther than excitatory feedback when all connections have the same spatial spread, because feedback inhibition to excitatory cells is disynaptic—via an excitatory synapse to an inhibitory cell, then via an inhibitory synapse to an excitatory cell. For example, if neurons connect to other cells whose preferred orientations differ by up to $\pm(\pi/4)$, then an excitatory cell can change the rate of an inhibitory cell with a $+(\pi/4)$ difference and that inhibitory cell inhibits a new excitatory cell with a $+(\pi/4)$ difference in preferred orientation from itself, so a $+(\pi/2)$ difference in preferred orientation from the original excitatory cell. In such a situation, activity in the original excitatory cell might increase the rate of all cells within $\pm(\pi/4)$ of its preferred orientation, but via the inhibitory cells, cause a decrease in rate of cells whose preferred orientation is between $+(\pi/4)$ and $+(\pi/2)$ (or $-(\pi/4)$ and $-(\pi/2)$).

With appropriate connections strengths, the result of the total network feedback is that cells with similar orientation preferences excite each other, while they inhibit cells with very different orientation preferences (see figure 6.19).

6.10 Ring Attractors for Spatial Memory and Head Direction

If the recurrent excitatory feedback is increased in a model for contrast-invariant orientation selectivity, then the bump of activity produced when the stimulus is on can remain after stimulus offset (figure 6.20). The activity in such a network can indicate both whether a stimulus was recently presented (a binary memory) and the angle corresponding to the stimulus (memory of a continuous quantity, called parametric memory).

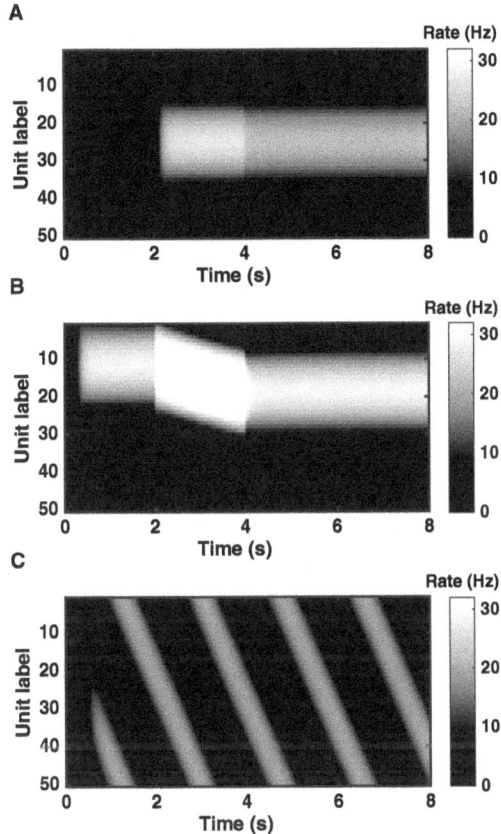

Figure 6.20
Stationary and nonstationary states in a ring model. (A) Both inactive and active states are stable. When a cosine stimulus with peak centered at unit 25 is applied at a time of 2 s and is removed at a time of 4 s, the model produces a bump of activity as in the orientation selectivity network. However, after the stimulus is removed, the bump of activity remains in place. (B) Only a stationary active "bump" state is stable. Increased excitatory feedback causes a bump to form in the absence of a stimulus. The same stimulus used in A causes the bump of activity to shift toward the peak of the stimulus. (C) Only a moving "bump" state is stable. Asymmetric inhibitory feedback suppresses one side of the bump more than the other, causing it to drift with constant velocity over time in the absence of any stimulus. Firing rates of the excitatory neurons are shown. This figure is produced by the online code ring_attractor.m.

Ring attractor A network in which neurons can be labeled by a variable, such as orientation, with circular symmetry, and thus arranges in a ring, in which the network possesses stable states with peak activity anywhere on the ring.

In figure 6.4 we saw the classes of stable states of the network progress from only an inactive state (figure 6.4A), to both an active and an inactive state (figure 6.4B), to only an active state (figure 6.4C) when the degree of excitatory feedback to a single unit was monotonically increased. The same progression through classes occurs in a ring network with spatially distributed excitatory feedback. For the ring network, all three of the classes of activity can provide useful functional roles.

First, we saw in the previous section that even if the inactive state is the only stable state following stimulus offset, the circuit can produce contrast-invariant responses during stimulus presentation. Second, a circuit with both active and inactive states can provide memory for the location (on a circle) of a stimulus while remaining inactive in the absence of a stimulus (figure 6.20A). Third, a circuit with no inactive state can provide important information to an animal about a property that should always exist, such as its head direction (figure 6.20B). Indeed, we shall see later in this section that a ring attractor circuit can provide both a memory of direction and update the direction by near-perfect integration of angular velocity if the latter is provided as an input.

Head-direction cell A neuron that is most active when an animal's head is facing in a particular direction, most likely produced by integration of angular velocity signals, since its tuning is present in the dark.

6.10.1 Dynamics of the Ring Attractor

The ring attractor is so called because of the ringlike symmetry to the system and because the active "bump" of activity that remains following stimulus offset has a stable shape that reappears—the network's activity is attracted back to it—after any temporary deviation. The position of the bump along the ring is marginally stable, as a low level of noise or a small asymmetric stimulus can cause it to drift gradually away from its initial position.[35–37] If the bump is initiated at a particular location (figure 6.20A) thereafter, the random drift causes the variance of its position to increase linearly in time. Such random walk–like behavior is typical of any line attractor (of which the ring-attractor is a special case).

The ring attractor was used to account for the visuospatial short-term memory[38] required in a task in which a monkey must recall the location of a stimulus—presented at a point on a circle—after a delay following stimulus offset. The monkey had to shift its eyes in a saccade to the remembered location following a cue at the end of the delay. Electrophysiology during the task revealed neurons in prefrontal cortex with firing rates in the delay period tuned to the prior stimulus location, just like neurons in a ring attractor.

As with the line attractor mentioned earlier in the chapter, the marginal stability of the ring attractor allows it to act as an integrator (figure 6.7). Any asymmetry in the connectivity of the circuit causes the bump of activity to move toward whichever side is receiving more excitation or less inhibition. Such behavior can lead to instability of a symmetric, so otherwise stationary bump, if the neurons have strong adaptation or synapses are strongly depressing. In such cases, once noise causes the bump to move in a given direction, then the trailing edge of the bump—where neurons were recently more active—has effectively fewer excitable neurons or weaker synapses than the leading edge of the bump, where neurons were recently inactive. Thus, the bump continues to move in one direction. Such behavior can be the cause of spontaneous waves in cortex.

More usefully, if the differences in excitability of the two sides of a bump can be controlled by an input signal, then the direction and velocity of the bump can be controlled. In this case, the bump can perform angular integration, moving more quickly in response to a large asymmetric input and more slowly in response to a weak asymmetric input. To control the level of asymmetry, the feedback should be via two populations one with a connection bias in one direction, the other with a connection bias in the opposing direction. Preferential input to one of these biased feedback populations causes the bump of activity to move in the corresponding direction.[39–41]

Simulations suggest that such angular integration is more robust in a ring model with inhibitory feedback connections (figure 6.17B) than with excitatory feedback connections (figure 6.17A). The range of velocities over which near-perfect integration arises is greatly enhanced in the former case (figure 6.21). Interestingly, the cells responsive to head direction (the integral of the head's angular velocity)[41] might arise in just such a ring attractor formed by inhibitory feedback.[39]

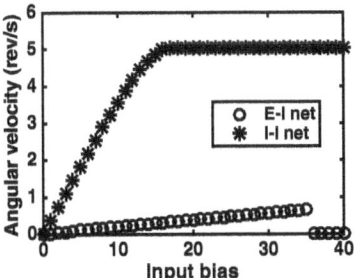

Figure 6.21
Angular integration in a ring attractor network. Angular velocity is approximately a linear function of input bias, so that position of the bump is approximately the integral of input bias over time. The ring network with only inhibitory neurons[39] (figure 6.17B) produces integration over a wider range of angular velocities than the standard excitatory-inhibitory ring network (figure 6.19). The code to produce this figure can be found online as `angular_integration.m`.

6.11   Tutorial 6.4: Orientation Selectivity in a Ring Model

Neuroscience goal: See how contrast-invariant gain can arise from different network configurations.

Computational goal: Simulate a large number of coupled ODEs, using arrays to store variables and their coupling constants.

In this tutorial, you will simulate a network of 100 firing-rate model units (50 excitatory and 50 inhibitory) to investigate how connections within the network shape neural responses to external input. We will compare a feedforward model with a recurrent circuit model (see figure 6.16). The tutorial follows the work of Ben-Yishai and Sompolinsky.[35]

Set up two arrays, with Nt rows and 50 columns, where Nt is the number of time points to be simulated. One array is for 50 excitatory cells and the other for 50 inhibitory cells. A simulation of 300 ms with a time step of 0.1 ms is sufficient.

Define three 50×50 connectivity matrices, each of which can be initialized with entries of zero. These are respectively W^{EE}, W^{EI}, and W^{IE}, for the connection strengths E-to-E, E-to-I, and I-to-E respectively (we ignore I-to-I connections in this tutorial). For each part of the tutorial the firing rates will be treated as a linear function of inputs, so their dynamics follow:

$$\tau_E \frac{dr_i^E}{dt} = -r_i^E + \frac{1}{N}\sum_j W_{ij}^{EE} r_j^E + \frac{1}{N}\sum_j W_{ij}^{IE} r_j^I + I_0^E + S_i^E; \quad \tau_I \frac{dr_i^I}{dt} = -r_i^I + \frac{1}{N}\sum_j W_{ij}^{EI} r_j^E + I_0^I + S_i^I$$

with the additional conditions $r_i^E > 0$ and $r_i^I > 0$ for all units and all times (rates cannot be negative).

In the preceding equations, I_0^E and I_0^I represent baseline input to excitatory and inhibitory units respectively. A negative value for these quantities is equivalent to a positive threshold in the f-I curve. The stimulus-dependent inputs, S_i^E and S_i^I, depend on the orientation preference of the unit labeled by i. We define the orientation preference of each excitatory unit as $\theta_i = \pi i / N$, where $N = 50$ is the total number of units of a given type, and use $\theta_{cue} = \pi / 2$ to denote the orientation of the stimulus. Given these definitions, we use:

$$S_i^E = A_E c[1 + \varepsilon \cdot \cos(2\theta_{cue} - 2\theta_i)], \quad S_i^I = A_I c[1 + \varepsilon \cdot \cos(2\theta_{cue} - 2\theta_i)],$$

with c being the contrast, which will vary from 0 to 1. $\varepsilon = 0.5$ represents the degree of modulation of input with orientation, in this case producing a threefold increase in input from the null direction (where $\theta_{cue} = \theta_i + \pi / 2$, so $\cos(2\theta_{cue} - 2\theta_i) = -1$) to the preferred direction (where $\theta_{cue} = \theta_i$, so $\cos(2\theta_{cue} - 2\theta_i) = +1$).

Other parameters depend on the network being simulated as follows:

Network A. $\tau_E = \tau_I = 10$ ms, $I_0^E = I_0^I = -10$, $A_E = 40$, $A_i^I = 40$, $W_{ij}^{EE} = W_{ij}^{EI} = W_{ij}^{IE} = 0$.

Network B. $\tau_E = \tau_I = 10$ ms, $I_0^E = I_0^I = -5$, $A_E = 40$, $A_i^I = 40$, $W_{ij}^{EE} = W_{ij}^{EI} = 0$,
$W_{ij}^{IE} = -[1 + \cos(\pi + 2\theta_i - 2\theta_j)]/N$.

Network C. $\tau_E = 50$ ms, $\tau_I = 5$ ms, $I_0^E = 2$, $I_0^I = 0.5$, $A_i^E = 100$,
$A_i^I = 0$, $W_{ij}^{EE} = 5[1 + \cos(2\theta_i - 2\theta_j)]/N$, $W_{ij}^{EI} = 3[1 + \cos(2\theta_i - 2\theta_j)]/N$,
$W_{ij}^{IE} = -4[1 + \cos(2\theta_i - 2\theta_j)]/N$.

Complete questions 1–4 using each of the networks A, B, and C.

1. With all firing rates initially at zero, simulate the network as a function of time, with the stimulus present the entire time, using contrasts of $c = 0, 0.25, 0.5, 0.75$, and 1.

Note: Firing rates can be stored as arrays, with each row representing a different time point and each column a different cell. This allows the sum over all excitatory units providing input to all excitatory cells within the network to be calculated using matrix multiplication as:

```
rE(i-1,:)*WEE
```

Each entry in the row vector produced by the preceding segment of code represents the total within-network excitatory input to a unit. The `i-1` in the segment of code represents the previous time point, because firing rates at the previous time point are used when calculating input at the current time point, `i`.

2. Plot, using separate figures for the excitatory and inhibitory units, firing rate as a function of time for those units with preferred orientations of $\theta_i = \theta_{cue} = \pi/2$ and $\theta_i = \pi$. Ensure the same figures are used for all five contrasts.

3. At the last time point simulated plot the firing rates of the population of excitatory and inhibitory units as a function of their index on two separate figures (one for excitatory units, the other for inhibitory units). Ensure the same figures are used for all contrasts.

4. Produce rescaled versions of the plots produced in 3 by dividing the firing rates of each unit by the mean firing rate averaged over all units for that stimulus contrast. Comment on any contrast-invariant scaling observed, or lack thereof—i.e., whether, when the response curves produced with different contrasts are rescaled by their mean response, they become identical to each other.

5. In addition to contrast invariance, further investigations showed that excitatory neurons received their greatest inhibitory input when the stimulus was at their preferred orientation, which is when they also received their greatest excitatory input. Which model(s) are compatible with this result?

6. Experiments typically involve the recording of a few neurons and acquire each neuron's tuning curve using a range of stimuli—in this case with stimuli of multiple orientations. How can we discuss the tuning curve (response of a single neuron to multiple orientations of stimuli) from our simulations when we only ever simulated stimuli with a single orientation?

Questions for Chapter 6

1. Connections from the thalamus to V1 are purely excitatory. Why would it be reasonable in a firing-rate model to allow a connection from a unit in the thalamus to a unit in V1 to have a negative strength?

2. What happens to (i) the typical time to threshold and (ii) the accuracy of responses in an integrator-type decision-making circuit if:

   a. Noise in the inputs is increased?

   b. Input synapses are strengthened?

   c. Thresholds of all units are increased?

3. A firing-rate model produces oscillations with a frequency of 20 Hz with neurons never firing at a rate above 15 Hz. Can such a model apply to real spiking neurons, and if it can, what would you expect to see in the neural spike trains?

4. The stationary bump state of a ring attractor can become unstable if synapses are depressing. Why does this happen, and what are the two possible stable bump states that are generated in this process?

7

An Introduction to Dynamical Systems

All living things and their component cells are dynamical systems. The molecules that form us are constantly in flux, changing in location and number throughout our bodies; cells, including neurons, are generated and die; connections between neurons form and disappear; activity in neural circuits rises and falls. Yet, beneath the fluidity and apparent tumult within the processes that comprise a living being, is a choreography that ensures stability of the key patterns that make us who we are. To gain a better understanding of this choreography we should first embrace the mathematical framework of dynamical systems. In this chapter, we summarize the properties and roles of dynamical systems used within this book, many of which we have discussed in earlier chapters.

For a more in-depth study of the role of dynamical systems in neuroscience, in particular at the single-neuron level, I highly recommend the book *Dynamical Systems in Neuroscience* by Eugene Izhikevich.[1] For a more pedagogical approach to the mathematics of dynamical systems in general, *Nonlinear Dynamics and Chaos* by Steven Strogatz[2] is exceptionally good.

7.1   What Is a Dynamical System?

In simplest terms a dynamical system is one in which the variables can change in time, i.e., are dynamic. Mathematically the term is typically reserved for the study of systems of nonlinear differential equations—such as those we have been using to describe neural activity.

As we have seen, a differential equation is needed to describe just about any biophysical process—rates of change of concentrations of molecules depend on concentrations of other molecules; rate of entry of ions into a cell depends on the fraction of open ion channels; rate of ion channel opening depends on the membrane potential, or temperature, or neurotransmitter release; rate of population growth of a species depends on numbers of that species, predators, and available food stocks. All of these processes are described via the rates of change of the corresponding variables. Moreover, the dependencies are nonlinear—thresholds, saturation, power law, and exponential dependences abound—except in very simple cases rarely seen in biology. Thus, nonlinear differential equations comprise the correct approach to modeling—and perhaps a key to the understanding of (see, e.g., ref. 3)—neural activity and, indeed, any biological system.

In the following sections, we explore the various behaviors of dynamical systems, predominantly using as a basis the firing rate model of neural activity (chapter 6). We shall see that even simple circuits can exhibit multiple, qualitatively distinct types of behavior, just as the membrane potential of a bistable neuron can oscillate or remain static. Moreover, quantitative changes in the values of parameters of a given circuit can produce dramatic, qualitative transitions in the circuit's behavior. The study of such transitions is the hallmark of courses in dynamical systems, whose key message applies to all biological systems: a significant change in behavior can be caused by a subtle, barely observable change in the system's parameters.

7.2   Single Variable Behavior and Fixed Points

Fixed point   A set of values of all variables of a system at which none of the variables change in time (the right-hand side of all ODEs evaluates to zero).

For a single variable, such as firing rate, r, the dynamical system can be expressed as:

$$\frac{dr}{dt} = F(r), \tag{7.1}$$

where $F(r)$ can be any, generally nonlinear, function. While we know how to solve such an equation computationally, its qualitative behavior can be understood simply by plotting $F(r)$ as a function of r. As we have seen, any value of r at which $F(r) = 0$ is a fixed point, since when set at that value, r does not change. Of importance is whether, if shifted slightly away from the fixed point, r moves further away or returns toward the fixed point. The former case corresponds to an unstable fixed point, the latter case corresponds to a stable fixed point. Stable fixed points are important, since they correspond to stable states or equilibria to which activity is drawn. These are usually dynamic equilibria at which the ongoing processes conspire to ensure the rate of increase of any variable is precisely countered by its rate of decrease. Unstable fixed points correspond to tipping points, or points of no return. The value of $F(r)$ either side of the fixed point tells us the direction in which r moves, so it tells us the fixed point's stability as well (figure 7.1).

7.2.1   Bifurcations

Bifurcation   A bifurcation occurs when a continuous change in a parameter at some point leads to a qualitative change in the behavior of a system, usually because the number and/or stability of fixed points changes at that value of the parameter.

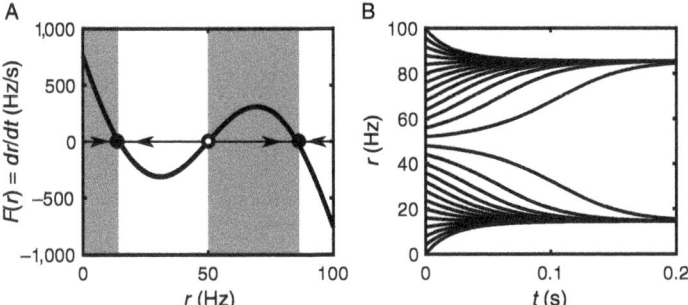

Figure 7.1
Stable and unstable fixed points in a single-variable firing rate model. (A) A plot of dr/dt against r, where

$$\frac{dr}{dt} = F(r) = \frac{-r + 100/\{1 + exp[-(Wr - I_{th})/I_\sigma]\}}{\tau}$$ is obtained from a firing-rate model with recurrent feedback of

strength W (see section 6.3). In this example, there are three fixed points (denoted by circles where $F(r) = 0$). Fixed points separate ranges of r where $dr/dt > 0$ (shaded gray) from ranges where $dr/dt < 0$ (unshaded). Arrows on the x-axis indicate the direction of change of r (rightward to increase r where $dr/dt > 0$ and leftward to decrease r where $dr/dt < 0$). The solid circles are stable fixed points, around which the direction of dr/dt is toward the fixed point, while the open circle at 50 Hz is an unstable fixed point, around which the direction of dr/dt is away from the fixed point. The system is bistable and can be a basic memory unit. (B) If the differential equation is simulated with multiple starting points, the trajectories of $r(t)$ move away from the unstable fixed point and toward one of the stable fixed points. Notice that since dr/dt depends only on r, if one were to take any horizontal line of fixed r, the gradient of the trajectories crossing that line would all be the same (fixed dr/dt at a fixed r). Parameters used are $\tau = 10$ ms, $W = 1$ nA/Hz, $I_{th} = 50$ nA, $I_\sigma = 20$ nA. This figure was generated by the online code `drdt_plot.m`.

Figure 7.1 was generated from a firing-rate model with fixed values for the feedback strength, W, and the firing threshold, I_{th} (see chapter 6). When either of these parameters is altered, the position of the fixed points changes and even the numbers of fixed points can change. In particular, if the feedback is too weak or the threshold is too high, then the state of high firing rate is no longer present. Conversely, if the feedback is too strong or the threshold is too low, then the state of low firing rate disappears.

For example, if we increase the value of excitatory feedback, W, from that of figure 7.1A (as would happen if synaptic strengths were increased), dr/dt would increase, shifting the curve up. As the curve of dr/dt versus r rises with such an increase of W, we see that the lower stable fixed point and the unstable fixed point approach each other (figure 7.2A, B) until they collide at a critical value of W (figure 7.2C). If W is further increased then only a high firing-rate stable fixed point remains (figure 7.2D). Conversely, if we were to decrease the value of W, then the higher stable fixed point and the unstable one would approach each other until they collide. If W were decreased further, then only a low firing-rate stable fixed point would remain.

We can combine these results into one figure to show how the fixed points change as we vary the feedback parameter, W. The resulting figure is a bifurcation plot (figure 7.3), in which the firing-rates given by the x-coordinates of the intercepts (the circles) in figure 7.2 are now plotted on the y-axis. As the parameter, W, is gradually varied along the x-axis, points arise where

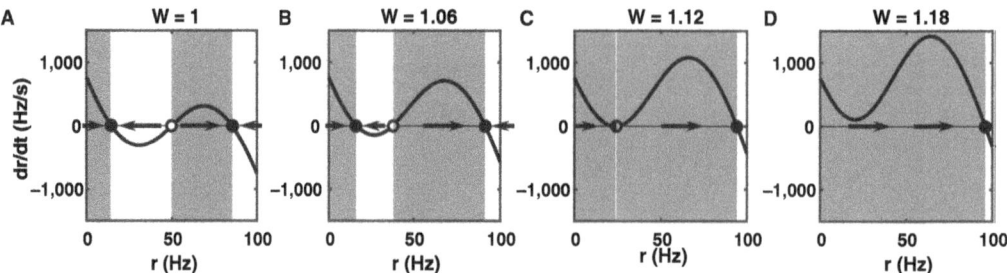

Figure 7.2
Increase of feedback strength causes loss of low firing-rate fixed point. (A) The bistable system of figure 7.1. (B) With increased feedback, W, dr/dt becomes more positive, so the curve shifts up and the lower fixed points move closer together. (C) At a critical value of feedback the lower two fixed points collide. (D) With even greater feedback, only one fixed point remains. These curves were produced using the online code `drdt_plot_manyW.m`.

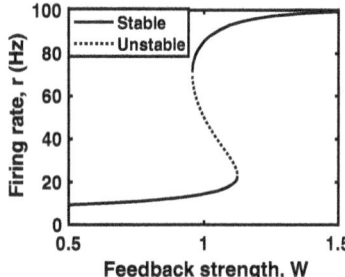

Figure 7.3
A bifurcation curve shows the stable and unstable states as a function of a control parameter. The feedback strength, W, controls whether the system has a single stable state of low firing rate (at low W), a single stable state of high firing rate (at high W), or two stable states with an intermediate unstable fixed point (at intermediate W). Two bifurcations occur with increasing W, one at the appearance, the other at the disappearance of the unstable fixed point. The x-axis crossing-points of the curves in figure 7.2 produce points on this curve: three points at feedback strengths of $W = 1$ and $W = 1.06$ (figure 7.2A, B); one point of high rate at $W = 1.18$ (figure 7.2D), with $W = 1.12$ (figure 7.2C) corresponding to the bifurcation point, at the value of W where the two lower fixed points collide. This figure was generated by the online code `bifurcation_varyW.m`.

there is an abrupt change in the number or the stability of fixed points—these are the points of bifurcation.

7.2.2   Requirement for Oscillations

Single variable dynamics of the form of equation 7.1 are quite constrained in their behaviors. For example, oscillations are not possible, because an oscillation requires the variable to change back and forth, so that at any value in its oscillating range, the variable's rate of change should be positive when on the increasing part of the cycle, but negative when on the decreasing part of the cycle. Such behavior is incompatible with equation 7.1, which provides a single value for the rate of change at each value of the variable, r.

However, we saw in chapter 2 that periodic activity is possible in a single-variable model such as the leaky integrate-and-fire model, if there is a mapping that instantaneously changes the value of the variable from one value to another. In these simple models of the neuron's membrane potential, the reset, by means of which the membrane potential instantaneously jumps from the high value (the threshold, V_{th}) to a lower value (the reset potential, V_{reset}) is essential to allow regular spiking. Such a mapping is akin to taking a sheet of paper with an x-axis drawn on it, then rolling it into a cylinder such that the highest x-value connects to the lowest x-value. Using a pencil to draw a line, if one is allowed to extend the line in only a single direction (e.g., if x increases only with time), the line will keep returning to itself periodically, so its x-coordinate will oscillate. Without such a mapping, two variables are needed to produce an oscillator.

7.3 Models with Two Variables

A rich variety of behaviors—including many important ones observed in real neurons and circuits—is possible in models of dynamical systems that possess two variables. The behavior of two variables can be plotted in two dimensions (one axis for each variable) so can be easily visualized. Because of these two features, it is very common for mathematicians to try to reduce more complicated models to two-variable ones.

For example, the Hodgkin-Huxley model (equation 4.9) can be reduced from four variables (V, m, h, n) to two variables (V, w). First, the need for a separate dynamical equation for the activation variable, m, is removed by assuming that it changes instantaneously with membrane potential, V. Second, sodium inactivation, h, and potassium activation, n, respond to changes in membrane potential with similar time constants, so that the value of one of them is highly predictive of the value of the other. Therefore, they can be combined into a single variable, w. The FitzHugh-Nagumo model,[4-6] which we shall study in section 7.6, is a simplified version of this two-variable reduction.

In general, a two-variable model requires two differential equations, each of which can depend on both variables. For example, with two connected firing-rate units, the general form is:

$$\frac{dr_1}{dt} = F(r_1, r_2)$$
$$\frac{dr_2}{dt} = G(r_1, r_2)$$

(7.2)

where $F(r_1, r_2)$ and $G(r_1, r_2)$ are typically two different functions, either because the two units represent different cell types or because the connections they receive are not identical. For example, in the decision-making circuit of section 6.5, where

$$\tau \frac{dr_1}{dt} = -r_1 + f(W_s r_1 + W_x r_2)$$
$$\tau \frac{dr_2}{dt} = -r_2 + f(W_s r_2 + W_x r_1)$$

(7.3)

and $f(I)$ is the single-neuron firing-rate response to input current, I, the two functions are $F(r_1,r_2)=[-r_1+f(W_sr_1+W_xr_2)]/\tau$ and $G(r_1,r_2)=[-r_2+f(W_sr_2+W_xr_1)]/\tau$.

Just as in single-variable models, the location and type of fixed points are key determinants of the behavior of a two-variable model. To see the fixed points and understand how they change with changes in parameters, it is useful to plot nullclines using phase-plane analysis, as will be explained in the next subsection.

Nullcline   A curve showing how the fixed point of one variable depends on the values of other variables in a system of coupled ODEs.

Phase plane   A plot on the axes of the two variables in a system of two coupled ODEs, which can show the position of fixed points, nullclines, and trajectories demonstrating how the variables of the system change together over time.

7.3.1   Nullclines and Phase-Plane Analysis

At a fixed point of a two-variable system, it is necessary that both differential equations produce zero rate of change, i.e., $F(r_1,r_2)=0$ and $G(r_1,r_2)=0$ in equation 7.2. To produce nullclines, we consider one of these equations at a time. By way of example, we will consider a decision-making circuit with threshold-linear units, in the absence of a stimulus:

$$\tau\frac{dr_1}{dt}=-r_1+W_sr_1+W_xr_2-\Theta$$

$$\tau\frac{dr_2}{dt}=-r_2+W_sr_2+W_xr_1-\Theta, \tag{7.4}$$

where $\Theta>0$ is the threshold, proportional to the amount of input current the neurons need to start firing spikes (as in tutorial 6.2). A negative threshold ($\Theta<0$) corresponds to a rate of spontaneous activity, the firing rate in the absence of input.

Considering the second of equation 7.4, we find that $dr_2/dt=0$ if

$$r_2=\frac{W_xr_1-\Theta}{1-W_s}, \tag{7.5}$$

which describes a straight line if we plot r_2 against r_1. This straight line forms one of the two nullclines of the system. It shows the values of r_2 where $dr_2/dt=0$ (i.e., fixed points of r_2) if r_1 were held constant.

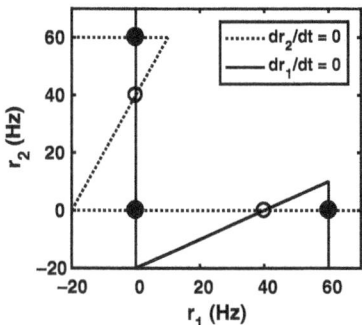

Figure 7.4
Nullclines showing the fixed points of a bistable system with threshold-linear firing-rate units. The nullcline for r_2 is shown as a dotted line, and for r_1 as a solid line. The nullclines cross at five fixed points, three of which are stable (solid circles) and two of which are unstable (open circles). The axes are extended to negative rates—which can never be reached in practice—so that the complete "Z" shape of the nullclines can be seen. The shapes of the curves are similar to reflections of figure 7.3 (reflected, because rate of one unit provides negative input to the other). In this case, the fixed points all lie on one of the axes. The online code `nullcline_bistable.m` was used to generate this figure. It can be altered so that the parameter-dependence of the nullclines and fixed points is easily studied.

We also can show that if firing rates are constrained to be nonnegative, then $dr_2/dt = 0$ if $r_2 = 0$ and $W_x r_1 - \Theta < 0$, since dr_2/dt is prevented from being negative in this situation. Similarly, if there is a maximum firing rate of r_{max}, then $dr_2/dt = 0$ if $r_2 = r_{max}$ and $-r_{max} + W_s r_{max} + W_x r_1 - \Theta > 0$, since dr_2/dt is prevented from being positive in this situation.

For concreteness, to find these fixed points, we will use the parameters of tutorial 6.2 (jumping mode): $r_{max} = 60$ Hz, $W_s = 1.05$, $W_x = -0.05$, and $\Theta = 4$ Hz. In this case, the dotted line of figure 7.4 is the resulting nullcline for r_2. Notice it has three parts, one following equation 7.5 in a diagonal direction, one horizontal at $r_2 = 0$ where the dynamics alone would reduce r_2 further, and one horizontal at $r_2 = r_{max}$, where the dynamics alone would increase r_2 further.

Similarly, by solving for how r_1 varies with r_2 when $dr_1/dt = 0$ in equation 7.4, we obtain a second nullcline, the solid curve in figure 7.4, whose diagonal portion follows the equation:

$$r_1 = \frac{W_x r_2 - \Theta}{1 - W_s}. \tag{7.6}$$

We then know that at the points where these two curves cross, both $dr_1/dt = 0$ and $dr_2/dt = 0$, so the system has a fixed point—a pair of values of (r_1, r_2) such that if the system were set at these rates, the rates would not change.

We can rearrange equation 7.6 to obtain $r_2 = ((1 - W_s)r_1 + \Theta)/W_x$. Then we can see that equation 7.6 produces the same line as equation 7.5 if the gradients are the same, such that $(1 - W_s)/W_x = W_x/(1 - W_s)$, and if the intercepts are the same, such that $\Theta/W_x = -\Theta/(1 - W_s)$. Therefore, if we set $W_x = -(1 - W_s)$ or $W_s - W_x = 1$, then the two lines are identical and the nullclines overlap along a straight line. This is identical to the requirement for an integrator in

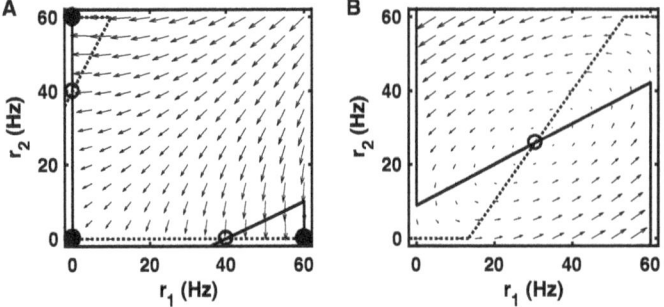

Figure 7.5
Vector fields indicate the dynamics of a system on a phase plane. (A) The multistable system shown in figure 7.4, with arrows indicating how the firing rates change together. Depending on the starting point, the system moves to one of the three stable fixed points (solid circles). (B) With adjustment of thresholds and connection strengths, with one excitatory unit of rate r_1, and one inhibitory unit of rate r_2, the nullclines cross at a single point. In this case an unstable fixed points is produced, around which the firing rates oscillate to produce an orbit in the phase plane. This circuit is equivalent to the one that produced PING oscillations (figures 6.11–6.12 and tutorial 6.3), although here the neural responses are threshold-linear rather than sigmoidal. In both panels, the nullcline for r_1, where $dr_1 / dt = 0$, is the solid line, while the nullcline for r_2, where $dr_2 / dt = 0$, is the dotted line. The code to produce this figure can be found online as vector_field.m.

a decision-making circuit (see text following equation 6.14) and we used this condition for one circuit in tutorial 7.2.

In addition to the nullclines, it can also be useful to plot arrows indicating the direction of change of firing rates (figure 7.5). Unlike the single-variable case depicted in figure 7.1, the arrows do not demonstrate with certainty whether a fixed point is stable, but they do provide a good indication. They can also suggest when oscillations occur, as oscillations appear as orbits around a fixed point when plotted as a function of two variables on a plane (figure 7.5B).

Orbit A closed trajectory, showing how the variables of the system can change together over time such that they return to their starting state, implying an oscillation of the system.

7.3.2 The Inhibition-Stabilized Network
Before looking at time-varying dynamical states, we consider here a stable fixed point that has unexpected, sometimes considered paradoxical properties in its response to external input.[7] Given the strength of feedback excitation and inhibition in many areas of the brain and careful modeling of the observed neural activity patterns in both visual cortex and hippocampus, there is reason to believe some neural circuits operate in an inhibition-stabilized regime.[8]

> **Inhibition-stabilized regime**   The set of parameters in which a circuit has sufficient excitatory feedback to destabilize any low firing-rate state, but with sufficient compensatory inhibitory feedback to stabilize the excitatory rate at a value that, without dynamic inhibitory feedback, would be an unstable fixed point intermediate between its low spontaneous rate and its high, saturated rate.

To be in the inhibition-stabilized regime (figure 7.6), one requirement is that the excitatory feedback to excitatory neurons (W^{EE} in figure 7.6A, B) is sufficiently strong that, in the absence of inhibition, the excitatory cells would not be able to fire stably at a low rate—the feedback is so

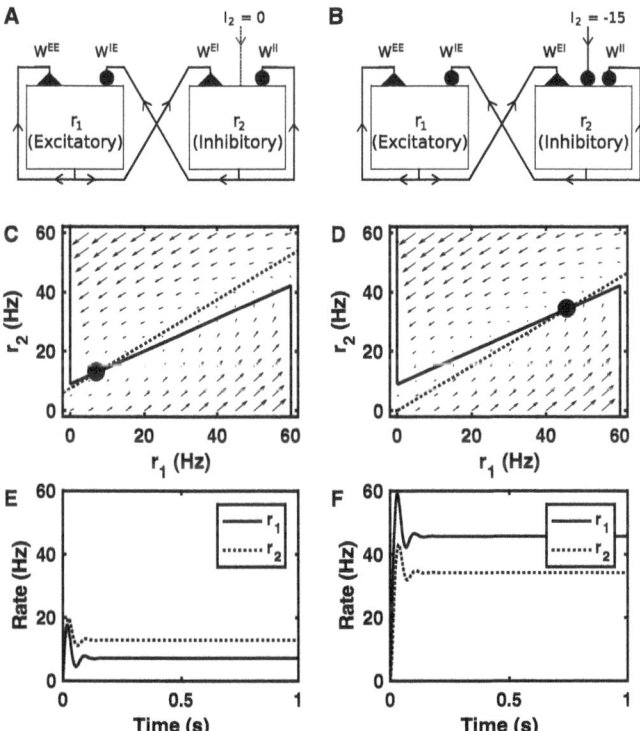

Figure 7.6
The surprising response to inputs of an inhibition-stabilized network. (A, B) The circuit consists of two units, one excitatory and one inhibitory, with strong excitatory (solid triangles) and inhibitory (solid circles) reciprocal connections. In A, there is no external input; in B, there is inhibitory input to the inhibitory unit. (C, D) The nullclines and vector fields are plotted as in figure 7.5. Notice the similarity to the oscillator of figure 7.5B, yet in this case the fixed point where the nullclines cross is stable. In D, the inhibitory input lowers the nullcline for r_2 (dotted), that is, reducing r_2 for any fixed level of r_1, but with the consequence that the crossing point of the two nullclines is raised, producing increased r_2 as well as increased r_1. (E) The firing rates settle at the fixed point in the absence of external input. (F) The firing rates settle to the higher fixed point (increased rate for both units) when the inhibitory unit is inhibited. The code to produce these results is `vector_field.m`. Parameters are $W^{EE} = 2.25$, $W^{EI} = 1.5$, $W^{IE} = -2.25$, $W^{II} = -1$, $\Theta_1 = -20$ Hz, $\Theta_2 = -15$ Hz, with $r^{(max)} = 60$ Hz for both units, $\tau = 10$ ms and otherwise linear f-I curves as in equation 7.4.

strong that each spike generates more than one extra spike, so the spike rate would increase until a maximum rate (or maximum level of feedback) is reached (as we saw in figure 7.2D).

The second requirement for a circuit to be in the inhibition-stabilized regime is that the inhibitory unit's nullcline (the dotted line of figure 7.6C, D) crosses the excitatory unit's nullcline on its unstable branch. To see this, consider the fixed point (the solid circle) on figure 7.6C. We see that if we move to the left then the arrows point leftward because $dr_1/dt < 0$, while if we move to the right then the arrows point rightward because $dr_1/dt > 0$. That is, considering only the x-direction the arrows point away from the diagonal solid line and toward the vertical solid lines at $r_1 = 0$ and at $r_1 = 60\,\mathrm{Hz}$. That is, if we do not take into account changes in inhibition, so fix r_2, the dynamics of r_1 alone corresponds to the situation in figure 7.1, with an intermediate, unstable fixed point.

The third and final requirement for the circuit to be in the inhibition stabilized regime is that the inhibitory feedback is sufficient to stabilize the fixed point in question. Without such stabilization, a system which satisfies the first two requirements is likely to oscillate—indeed, such is the case for the circuit shown in figure 7.5B, which fulfills the first two requirements, but not the third. The necessary stabilization can arise with nullclines identical to those of figure 7.5B, if the time constant for inhibitory response is reduced to be faster than that for excitatory feedback. Alternatively, stabilization can arise with different parameters, so that the two nullclines cross in a manner that is qualitatively the same, but with different gradients (figure 7.6C, D). If the fixed point is stabilized, then its position indicates the steady state of the system and an input-driven shift in the fixed point determines the system's response to inputs.

The response of an inhibition-stabilized network to input can be puzzling at face value. In particular, the first so-called paradoxical effect is that the stable firing rate of the inhibitory neurons changes in the opposite direction to that produced by any direct input to them. For example, in figure 7.6, we show that the circuit, in the absence of input, has a stable set of low firing rates. However, if the inhibitory neurons are provided with inhibitory input—which would normally decrease their firing rate—the network feedback causes both the excitatory and the inhibitory neurons to increase their rate. Thus, the response of inhibitory neurons to inhibitory input is an increase of rate and conversely, if they were provided with excitatory input, their rate would decrease.

Such puzzling behavior can be explained graphically in figure 7.6. If we compare the nullcline for the inhibitory unit, which describes its fixed points if feedback excitation were held fixed, we see that when inhibitory input is added (in figure 7.6D versus figure 7.6C) the nullcline moves down, to lower r_2. However, as it does so, the crossing point of the two nullclines (i.e., the fixed point) moves up, to higher r_2, as well as higher r_1.

The intuitive (rather than graphical) explanation is that the dominant input to the inhibitory neurons is from the local excitatory neurons in the feedback circuit. If the direct external input causes the inhibitory neurons to reduce their firing rate, the local excitatory cells overrespond to reduced inhibition by firing a lot more. The increased excitatory activity is more than sufficient to raise the inhibitory rate beyond its initial level.

The second so-called paradox then arises: Following this increase in excitation and subsequent increase in inhibition, the excitatory units are firing at a higher rate while receiving more inhibition than before. So why does their rate not decrease again? The answer to this is that they are on the unstable branch of their nullcline, where the increased direct excitatory feedback they are receiving at high excitatory rate compensates for the increased indirect inhibitory feedback they receive at this rate.

7.3.3   How Inhibitory Feedback to Inhibitory Neurons Impacts Stability of States

For excitatory units, any excitatory feedback increases the effective time constant for decay of activity in the circuit (see the text following equation 6.14). The converse for inhibitory units is that inhibitory feedback causes a more rapid decay of activity following a transient increase, corresponding to a decrease in the effective time constant. Recall also that slow inhibitory feedback is more likely to engender oscillations, whereas fast inhibitory feedback is more likely to clamp down and stabilize any set of firing rates. The net result in a circuit with reciprocal excitatory and inhibitory feedback is that the stronger the inhibitory-to-inhibitory connections, the more likely a stabilized state arises, whereas the weaker those connections, the more likely oscillations or bistability arise.

7.4   Tutorial 7.1: The Inhibition-Stabilized Circuit

Neuroscience goals: Understand how constant input current can shift a circuit into the inhibition-stabilized regime; understand how bistability can arise between states of low firing rate.

Computational goals: Gain more experience at connecting circuits and simulating firing-rate models with different types of f-I curve; use the sign function and ensure bounds in rate.

In this tutorial, you will produce a model of an excitatory unit coupled to an inhibitory unit using a firing-rate model for each unit's activity. The excitatory unit will have a quadratic firing rate curve, which has the interesting property that its gradient is low at low rates, meaning the effective excitatory feedback can be relatively weak and lead to stability, whereas the gradient increases at high rates, increasing the effective excitatory feedback to a point of instability. Thus, this circuit requires inhibition for stability at high rates—where it can be in an inhibition-stabilized regime—but not at low rates. Such a situation appears to be the case in the circuitry of visual cortex.[9] You will also see how the dynamical regime of the circuit depends on the time constants. In part B, you will couple two such excitatory-inhibitory pairs to produce a bistable circuit with distinct states of low activity.

Part A

Simulate the activity of an excitatory unit of rate r_E, coupled to an inhibitory unit of rate r_I (as in figure 7.6A). The firing rates follow respectively:

$$\tau_E \frac{dr_E}{dt} = -r_E + \alpha_E (I_E - \Theta_E)^2 \cdot sign(I_E - \Theta_E)$$

$$\tau_I \frac{dr_I}{dt} = -r_I + \alpha_I (I_I - \Theta_I),$$

with the conditions $0 \le r_E \le r_{max}$, $0 \le r_I \le r_{max}$, which you must enforce. The $sign(\ )$ function (use $sign$ in MATLAB) returns the sign of the quantity within parentheses as ± 1, and the total currents to each unit are given by:

$$I_E = W^{EE} r_E + W^{IE} r_I + I_E^{(App)}$$

$$I_I = W^{EI} r_E + W^{II} r_I + I_I^{(App)}.$$

The f-I curves and the connection strengths are fixed through all questions in part A, so the following parameters should be set: maximum firing rate, $r_{max} = 100$ Hz; threshold of excitatory (E-) cells $\Theta_E = -5$ (so their spontaneous rate is 5 Hz); threshold of inhibitory (I-) cells, $\Theta_I = 0$; gain of E-cells, $\alpha_E = 0.05$; gain of I-cells, $\alpha_I = 1$; E-to-E connection strength, $W^{EE} = 2$; E-to-I connection strength $W^{EI} = 2.5$; I-to-E connection strength, $W^{IE} = -2.5$; and I-to-I connection strength, $W^{II} = -2$.

In each question, you will simulate the system with given baseline currents (with units in Hz, based on the firing-rate change they cause), which are incorporated as $I_E^{(App)}$ and $I_I^{(App)}$. In a total simulation of 3 s, you should apply an extra current to increase $I_I^{(App)}$ by 20 in the middle of the simulation with onset at 1 s and offset at 2 s. That is $I_E^{(App)}(t) = I_E^{(base)}$ and $I_I^{(App)}(t) = I_I^{(base)} + I_I^{(stim)}(t)$ where $I_I^{(stim)}(t) = 20$ if $1 < t \le 2$, otherwise $I_I^{(stim)}(t) = 0$.

In each question, explain the system's behavior before, during, and after the extra positive applied current to the inhibitory cells.

1. Simulate the system in a default condition with parameters $I_E^{(base)} = 0$, $I_I^{(base)} = 0$, $\tau_E = \tau_I = 5$ ms.

2. Simulate the system with increased baseline current, $I_E^{(base)} = 25$, $I_I^{(base)} = 15$, $\tau_E = \tau_I = 5$ ms. (During the stimulus $I_I^{(App)}$ is still increased by 20, so becomes 35).

3–4. Respectively repeat 1 and 2 but with altered time constants such that $\tau_E = 2$ ms and $\tau_I = 10$ ms.

Be sure to comment on any differences in the results for questions 1–4.

Part B

Duplicate the circuit of part A so that there are two independent pairs, each pair comprising two units, one excitatory and one inhibitory, with the dynamics provided in part A. Add two connections between the pairs, from the excitatory unit of each pair to the inhibitory unit of the other pair, with a value $W_{EI-X} = 1.75$ (keep all within-pair connections unchanged from part A).

5. Simulate the system for 3 s, applying a pulse of additional excitation, $I_I^{(App)} = 10$, to the first inhibitory unit for a duration of 100 ms at the 1-s time point, and applying such a pulse of excitation to the second inhibitory unit for a duration of 100 ms at the 2-s time point.

For this simulation set $I_E^{(base)} = 25$ and $I_I^{(base)} = 20$, with $\tau_E = \tau_I = 5$ ms.

Explain the resulting activity of the network. How many stable states do you believe the system possesses? Describe each of them (giving the set of firing rates of any fixed point). Test that each is stable by, for example, perturbing them by a small amount with an applied current or adding a small amount of noise.

7.5 Attractor State Itinerary

> **Attractor state itinerancy**   The process of a system changing from the vicinity of one fixed point to another, typically with dwelling periods in the vicinity of each fixed point being longer than the times taken to transition between them.

Neural activity in vivo is notoriously variable, which is why many trials with identical stimuli are needed to build up a reliable picture of the neural responses[10]. Such variability appears to be at odds with any theory of information processing based on attractor states, in particular the fixed-point attractors discussed so far. However, if neural activity "jumps" between different, distinct attractor states—a process called attractor state itinerancy[11]—the spike times of single neurons would be highly variable. Moreover, if the timing of such jumps were unreliable, varying from trial to trial, then the rapid transitions between states may not be revealed when spike trains are averaged across trials according to standard procedures. In chapter 9 we consider methods that do not rely upon trial-averaging for analysis of neural activity patterns, so could reveal any such attractor state itinerancy. Here we consider a simple model of one of the more likely occurrences of attractor state itinerancy.

7.5.1 Bistable Percepts
As we saw in section 5.7, the Necker cube (figure 7.7A) is one of many examples of visual image that can be perceived in more than one way (see also figure 5.8). If we stare at the Necker cube, initially one of the squares appears to be the closest face. However, after a few seconds the cube appears to switch orientation, with what was previously a rear face becoming the closest face. The times between such transitions are variable and their distribution enables us to gain some insights into the underlying neural processes that cause such transitions.[13]

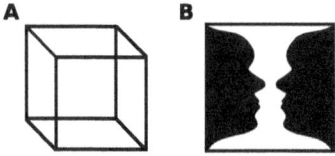

Figure 7.7
Alternating visual percepts are produced by ambiguous images. (A) The Necker cube can be perceived with either the upper square or the lower square front-facing. (B) Either a white vase or two black faces are perceived (first created by the Danish psychologist Edgar Rubin in 1915).[12] If we gaze at either image, our percepts alternate.

Perceptual rivalry   A phenomenon in which a single stimulus can be perceived in more than one manner, causing a switching back and forth between these different percepts.

Studies of transitions between dominant percepts—or perceptual rivalry—can arise when different images are presented to the two eyes (binocular rivalry)[14,15] or when a subject views moving oriented plaids, where two sets of parallel stripes move across each other at an angle.[16,17] In the case of moving plaids, percepts of up-down motion, or left-right motion, or a coherent motion of diamonds formed by combining the two sets are all possible. The widths, contrasts, and orientations relative to direction-of-motion, all contribute to determining the distribution of durations of each percept in between transitions.

Auditory stimuli can also provide bistable percepts[18,19]—sequences of tones of different frequencies can be perceived as a single stream or as distinct sources. It would be interesting to find such stimuli in other modalities, such as smell, taste, or touch.

It is relatively easy to produce ambiguous percepts in the visual domain because we perceive a three-dimensional world, while our eyes only receive a two-dimensional projection of the world around us. It is perhaps the combination of such inherent ambiguity with the greater extent of model building of the visual domain internally within our brains—and the concomitant attractor states underlying such internal models—that renders visual sensory inputs the most likely to produce attractor state itinerancy.

7.5.2   Noise-Driven Transitions in a Bistable System

If a dynamical system, such as a neural circuit, possesses two fixed points, a transient external input can cause a transition from one fixed point to another (figure 6.4B). Random fluctuations of an input that is in the form of zero-mean white noise can cause transitions in a similar manner. In the latter case then the time spent in one state before a sufficiently large noise-fluctuation knocks the system into the other state is highly unpredictable and follows a near-exponential distribution (figure 7.8B, E). In fact, in the limit that the time between transitions is much longer than the time to actually make the transition (so most of the time is spent "waiting" for a large noise

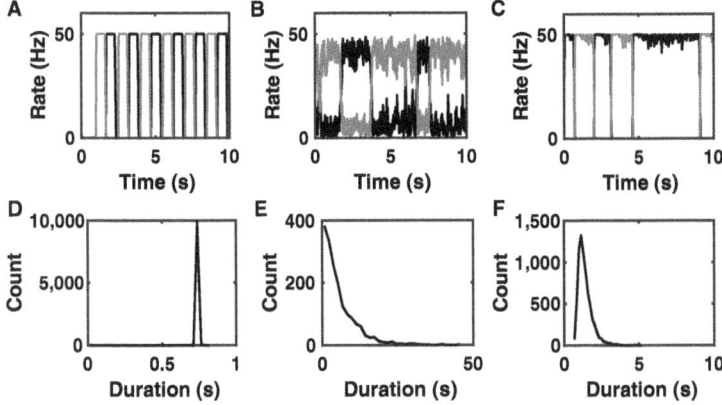

Figure 7.8
Transitions between attractor states in a bistable system. (A) An adaptation current is added to the winner-takes-all network of figure 6.8. The time for the adaptation current to build up in the highly active unit depends on that unit's firing rate, while the time for the adaptation to decay in the silent unit is equal to the time constant (of 250 ms in this example). The resulting switching of the dominant unit is periodic, with near-identical state durations (D). (B) Instead of an adaptation current, external noise is added to both units in a bistable winner-takes-all network. State durations (times between transitions) are highly variable, producing an exponential distribution (E). (C) A combination of weaker adaptation current and weaker noise produces variable state durations. Very brief state durations are rare because the adaptation must build up enough for the noise to have a significant chance of causing a transition. The resulting shape of the distribution of state durations (F) matches that of percept durations during perceptual rivalry tasks.[20] (D–F) The distribution of the number of occurrences of different state durations over, in each panel, a 10,000-s trial of the activity shown in A–C respectively. The code to generate this figure is available online as `bistable_percept.m`.

fluctuation) the state transitions follow a Poisson process, which has an exponential distribution of intervals between transitions (section 3.3.3). In general, an analysis of the distributions of state durations (figure 7.8D–F) can provide strong evidence for a particular architecture of the underlying neural circuitry.[21]

7.6   Quasistability and Relaxation Oscillators: The FitzHugh-Nagumo Model

If one variable of the system changes slowly enough compared to other variables, the system can be analyzed with the slow variable treated as a fixed parameter that determines the available states of the system. Once the available states are found assuming a fixed slow parameter, then the slow variable is allowed to gradually vary. As the slow variable gradually varies, so too do the available states, until a bifurcation point is reached where the system's current activity state loses stability.

Quasistable state   A state of a system, which is almost stable and appears stable for a duration longer than the system's fastest timescales, but eventually proves to be unstable.

> **Relaxation oscillator**   A system that regularly switches between different quasistable states, with long durations within each state compared to the timescale of more rapid jumps between those states.

If multiple attractor states are present in the system with the slow variable fixed then interesting behavior resembling attractor state itinerancy can arise when the slow dynamics of that variable are included. For example, the attractor states may slowly change until the one corresponding to the system's activity becomes unstable—in which case the state was "quasistable," not completely stable. The activity will then rapidly change to reach a new quasistable attractor state. The time between the system's transitions from one attractor state to another is the time taken for the attractor state to become unstable, which depends on how rapidly the slow variable changes. We saw an example of such behavior in figure 7.8A, where the slow adaptation variable changes more than an order of magnitude more slowly than neural firing rates. The resulting behavior in a two-state system can be periodic—as it is in figure 7.8A—in which case the system is a type of oscillator called a relaxation oscillator.

One of the best-known examples of a relaxation oscillator is the FitzHugh-Nagumo model,[4–6] which is a two-variable simplification of the Hodgkin-Huxley model capable of reproducing and explaining many of its type-II properties (section 4.2.3).

The two variables in the FitzHugh-Nagumo model are the membrane potential, V, and an adaptation variable, which is given the symbol w. The key ingredient of the model is the cubic dependence of the rate of change of membrane potential, dV/dt, on membrane potential, V. Such a cubic dependence matches the shapes of the dr/dt equation in figures 7.1 and 7.2, which we saw can lead to bistability. As in other single neuron models, dV/dt increases with applied current, I, which shifts up the curve of dV/dt versus V, and decreases with adaptation, which shifts down the curve of dV/dt versus V. This means that at fixed high levels of adaptation only lower values of membrane potential can be stable, at fixed low levels of adaptation only higher values of membrane potential can be stable, whereas at fixed intermediate levels of adaptation both lower and higher values of membrane potential can be stable (see figure 7.9).

The ODEs describing a general form of the FitzHugh-Nagumo model are:

$$\tau_V \frac{dV}{dt} = -\beta V (V - V_1)(V - V_2) - w + I^{(app)}$$

$$\tau_w \frac{dw}{dt} = V - V_0 - w, \tag{7.7}$$

where V_1 and V_2 determine the range of membrane potential variation, V_0 is the membrane potential above which the adaptation variable rises above zero, β determines how much adaptation is needed to cause a jump in V, τ_V is a very short time constant that determines the rate of rapid

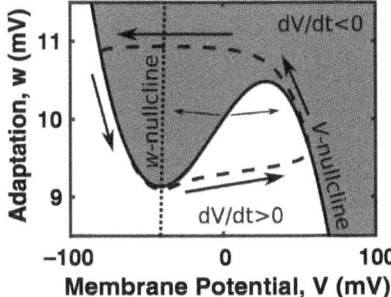

Figure 7.9
The FitzHugh Nagumo model. The cubic V-nullcline (solid line) separates regions where $dV / dt > 0$ (unshaded) from regions where $dV / dt < 0$ (shaded gray). When the system is oscillating, (dashed trajectory) it changes slowly when its variables lie on or near the V-nullcline, but it jumps rapidly (horizontal arrows) between the high-V and low-V branches of the nullcline when the system reaches the bifurcation points (maxima and minima of the V-nullcline). In this manner, the membrane potential cycles around the range of adaptation, w, where, if w were fixed, V would be bistable. This figure was produced by the available online code FHNmodel.m.

upswing and reset of membrane potential during a spike, τ_w is a much longer time constant, which determines the rate of recovery between spikes and the spike-width, and $I^{(app)}$ is the applied current in units of the voltage change it would cause in the absence of the spike mechanism.

The separation of timescales ($\tau_V \ll \tau_w$) ensures that the system spends most of its time on the V-nullcline, given by $dV / dt = 0$. This cubic nullcline, derived from equataion 7.7, is given by the equation

$$w = -\beta V (V - V_1)(V - V_2) + I^{(app)}, \tag{7.8}$$

and plotted in figure 7.9. Whenever w is less than this value (below the curve in figure 7.9, unshaded region) then equation 7.7 shows us that $dV / dt > 0$, so the membrane potential increases rapidly and the trajectory moves to the right until it hits the V-nullcline, where equation 7.8 is satisfied.

When equation 7.8 is satisfied, so that the system rests on the V-nullcline, then according to equation 7.7, the adaptation variable slowly changes, increasing if $w < V - V_0$ while to the right of the w-nullcline and decreasing if $w > V - V_0$ (while to the left of the w-nullcline). Therefore, with the parameters that produce figure 7.9, when the membrane potential is at its high level, adaptation increases until it is so high that the high-V branch of the V-nullcline disappears. Once there is no longer a stable state of high membrane potential, the membrane potential rapidly drops until the system is on the low-V branch of the V-nullcline. When the system is on the low-V branch of the V-nullcline, the adaptation variable decreases, until the membrane potential no longer has a stable low value, in which case the membrane potential rapidly jumps to the high-V branch of its nullcline to initiate a spike. The process repeats to produce oscillations (figure 7.10B).

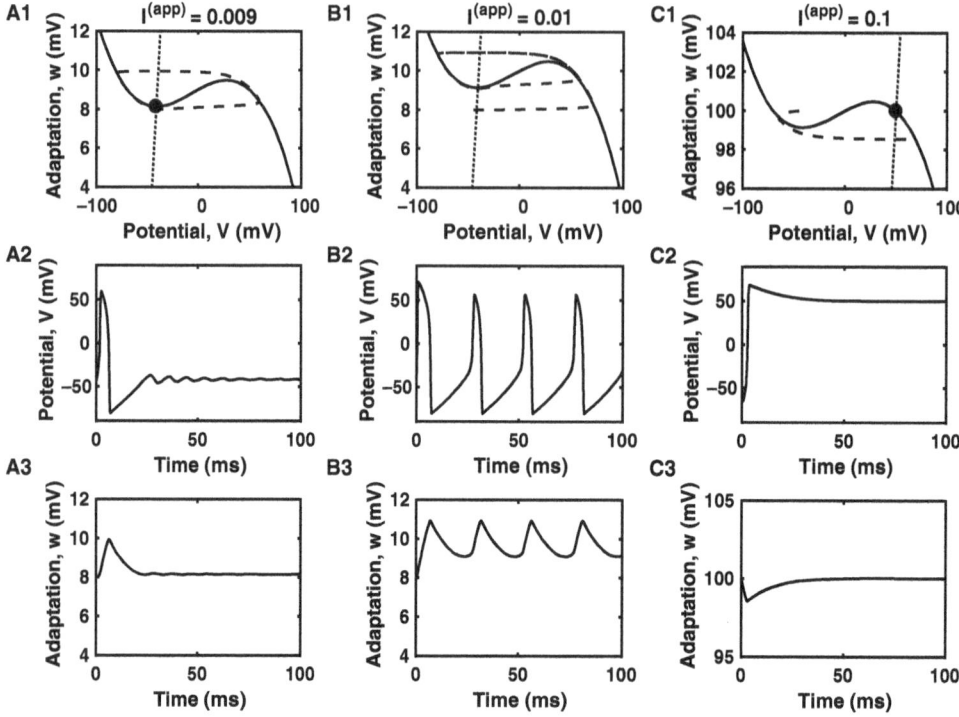

Figure 7.10
The FitzHugh-Nagumo model behaves like a type-II neuron. (A) With small applied current, $I^{(app)} = 0.009$, a single spike occurs, followed by subthreshold oscillations. (A1) Nullclines are shown with $dV / dt = 0$ solid and $dw / dt = 0$ dotted. The trajectory of $w(t)$ versus $V(t)$, dashed, depicts the spike as a single excursion jumping between stable sections of the V-nullcline, but ultimately resting at the stable fixed point (solid circle) where the two nullclines intersect. (A2) Membrane potential versus time and (A3) adaptation variable versus time for the same applied current of $I^{(app)} = 0.009$. (B) With a larger applied current, $I^{(app)} = 0.01$, the V-nullcline shifts up to higher values of w (B1) and regular spiking occurs (B2), with repeated loops between the higher and lower branches of the V-nullcline followed. Axes of B1–B3 are identical to those of A1–A3. (C) With very large applied current, $I^{(app)} = 0.1$, the w-nullcline intersects the V-nullcline on its upper branch, producing a stable fixed point at high membrane potential. Spike-like oscillations are no longer possible. Axes in C1–C3 are shifted to higher values of w. The figures depict solutions of equation 7.7, with parameters $V_0 = -50$ mV, $V_1 = -70$ mV, $V_2 = 50$ mV, $\beta = 8$ mV^{-2}, $\tau_V = 0.01$ ms, $\tau_w = 200$ ms, and the applied currents shown. This figure was produced by the available online code FHNmodel.m.

7.7   Heteroclinic Sequences

Saddle point   A fixed point, which is stable in at least one direction but also unstable in at least one direction. Given the impossibility in practice of having no motion in the unstable direction, saddle points are overall unstable, but they can be approached before being departed from. The system can reside near the saddle point for a long time.

Heteroclinic sequence   A trajectory of a dynamical system that passes from one saddle point to another, typically with long times of relative constancy near each saddle point interspersed with rapid changes in the variables as the system processes between saddle points.

We have so far considered fixed points that are either stable (so produce attractor states) or unstable. We have also considered the marginal stability of a line attractor (section 6.3.5). In systems with more than one dimension (i.e., in any system with multiple variables) unstable fixed points can be saddle points—there can be directions from which the network's activity moves toward the fixed point, as well as directions in which the network's activity moves away from the fixed point. That is, depending on the set of values of all variables, the activity approaches or departs from a fixed point. The unstable fixed points of figure 7.5A are examples of saddle points, with a clearer example provided in figure 7.11.

High-dimensional systems, like those comprising many neurons or units, with a great deal of heterogeneity in their interconnections, are likely to possess many saddle points.[22,23] In such cases the system's activity can process from the vicinity of one saddle point to another, producing a heteroclinic sequence.[24] If the activity returns to the vicinity of the original saddle point and then repeats, a heteroclinic orbit is produced.

A heteroclinic sequence can appear to have near-stationary activity while in the vicinity of a fixed point (cf. figure 7.11D, E). Moreover, the time spent near the fixed point depends on how closely the activity approaches the fixed point (as without noise, if the system ever reached the fixed point precisely, it would remain there forever). Therefore, while multiple trials with small differences in initial conditions can lead to a reproducible sequence of states, each state being the activity at a fixed point, the time spent in those states can vary. In this sense, the heteroclinic sequence can resemble a series of noise-driven transitions between discrete attractor states.[11]

7.8   Chaos

Perhaps the best-known example of a chaotic system is the atmosphere, with its weather patterns being notoriously unpredictable beyond a few days in many parts of the world. Indeed, one of

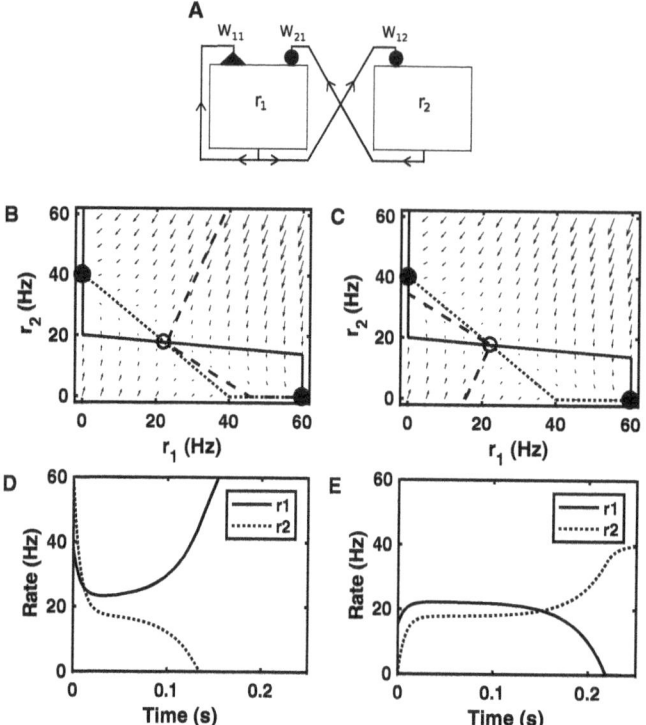

Figure 7.11
Example of a saddle point in a circuit with cross-inhibition. (A) Circuit architecture with cross-inhibition between two units similar to that of a decision-making network (section 6.5) (solid circles = inhibitory connections; solid triangle = excitatory connection). (B, C) Nullclines cross in three places with stable fixed points shown as solid circles and the intermediate unstable fixed point, which is a saddle point, as an open circle. Arrows indicate direction of change of the system's firing rates. Dashed lines are trajectories, which initially approach the saddle point then deviate to one of the stable fixed points. (D, E) The firing rate curves that correspond to the trajectories in B and C respectively. Note the period of slower rate of change and the nonmonotonic variation of one of the firing rates, a hallmark of approach toward and then departure from a saddle point. The code to produce these results is `vector_field_saddle.m`. Parameters are $W_{11} = 0.95$, $W_{12} = -1$, $W_{21} = -0.5$, $\Theta_1 = -10$ Hz, $\Theta_2 = -40$ Hz, with $r^{(max)} = 60$ Hz for both units, $\tau = 10$ ms and otherwise linear f-I curves as in equation 7.4.

the criterion of chaos is epitomized in the "butterfly effect," so called because the tiniest perturbation, such as the flutter of a butterfly's wings, could make the difference between a hurricane hitting one continental region or another a week later. Mathematically, chaotic systems have such strong sensitivity to initial conditions that the impact of a tiny effect grows exponentially over time. The high variability of neural firing in vivo—a variability that can be reduced by input from a stimulus[25,26]—is suggestive that the cortex operates in a chaotic state.

In an unstable single-variable system—such as one of exponential growth—the effect of a small change at one point in time can be exponentially magnified as time progresses, but such systems are not chaotic. Therefore, a second criterion is needed to define a chaotic system, namely the requirement of mixing of trajectories. For example, if the system is changed by a small increase in one variable, then for the system to be chaotic, as that variable changes through time at some point it must become less than what it would have been without the initial increase. For this condition to be satisfied, it turns out that chaotic systems must be nonlinear and contain at least three variables.[2]

Chaotic system   A system whose behavior is extremely sensitive to initial conditions, with diverging trajectories that remain bounded and cross over each other in any 2D projection.

Edge of chaos   A system operating with a set of parameters, which if adjusted in one direction cause the system to be chaotic, while if adjusted in the other direction cause the system to have stable states. The particular set of parameters can be called "critical," and the system may exhibit criticality.

Criticality   The property of a system with fluctuations on all length scales and time scales, typified by power-law exponents in the distributions of measured properties, such that measurements of the system at different length scales or time scales generate qualitatively identical results to each other.

7.8.1   Chaotic Systems and Lack of Predictability

The exquisite sensitivity to initial conditions of a chaotic system leads to its behavior, just like the weather, being unpredictable beyond a small period of time. It may seem that such unpredictability should be avoided in neural circuits that control our behavior and responses to the environment—if we see a red stoplight ahead of us when driving, we would hope that our foot reliably presses the brake pedal. However, initially chaotic systems can be trained to produce reliable responses (i.e., with small changes in connections they are shifted out of the chaotic regime), so long as they are not initially "too chaotic."[27,28] It is likely that, with their vast repertoire of internal dynamics, when operating near the "edge of chaos,"[29] chaotic systems are the most versatile when it comes to training a circuit to produce a novel, arbitrary response.[30]

As a source of unpredictability, moreover, chaos could be beneficial. Just as evolution relies on random variation to produces differences in genetic expression that occasionally might be favorable (and then propagated), so too can random variability in our responses to the environment allow us to explore alternative behaviors. If those alternative behaviors prove to be beneficial, we would want them to be reinforced so that we are more likely to repeat them in similar circumstances. Such reinforcement is an aspect of learning that we will consider in chapter 8, but for here the key point is that random variation of behavior or strategy when faced with a decision or task is an important step in achieving an optimal response. In this context, the random variation is known as exploration.

Unpredictability is also beneficial in the competitive systems explored by game theory.[31] For example, if you are a potential prey, then if your actions are completely predictable ahead of time, a predator can lie in wait and obtain an easy meal. If our actions, at least at the level of selecting alternatives, were based on the state of a chaotic system, we would be much harder to catch.

While the unpredictable behavior of a chaotic neural circuit may therefore be beneficial, such unpredictability need not arise from the sort of chaos that we consider here.

Neural activity in vivo is indeed unpredictable on a trial-by-trial basis, particularly at the low level of individual neural spikes, as would be expected in a chaotic system. However, such unpredictability need not arise from the sort of chaos that we consider in this book—that is, chaos arising from the coupled differential equations reflecting neural firing rates, or the interaction between membrane potentials and conductance of various channels. Rather, it is also possible that the observed variability is accounted for by a combination of environmental fluctuations and microscopic noise sources within the brain.

Microscopic noise arises from the motion of molecules—a motion, which in of itself is chaotic—driven by thermal energy. Microscopic noise may seem a very unlikely source of variability in the behavior of an animal weighing over 100 pounds, given the law of large numbers: random variations at the microscopic level tend to cancel out on a large scale, so rocks do not suddenly roll uphill (as they might if all of the air molecules hitting them ever did so in a concerted fashion). Yet neural systems appear to be fashioned so that there are many "points of no return" or "all-or-none" effects, from vesicle release to a neural action potential to a transition between attractor states of neural activity. These "all-or-none" effects can occur in subsystems with few molecules (e.g., the number of proteins that must be bound near an axon terminal to ensure a vesicle's release) or few components at a larger scale (e.g., the number of release-ready vesicles at a synapse). Each "all-or-none" event amplifies noise at a given scale to a larger scale. All of these amplifications can combine in an avalanche-like manner to allow a tiny fluctuation at the microscopic level to increase to a level that produces a change in behavior.

Deterministic   A system in which the dynamics are fully determined given the differential equations it follows, combines with its initial conditions—i.e., no randomness is added.

In many of our prior simulations, we have included such microscopic noise via random numbers. By contrast, when we discuss chaos in a neural system, we typically ignore such low-level noise and study the random-looking behavior that arises from deterministic coupled differential equations. The subtlety of deterministic chaos is that although at each timestep the values of variables are completely determined by their values in the prior timestep (as in all of the ODEs we have solved without an additional random noise term) the values of the variables many timesteps ahead cannot be determined with any reliability if their current values are not know with infinite precision—a precision that is practically impossible in all scenarios outside of a computer or mathematical equation.

7.8.2 Examples of Chaotic Neural Circuits

We will consider first an example of a chaotic three-variable system. The system produces the simplest type of chaos, based on near-oscillatory behavior of three firing-rate model units. Later in the section we will consider a system with many more firing-rate units whose chaotic behavior is high-dimensional, so that it appears more like the "chaos" of everyday use, containing no obvious patterns.

Chaotic systems with few variables can arise when oscillators become unstable through a parameter shift (e.g., if a modulator alters a conductance, or if a synaptic connection changes) or through an interaction with another oscillator. Remnants of the underlying oscillator are visible in the behavior of these low-dimensional chaotic systems, as is the case of the model neural circuit shown in figure 7.12. Office toys, sometimes called kinetic sculptures, with a driven magnetic pendulum connected to a second arm that is able to rotate in either direction have similar behavior (e.g., those sold as "Mars" or "Jupiter") with the direction of motion of the central arm becoming rapidly unpredictable.

In figure 7.12 we show that a connected circuit of three firing-rate units—in this example, with firing-rate curves that are linear, combined with a threshold and a maximum rate—can be chaotic. The structure of the circuit is such that units 1 and 2 alone produce an oscillator based on self-excitation within unit 1 that receives inhibitory feedback via unit 2. Units 3 and 2 together produce a similar, though slightly less excitable, oscillator. No noise source is added to the system, but the times of switches—when the dominant oscillation of units 1 and 2 becomes temporarily suppressed because unit 3 produces a burst of activity—become rapidly unpredictable. Moreover, an imperceptible change in initial conditions (compare panel C with B) produces a change in activity that diverges more and more from the unperturbed activity as time progresses until the two activity traces (the one with and the one without the perturbation) are completely uncorrelated.

A system with many more units becomes chaotic if units receive balanced input—in firing rate models this corresponds to a mean current of zero as excitation is matched on average by inhibition—while connection strengths between units exceed a threshold.[32] Indeed, interest in chaotic behavior of neural circuits was spurred in the late 1980s, by a paper by Sompolinsky and colleagues,[33] showing this result for an infinite system via mathematical proof.

A

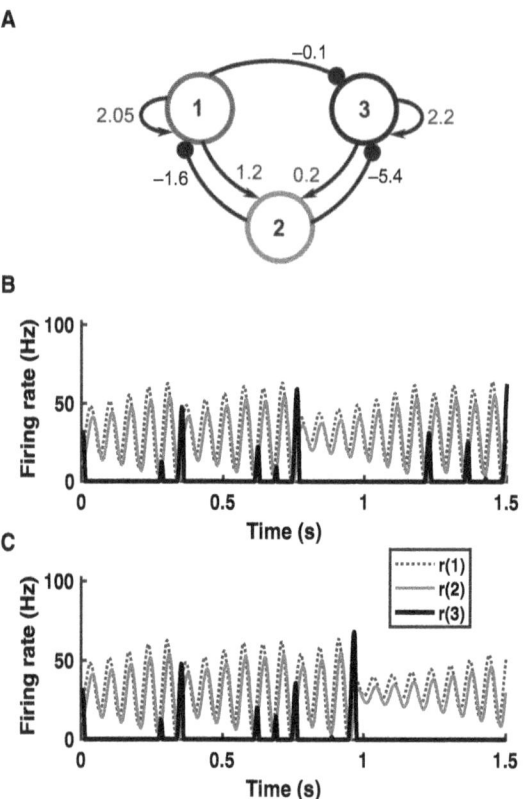

B

C

Figure 7.12
Chaotic behavior of a circuit of three firing-rate model units. (A) Circuit architecture is based on two coupled oscilla-tors, with units 1 and 2 forming one excitatory-inhibitory oscillator and units 3 and 2 forming a separate one. Arrows represent excitatory connections, solid balls represent inhibitory connections. (B) Oscillations of units 1 and 2 grow in amplitude, but are reset by irregular, transient bursts of firing of unit 3. (C) The initial conditions are altered by less than one part in 1,000—the initial rate of unit 1 is shifted from 15 Hz in (B) to 15.01 Hz in C. The system begins to fol-low almost the same trajectory as in B, but gradually diverges. Once a time of 1 second has passed, the system's state in C is completely uncorrelated with its state at that time in B, and the time of bursts of activity of unit 3 in C cannot be predicted from the prior simulation in B. The code producing this figure is available online as chaotic_3units.m.

Lyapunov exponent   A measure of the tendency for trajectories of a system to approach each other (a negative exponent, suggestive of attractor states) or diverge from each other (a positive exponent, suggestive of chaos).

The behavior of an example network with 200 excitatory and 200 inhibitory coupled units is shown in figure 7.13. The sensitivity to initial conditions of this network is so extreme that if the rate of a single unit is altered by only 10^{-14} Hz, a large-scale change in activity arises within a few hundred milliseconds. The lower set of panels (figure 7.13B1–3) depict the effect of such a microscopic deviation (which is physically implausible in terms of spiking activity). If we measure the mean across all 200 excitatory cells of the absolute change in their firing rates caused by the tiny perturbation, we see that the change grows exponentially until the system size is reached (figure 7.14). Such exponential growth is a hallmark of chaos and the rate of growth, known as the Lyapunov exponent, is positive for a chaotic system—whereas for a system of attractors it would be negative, indicating exponential decay to stable fixed points.

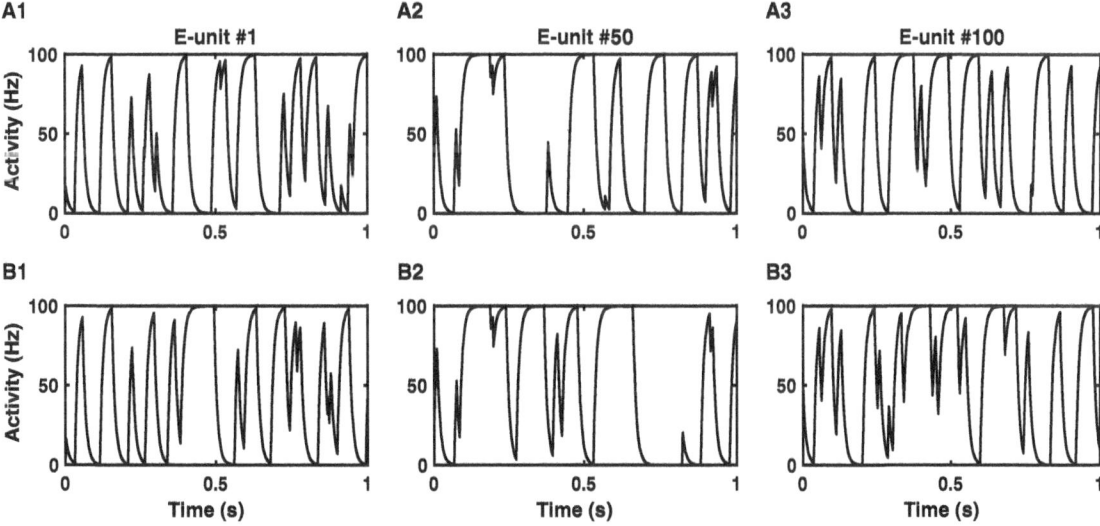

Figure 7.13
Chaotic neural activity in high dimensions. Firing rate as a function of time is shown for three excitatory units selected from a balanced network of 200 excitatory and 200 inhibitory units with sigmoidal firing rate curves and random, sparse connections between units. Unlike figure 7.12, no vestige of regular oscillation remains in the chaotic activity. (A) Unperturbed activity with initial rates evenly spaced in the range from 0 to 100 Hz. (B) Perturbed activity with the initial rate of a single unit shifted from 0 to 10^{-14} Hz, while the initial rates of other units remain identical to those in A. The activity in B initially appears unchanged from that of A, but deviations are visible after a few hundred milliseconds. The timescale of divergence is independent of the particular unit that is perturbed initially. The code producing this figure is available online as `highD_chaos.m`.

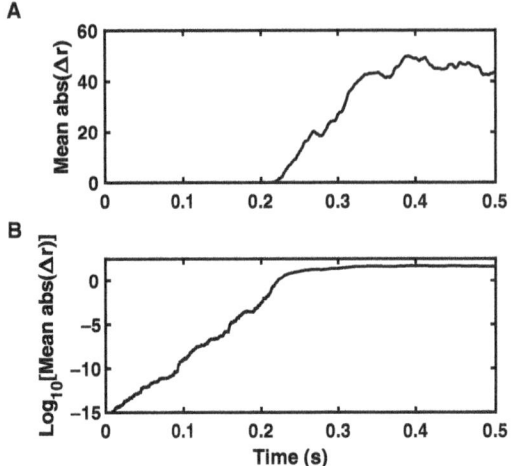

Figure 7.14
The divergence of firing rates grows exponentially following a small perturbation in a chaotic system. (A) The mean of the absolute difference in firing rates of all excitatory units of the network used to produce figure 7.13, is plotted as a function of time. The initial perturbation of 10^{-14} Hz in one unit out of 200 corresponds to a mean change of 5×10^{-16} Hz across all excitatory units. The change in network activity appears negligible until after 200 milliseconds the exponential growth brings the change up to the size of the system. (B) The logarithm (base 10) of the plot in A reveals the exponential period of growth as a straight line. The code producing this figure is available online as `highD_chaos.m`.

7.9 Criticality

Criticality is a phenomenon that has been studied intensely and documented abundantly in the field of physics[34,35] prior to its more recent observation and study in neural circuits.[36,37] The main hallmark of criticality is one of correlations that arise on many scales of length and time. That is, if the system is changed in one place, that change may have a very local effect, or it may propagate to alter the entire system (see figure 7.15).

As suggested by figure 7.15, criticality is closely related to the need for the brain to reside in a state in which some but not all neurons fire. If the mean strength of connections between neurons is too weak, or if neurons are not excitable enough, any initial pulse of activity quickly dissipates in neural circuits (figure 7.15A). Conversely, if connections are too strong and/or neurons are too excitable, any small input leads to runaway excitation with the majority of neurons eventually firing synchronously (figure 7.15C), which is a symptom of epilepsy. Between these two types of activity, just like at the "edge of chaos," the brain can maintain activity in a fraction of neurons that most of the time neither grows excessively nor disappears (figure 7.15B). Typically to achieve such a state, one or more homeostatic processes is required. We will consider such processes in chapter 8.

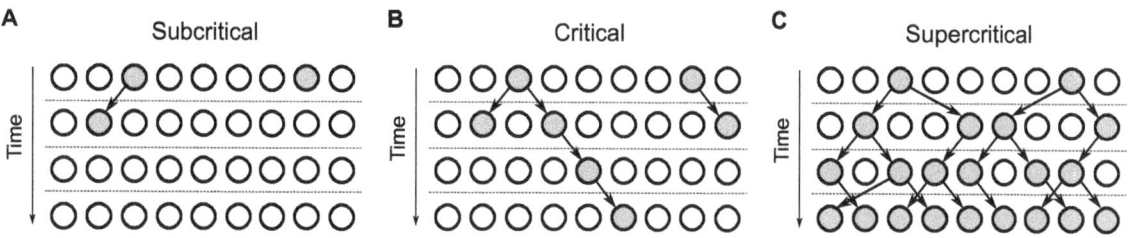

Figure 7.15
Propagation of activity as "avalanches" indicates whether a network is in the critical state. (A) In a subcritical network, the effect of each spike is to produce, on average, less than one extra spike in all other neurons, so avalanches typically decay. Two avalanches are shown, one of size 2 and duration 2 (on the left), the other of size 1 and duration 1 (on the right). (B) In the critical state, each spike causes, on average, exactly one extra spike in another neuron, so avalanches can propagate or decay quickly. Two avalanches are shown, one of size 5 and duration 4 (on the left) and one of size 2 and duration 2 (on the right). (C) In a supercritical network, each spike causes on average more than one extra spike in other neurons, so avalanches commonly propagate and can extend to the full size of the system. The two avalanches initiated in the first timestep merge and quickly reach the system's size.

Homeostasis   A point of balance that is a long-term equilibrium for the system, dependent on homeostatic processes, which produce such a "Goldilocks effect" of being neither "too hot" nor "too cold."

It may be intuitive that a system balanced between dissipation and explosion of activity has computational advantages. In particular, the range and number of distinct responses to inputs is greater in a system at a "sweet spot," where activity can both increase or decrease overall, or simply shift between cells, compared to a system in which most cells respond with high activity or remain silent when receiving input. Critical systems, in which responses can range from the tiny to the macroscopic, provide the largest repertoire of possible responses to initial perturbations. Such a large repertoire of responses is an indication that a system in the critical state, which can be at the "edge of chaos,"[38] has a high capacity for information processing.

7.9.1   Power-Law Distributions
One hallmark of criticality is a power law in the distribution of sizes and durations of avalanches. An avalanche can be defined as a sequence of time bins across which activity exceeds a defined threshold. For example, when a subset of neurons is recorded, the observation of one or more spikes in a series of successive time bins constitutes an avalanche. Alternatively, activity can be measured at a larger scale, either through the local field potential (arising from the shifts in ions averaged across many neurons) or functional magnetic resonance imaging (fMRI, which measures change in blood flow to small volumes called voxels in response to activity of many neurons there). In this case, a threshold level is set and the spatial data is transformed into a set

of entries of 1 or 0, with each entry indicating whether the activity level is above or below the threshold at a given location and in a given time bin.

Avalanche   A measure of neural activity that exceeds a given threshold over multiple successive time bins.

The duration of an avalanche is then the stretch of time—or number of successive time bins—without loss of activity. The size of an avalanche is the number of total spikes (or distinct activations of sites) during the avalanche. The size of an avalanche can be greater than the system's size (the number of recorded sites) because reactivation of cells is possible (as in figure 7.13C). In the critical state, plots of the number of avalanches as a function of their size, the number of avalanches as a function of their duration, and the mean sizes of avalanches as a function of their duration all are best fit by power laws. Given the characteristic property of logarithms, $log(x^\alpha) = \alpha log(x)$, the best way to observe a power-law distribution is to plot the logarithm of the number of occurrences versus the logarithm of the value being measured and look for a straight line.

For example, if the number of avalanches, N, with a given number of spikes, s, follows a power law so that $N = ks^{-\beta}$ then $log(N) = log(k) - \beta log(s)$ so a plot of $log(N)$ against $log(s)$ is a straight line with a gradient of $-\beta$. Examples of such plots are shown in figure 7.16.

Observation of power-law distributions is not sufficient to indicate the presence of criticality. Indeed, many noncritical systems can produce such power laws.[39] Moreover, it can be very difficult to show that a distribution of avalanches follows a power law, because the range of sizes or durations of avalanches should range over three orders of magnitude for other distributions to be ruled out. It takes a long time to obtain sufficient statistics to compare the relative abundances of avalanches across such a wide range, in particular because the large ones are so rare.

7.9.2   Requirements for Criticality

When analyzing the distributions of avalanches, three power laws should be found if the system is critical. The first is the number of avalanches of a given duration as a function of duration (figure 7.16A). The second is the number of avalanches of a given size as a function of size (figure 7.16B). The third is the mean size of an avalanche of a given duration as a function of duration (figure 7.16C). These power laws have exponents that are defined respectively in the following equations:

$$f_d(T) \propto T^{-\alpha} \tag{7.9}$$

$$f_S(S) \propto S^{-\tau} \tag{7.10}$$

$$\langle S \rangle(T) \propto T^{\frac{1}{\sigma \nu z}} \tag{7.11}$$

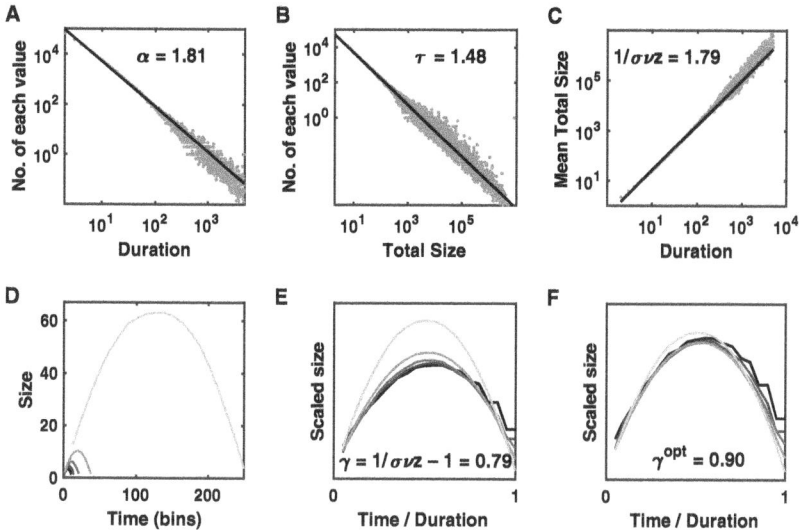

Figure 7.16

Analysis of avalanche data from a simple "birth-death" model of neural activity. (A) A log-log plot of the number of avalanches as a function of avalanche duration. Gradient of the straight line yields α, the exponent of the power law. (B) A log-log plot of the number of avalanches as a function of the total size of the avalanche. Gradient of the straight line yields τ, the exponent of the power law. (C) A log-log plot of the mean size of avalanches as a function of their duration (i.e., x-values of panel B, plotted against x-values of panel A). Gradient of the straight line yields $1/\sigma \nu z$, the exponent of the power law. However, in this case the scaling relation of a critical system is not quite satisfied, as $1/\sigma \nu z = 1.79 \neq (\alpha - 1)(\tau - 1) = 1.69$. (D) The instantaneous size as a function of time is calculated for all avalanches, which are combined according to duration into five distinct groups. (E) The shape of the profile is similar for all avalanches, as seen by replotting the data shown in D, with the timescale normalized by the duration, T, (i.e., $t \rightarrow t/T$) and after the size of each avalanche is scaled by a power-law of the duration (i.e., $s \rightarrow s/T^\gamma$), in which $\gamma = (1/\sigma \nu z) - 1 = 0.79$. (F) The scale-free behavior is seen more clearly by using a fitted value of γ, $\gamma^{opt} = 0.90$, obtained from the mean value of each curve. The code that generated this figure is available online as `avalanche_data.m`.

where T is the duration of an avalanche, f_d is the number of a given duration, s is the size of an avalanche, f_S is the number of a given size, and $\langle S \rangle (T)$ is the mean size of a given duration. The three exponents, τ, α, and $1/\sigma \nu z$ (the last one appears to have three variables within it, based on the theory of critical systems, but is treated as a single exponent) should be measured separately from fits to log plots of the avalanche distributions.

If the system is critical, then its activity should be scale-free, meaning the distribution of trajectories (number of spikes per time bin versus time) of avalanches of different sizes can be multiplicatively scaled to appear identical (see figure 7.16D, E). Such scale-free activity does not mean the system is a scale-free network, which means the connectivity pattern is scale-free—critical networks can have local connectivity. Scale-free activity in avalanches can be tested in two ways.

Scale-free activity Activity that cannot be ascribed a particular scale of length or time over which correlations decay.

Scale-free network A network whose connections are not confined to or dominated by a particular length scale.

Universal curve A curve onto which multiple datasets fall, if each dataset has its x-coordinate and y-coordinate scaled by separate appropriate values that are related to each other.

First, the mean instantaneous size of all avalanches of a given duration can be plotted as a function of time (figure 7.16D). All avalanches begin with an increase of activity from zero and they finish when the activity returns to zero. If the activity is scale-free, the shape of the curve describing how the mean activity increases then falls back toward zero is independent of the total duration of the set of avalanches being averaged. That is to say a universal curve exists, so that if the curves produced from different durations are scaled appropriately, they will lie on top of each other if the system is critical and therefore scale-free.

To see whether avalanches produce a universal curve, the time index should be scaled by (i.e., normalized by or divided by) the total duration of the avalanche. So, we recalculate time for each avalanche as the fraction of the avalanche's duration. The appropriate scaling of instantaneous size as a function of time can be calculated from the relationship between the total sizes of avalanches, $S(T)$, and their duration, T.

To see this, let us assume that the mean peak size of avalanches scales according to the duration of avalanches by a factor of T^γ, where $\gamma > 0$ (longer avalanches have greater peak instantaneous size). If the curve is universal then the mean size at any fraction of the duration of the avalanche also scales by T^γ. Now the total size of the avalanche is the sum of the instantaneous size across all time bins. Even if the instantaneous size did not change, longer avalanches would have greater total size by a factor of T (for example, if each avalanche comprises a series of time bins with one neuron/site active in each time bin, then the total size of each avalanche would be the number of time bins, T). Therefore, if the instantaneous size also increases, in this case by a factor of T^γ, the total size of the avalanche across all time bins is proportional to $T^{\gamma+1}$. However, we have already found the exponent for the total mean size of avalanches, $S(T) \propto T^{\frac{1}{\sigma \upsilon z}}$ so we can identify $1/\sigma \upsilon z$ as equal to $\gamma + 1$, or equivalently obtain γ as:

$$\gamma = \frac{1}{\sigma \upsilon z} - 1. \tag{7.12}$$

This result is proven in the appendix.

In figure 7.16E, we replot figure 7.16D, but after dividing all size values by $T^{\frac{1}{\sigma v z}-1}$ and after dividing all time values by T for each avalanche of duration T. As can be seen, the curves are similar, but with a systematic shift mainly because the total size does not scale exactly as a power law (in figure 7.16C the curve is slightly supralinear).

A second, quicker test (but a less rigorous one) of scale-free behavior relies upon comparison of the three exponents calculated in equations 7.9–7.11. If avalanches are scale-free in the manner described in the previous paragraphs, then the exponents are not independent, but satisfy the equation:

$$\frac{\alpha-1}{\tau-1}=\frac{1}{\sigma v z}, \tag{7.13}$$

which is known as the scaling relation. The result is derived in the appendix. The breakdown of scaling observed in figure 7.16E means that equation 7.13 does not quite hold for the exponents calculated for such a simplified, near-critical system.

7.9.3 A Simplified Avalanche Model with a Subset of the Features of Criticality

The data used to obtain the power-law curves in figure 7.16 are obtained from the birth-death process described below (see the online code `avalanche_data.m`). Each avalanche is initiated with a single active unit. At each timestep thereafter that active unit becomes inactive (a "death") but gives rise to a number of new active units. The number of new active units ("births") is probabilistically chosen from the Poisson distribution with a mean of exactly 1. The process is repeated on the next time step, with a random number of "births" produced by each active unit so that the number of active units in the next time step is the sum of all these "births." That is, the number of active units, N_{i+1}, in timestep $i+1$, depends on the number of active units, N_i, in time step i, as:

$$N_{i+1} = \sum_{n=1}^{N_i} B_n \tag{7.14}$$

where each value, B_n, (the random number of newly activated units or "births" produced by each previously active unit) is the realized value of a Poisson process. That is, each B_n can be any nonnegative integer, chosen with probability $P(B_n)$ given by:

$$P(B_n)=\frac{e^{-1}}{B_n!}. \tag{7.15}$$

If a time bin, i, is reached, where all the randomly chosen values of B_n are zero, so that $N_{i+1}=0$, then the duration of the avalanche is recorded as $T=i$ and the total size is recorded as $S = \sum_{i=1}^{T} N_i$.

The avalanches produced by this model do exhibit power-law (or nearly power-law) statistics (figure 7.16A–C). Moreover, the mean shapes of the avalanches are approximately universal (figure 7.16F), meaning the system is close to criticality. The expected scaling relationships are not quite achieved, mainly because of deviations at the level of very small avalanches, where the discrete time bins and discrete numbers of active units limit scale-free behavior. Since the smallest avalanches are the most numerous, they can play a dominant role when fitting the exponents.

More generally, scale-free behavior is limited for small avalanches by the discrete counting of neurons, and for large avalanches by the finite number of neurons being recorded. Given these restrictions, observation of scale-free behavior in neural systems is rarely as clear as that depicted by data such as these, obtained from computer simulations (albeit time-consuming ones) of a simple model.

7.10 Tutorial 7.2: Diverse Dynamical Systems from Similar Circuit Architectures

Neuroscience goals: Learn how near-identical circuits can produce very different types of activity; learn how highly simplified neural circuits can nevertheless produce complex patterns of activity.

Computational goals: Gain more practice at using matrices and arrays to simulate coupled units; understand the idea of using a single code or function to simulate multiple types of circuit by changing particular parameters.

For each question you will simulate a two-unit or three-unit circuit (i.e., $N_{units} = 2$ or 3), with connection matrices defined as

$$\underline{\underline{W}} = \begin{pmatrix} W_{11} & W_{12} \\ W_{21} & W_{22} \end{pmatrix}$$

or

$$\underline{\underline{W}} = \begin{pmatrix} W_{11} & W_{12} & W_{13} \\ W_{21} & W_{22} & W_{23} \\ W_{31} & W_{32} & W_{33} \end{pmatrix},$$

respectively, where W_{ji} is the connection strength from unit j to unit i. Each unit, i, responds to a total input current, I_i, which is given by:

$$I_i = I_i^{(app)} + \sum_{j=1}^{N_{units}} W_{ji} r_j,$$

where $I_i^{(app)}$ is an external, applied current. Each unit responds with a threshold-linear firing-rate curve, so that

$$\tau \frac{dr_i}{dt} = -r_i + I_i - \Theta_i,$$

with time constant, $\tau = 10$ ms, and where Θ_i is the threshold for unit i. If the threshold is negative, it indicates the level of spontaneous activity in the unit in the absence of input current.

To ensure rates are positive and do not exceed a maximum, we include the further conditions: $0 \le r_i \le r_{max}$, where $r_{max} = 100$ Hz.

The values of connection strengths and thresholds will vary across the nine circuits you will simulate. For each circuit, determine what sort of dynamical system is produced. For example, find the number of distinct states within the circuit or whether the system is chaotic or possesses a heteroclinic sequence. You should use pulses of applied current or changes in the initial conditions to aid your analysis.

Try to write the code with only a single loop for the time simulation written in the code and a separate loop to run through the nine questions with different parameters. You can use the commands `switch` and `case` in MATLAB to update the parameters for each question.

1. $\underline{\underline{W}} = \begin{pmatrix} 0.6 & 1 \\ -0.2 & 0 \end{pmatrix}$, $\Theta_1 = -5$, $\Theta_2 = -10$.

2. $\underline{\underline{W}} = \begin{pmatrix} 1.2 & -0.3 \\ -0.2 & 1.1 \end{pmatrix}$, $\Theta_1 = 10$, $\Theta_2 = 5$.

3. $\underline{\underline{W}} = \begin{pmatrix} 2.5 & 2 \\ -3.0 & -2 \end{pmatrix}$, $\Theta_1 = -10$, $\Theta_2 = 0$

4. $\underline{\underline{W}} = \begin{pmatrix} 0.8 & -0.2 \\ -0.4 & 0.6 \end{pmatrix}$, $\Theta_1 = -10$, $\Theta_2 = -10$.

5. $\underline{\underline{W}} = \begin{pmatrix} 2 & 1 \\ -1.5 & 0 \end{pmatrix}$, $\Theta_1 = 0$, $\Theta_2 = 20$.

6. $\underline{\underline{W}} = \begin{pmatrix} 1.5 & 0 & 1 \\ 0 & 2 & 1 \\ -2.5 & -3 & -1 \end{pmatrix}$, $\Theta_1 = -10$, $\Theta_2 = -5$, $\Theta_3 = 5$.

7. $\underline{\underline{W}} = \begin{pmatrix} 2.2 & -0.5 & 0.9 \\ -0.7 & 2 & 1.2 \\ -1.6 & -1.2 & 0 \end{pmatrix}$, $\Theta_1 = -18$, $\Theta_2 = -15$, $\Theta_3 = 0$.

8. $W = \begin{pmatrix} 2.05 & -0.2 & 1.2 \\ -0.05 & 2.1 & 0.5 \\ -1.6 & -4 & 0 \end{pmatrix}$, $\Theta_1 = -10$, $\Theta_2 = -20$, $\Theta_3 = 10$.

9. $W = \begin{pmatrix} 0.98 & -0.015 & -0.01 \\ 0 & 0.99 & -0.02 \\ -0.02 & 0.005 & 1.01 \end{pmatrix}$, $\Theta_1 = -2$, $\Theta_2 = -1$, $\Theta_3 = -1$.

Questions for Chapter 7

1. Which combination of changes in parameters—either increasing or decreasing the feedback connection strength, while either increasing or decreasing the threshold of a unit—can have the effect of enhancing the overall stability of a bistable unit with positive feedback?

2. Two nullclines cross at five different points. What is the greatest number of stable fixed points the corresponding system can possess?

3. In an inhibition-stabilized circuit, the inhibitory neurons gain some extra excitatory input to reach a new stable state. In what direction do the following change (if they change at all)?

   a. The firing rate of inhibitory cells.

   b. The firing rate of excitatory cells.

   c. The total inhibitory input to excitatory cells.

   d. The total excitatory input to excitatory cells.

   e. The total inhibitory input to inhibitory cells.

   f. The total excitatory input to inhibitory cells.

4. Suggest one way in which chaotic activity would be beneficial for the brain and one way in which it would be detrimental.

7.11   Appendix: Proof of the Scaling Relationship for Avalanche Sizes

1. Proof of equation 7.12, that $1/\sigma v z = \gamma + 1$.

   If avalanches are scale-free, then the instantaneous mean size, s, as a function of time, t, for avalanches of total duration, T, can be written as:

$$s(t,T) = s_0\left(\frac{t}{T}T_0, T_0\right) \cdot \left(\frac{T}{T_0}\right)^{\gamma} \tag{7.16}$$

where s_0 is a universal function, describing the mean shape of avalanches, and we set the "standard" avalanche to have a duration of T_0. The change in variable of the instantaneous time t, into $(t/T)T_0$, is done so that when t is a given fraction of T for any example avalanche, the equivalent time on the standard avalanche should be the same fraction of T_0, that is, $t/T = ((t/T)T_0)/T_0$.

The mean total size of avalanches of length T is then found by integrating over all instantaneous sizes, using equation 7.16:

$$
\begin{aligned}
S(T) &= \int_0^T s(t,T)\,dt \\
&= \left(\frac{T}{T_0}\right)^\gamma \int_0^{T_0} s_0(t',T_0)\,dt'\left(\frac{T}{T_0}\right) \\
&= \int_0^T s_0\left(\frac{t}{T}T_0,T_0\right)\cdot\left(\frac{T}{T_0}\right)^\gamma dt \\
&= \left(\frac{T}{T_0}\right)^{\gamma+1} S(T_0).
\end{aligned}
\tag{7.17}
$$

where in the second line we substituted $t' = \dfrac{t}{T}T_0$, so $dt = dt'\left(\dfrac{T}{T_0}\right)$.

Therefore, the total size of avalanches of duration T would scale as $T^{\gamma+1}$, so we can make the identity $1/\sigma vz = \gamma + 1$ (equation 7.12), since $1/\sigma vz$ is already defined in equation 7.11 as the exponent in the power-law relationship between mean total size and duration of avalanches.

2. Proof of equation 7.13, that $1/\sigma vz = (\alpha-1)/(\tau-1)$.

We will calculate the distribution of total sizes of avalanches, assuming three properties that are true for critical systems:

Property A. The distribution of avalanche durations is a power law with exponent $-\alpha$;

Property B. The probability of an avalanche of a given size, if we know the duration, is scale-free (so only depends on the mean size of that duration);

Property C. The mean size of a given duration is also a power law, with exponent $1/\sigma vz$.

We can begin by writing the relationship between distributions of total size, $P(S)$, and duration, $P(T)$, as:

$$
P(S) = \int P(T)P(S|T)\,dT,
\tag{7.18}
$$

where $P(S|T)$ is the probability of the total size being S when the duration is T.

The key step is to notice then that if the shape of the distribution of sizes as a function of duration is scale-free (property B), with a mean size that increases with duration as $T^{\frac{1}{\sigma vz}}$ (property C), then the spread of sizes of the distribution scales by the same factor as its mean, $T^{\frac{1}{\sigma vz}}$. Given the area of the distribution must be fixed at 1 (it is a probability distribution, so the area is the total probability of an avalanche having some value of size) then its height scales down by the increase in width, which introduces a factor of $T^{-\frac{1}{\sigma vz}}$. We can then write:

$$P(S \mid T) = T^{-\frac{1}{\sigma vz}} f\left(\frac{S}{T^{\frac{1}{\sigma vz}}}\right),$$

(7.19)

where f is an arbitrary function, which we do not need to know to find the scaling relation (just as we do not need to know the shape of $P(S \mid T)$).

We now use $P(T) = AT^{-\alpha}$ (property A) and substitute equation 7.19 into equation 7.18 to find

$$P(S) = \int AT^{-\alpha} T^{-\frac{1}{\sigma vz}} f\left(\frac{S}{T^{\frac{1}{\sigma vz}}}\right) dT.$$

(7.20)

We are only concerned with how this result depends on size, S, so we do not need to evaluate the integral exactly (which we cannot do anyway, while the function f is undefined). We proceed by making the substitution $x = S \Big/ \left(T^{\frac{1}{\sigma vz}}\right) = ST^{-\frac{1}{\sigma vz}}$, so $T = S^{\sigma vz} x^{-\sigma vz}$ and $dT = -\sigma vz S^{\sigma vz} x^{-\sigma vz - 1}$, which leads to a rewriting of equation 7.20 as:

$$P(S) = A\sigma vz S^{\sigma vz} S^{-1} S^{-\alpha \cdot \sigma vz} \int f(x) dx.$$

(7.21)

The term with the integral now produces a constant that is independent of S or T, and which we do not need to know. The preceding terms show us that the probability distribution is a power law in S, with an exponent of $\sigma vz - 1 - \alpha \cdot \sigma vz$. Comparing this form with $P(S) \propto S^{-\tau}$ allows us to equate the exponents, so that:

$$-\tau = \sigma vz - 1 - \alpha \cdot \sigma vz,$$

(7.22)

which can be rearranged to show the relationship (equation 7.13) between the scaling exponents that we aimed to prove:

$$\frac{1}{\sigma vz} = \frac{\alpha - 1}{\tau - 1}.$$

(7.23)

8

Learning and Synaptic Plasticity

Our brains enable us to produce models of our surrounding environment and to predict the effect of our actions on that environment. In particular, they allow us to learn from experience so that, in an ever-changing environment, we can be more successful at our endeavors than any organism whose behaviors and responses are genetically preprogrammed before birth. The initial building of these models and the lifelong process of updating them in response to new information rely on changes in neural circuits within our brains. That is, synaptic connections between neurons change in their pattern and in their strengths through development and through lifelong learning experiences. Such changes in connections are known as synaptic plasticity.

8.1 Hebbian Plasticity

In his 1949 book *The Organization of Behavior*,[1] Donald Hebb wrote the following prescient sentence:

When an axon of cell A is near enough to excite a cell B and repeatedly or persistently takes part in firing it, some growth process or metabolic change takes place in one of both cells such that A's efficiency, as one of the cells firing B, is increased.

The sentence, known as Hebb's postulate or Hebb's rule, was suggested as a method for producing the circuitry needed for persistent activity via a reverberatory loop of excitatory feedback (chapter 6, figures 6.3–6.5). While Hebb's rule contains a causal direction, so that cell A should fire before cell B for the connection strength to change, in this chapter the term "Hebbian plasticity" will be used if coactivity of two excitatory neurons leads to an increase in connection strength in either direction.

Hebbian plasticity A change in synaptic strength produced by a positive correlation, which should be causal, between presynaptic spikes and postsynaptic spikes.

Experimental support for Hebbian plasticity has accrued since 1966,[2] when long-term potentiation between excitatory cells was first discovered by Lømo and then later characterized by Bliss and Lømo[3] using hippocampal slices from rabbits. Later, in 1997,[4,5] the observation of spike-timing dependent plasticity provided evidence for the causal aspect of Hebb's rule. The biophysical mechanisms underlying Hebbian plasticity, which manifests as an increase in the strength of synaptic connection between the two neurons, have been gradually revealed over time thanks to numerous painstaking studies.

The particular mechanisms for Hebbian plasticity vary across cell types and brain areas, but all depend on a transient increase in calcium concentration within the postsynaptic cell in the vicinity of the synapse.[6–9] Such an increase can arise via influx through calcium channels[10] (section 4.5), NMDA receptors[11] (section 5.1.2), or efflux from internal stores.[12]

The NMDA mechanism is most easily understood.[13] A presynaptic spike causes glutamate to be released so that it binds to the postsynaptic receptors where it can remain for many tens of milliseconds. If, during that period, the postsynaptic membrane potential rises to a high level—as it could if the postsynaptic cell fires a spike or burst of spikes—then the NMDA channel opens and admits calcium ions. Importantly, this mechanism is directional—if the postsynaptic cell is active before the presynaptic cell, the high membrane potential—which returns to baseline after a few milliseconds—never coincides with the glutamate binding, so the NMDA channels do not open. We simulate this process in section 8.3.6.

Long-term potentiation (LTP)   An increase in synaptic strength that persists for longer than a few tens of minutes.

Long-term depression (LTD)   A decrease in synaptic strength that persists for longer than a few tens of minutes.

It is important to note that Hebbian plasticity on its own is destabilizing for two reasons. First, Hebb's postulate only provides a rule for an increase in the strength of synapses (as in figure 8.1A)—if the only possibility were for synapses to strengthen, they would all strengthen by chance eventually, leading to runaway excitation in the brain. So, an opposing mechanism is required to allow synapses to reduce their strength. Indeed, in parallel with the work on long-term potentiation, rules and mechanisms for long-term depression have also been elucidated.[14,15]

Hebb, of course, was aware of such a need. In order for his hypothesized reverberatory loop of excitation to be contained within a specific subset of cells, it was important that those cells inhibit, rather than excite other cells outside of the subset. Therefore, the active cells should

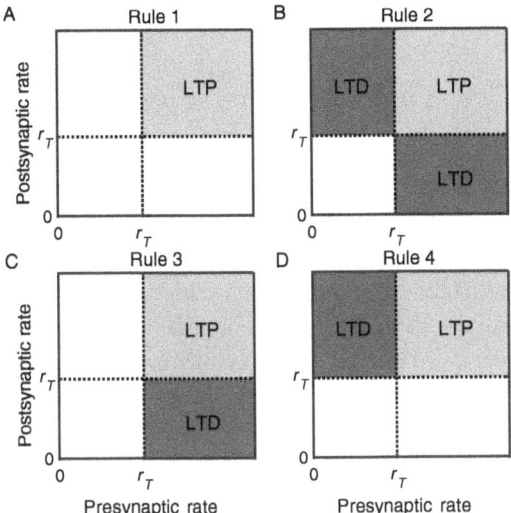

Figure 8.1
Alternative forms of Hebbian plasticity. (A) Potentiation only (symmetric). If both the presynaptic and the postsynaptic cells' activities are above-threshold, then connections between the two cells strengthen. Otherwise no change in synaptic weight occurs. To prevent runaway excitation a global normalization is needed. (Rule 1, tutorial 8.1) (B) Potentiation and depression (symmetric). Potentiation occurs as in A, but if one neuron's activity is above-threshold while the other neuron's activity is below-threshold, then any connection between the two cells weakens. If both neurons have low activity there is no change (rule 2, tutorial 8.1). (C, D) Potentiation and depression (asymmetric). Potentiation occurs as in A. Depression occurs only if the presynaptic cell's activity is above-threshold while the postsynaptic cell's activity is below-threshold in C, while depression occurs only if the postsynaptic cell's activity is above-threshold while the presynaptic cell's activity is below threshold in D. This figure is produced by the online code `plasticity_four_rules.m`.

weaken their excitatory connections to less active excitatory cells, whose occasional, spontaneous spikes would otherwise cause synaptic strengthening. The observed data on long-term depression (not to be confused with short-term depression of section 5.3.1) follow this rule qualitatively—particular protocols for inducing long-term depression include weak depolarization of the postsynaptic cell (to mimic its being only weakly excited) while causing the presynaptic cell to fire spikes,[16] or low frequency stimulation of presynaptic axons.[17]

Even with such an opposing form of synaptic plasticity, the two mechanisms together are still destabilizing because they are each a form of positive feedback: The change produced enhances the likelihood of further change. That is, potentiated synapses are likely to cause even more activity in the postsynaptic cell following presynaptic activity, resulting in further potentiation. Similarly, depression of excitatory synapses increases the likelihood of low activity in the postsynaptic cell at the time of presynaptic activity, resulting in increased likelihood of further depression. The end point of these processes alone, is for a subset of synapses to ever increase in strength up to whatever their biologically constrained maximum might be, while a complementary subset of synapses ever decreases in strength until the synapses are eliminated. Later in this chapter, in section 8.6, we consider homeostatic processes that counter such synaptic extremism and allow for the observed broad ranges of synaptic strengths.

8.1.1 Modeling Hebbian Plasticity

We will first simulate Hebbian plasticity using a firing rate model—we will explore simulations with spiking neurons in section 8.3, when we investigate the consequences of spike-timing dependent plasticity. In the simpler method here, we ignore the causal aspect of Hebb's rule (i.e., the order of firing does not matter) and determine the change in connection strength between two units via the covariation of their firing rates. In particular, if the two units have high activity at the same time then we strengthen the connection between them. This can be written using a binary function for the rate of change of connection strength, so that plasticity is either on or off, as:

$$\tau \frac{dW_{ij}}{dt} = \Theta(r_i - r_T) \cdot \Theta(r_j - r_T) \tag{8.1}$$

where $\Theta(x)$ is the Heaviside function, defined as $\Theta(x) = 1$ if $x > 0$ and $\Theta(x) = 0$ if $x < 0$; W_{ij} is the strength of connection from unit i to unit j; τ sets the rate of change; and r_T is the threshold rate for inducing Hebbian plasticity (see figure 8.1A).

We shall investigate the impact of such a rule, in combination with a rule for reducing synaptic strength in tutorial 8.1. Other rules are possible, in particular the rate of change of synaptic strength may be a continuous, graded function of firing rates rather than a binary function. A rule that incorporates such rate-dependence and includes depression as well as potentiation is:

$$\tau \frac{dW_{ij}}{dt} = r_i r_j (r_j - r_T). \tag{8.2}$$

In equation 8.2, r_i, the rate of the presynaptic cell, simply scales the rate of plasticity. However, r_j, the rate of the postsynaptic cell, both scales the rate of plasticity and alters the sign of plasticity, since the rule yields synaptic depression if $r_j < r_T$ and synaptic potentiation if $r_j > r_T$ (figure 8.2). This formula can also be derived from plasticity rules in spiking neurons (see equation 8.7 in section 8.3.4, and the corresponding derivation in appendix B of this chapter).

Pattern completion   A network produces a correct pattern based on prior examples when provided with a partial or corrupted input of that pattern, just as yu cn dcphr thse wrds.

Pattern separation   A network produces different responses for different patterns, so the particular pattern that is completed depends on which one the input most closely resembles.

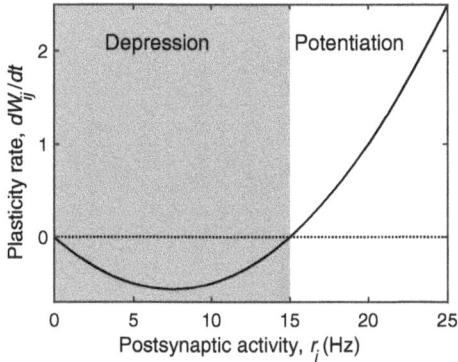

Figure 8.2
Plasticity rule with quadratic dependence on postsynaptic firing rate. The crossover from synaptic depression to potentiation occurs when the postsynaptic firing rate increases above the threshold rate, which is 15 Hz in this example. In some models the threshold rate can vary slowly over time, increasing if the postsynaptic cell has a long-term high firing rate and decreasing if the postsynaptic cell has a long-term low firing rate. Such metaplasticity, as proposed in the BCM model,[48] was an early suggestion of homeostasis (see section 8.6) to counter the instabilities of Hebbian plasticity. This figure is produced by the online code `quadratic_plasticity_figure.m`.

8.2   Tutorial 8.1: Pattern Completion and Pattern Separation via Hebbian Learning

Neuroscience goals: Produce a powerful model of associative learning in the brain; gain awareness of how different synaptic plasticity parameters impact such learning.

Computational goals: Convert matrices to vectors and back; implement matrix multiplication to calculate the results of multiple integrals.

In this tutorial, you will train a circuit undergoing Hebbian synaptic plasticity with many approximate examples of four distinct input patterns, then test whether the circuit reproduces the desired pattern when presented with new approximations to each pattern. The system is a type of autoassociative network,[18] the best-known of which is the Hopfield network.[19] These networks have the property of pattern completion (retrieve a complete pattern given partial input) and pattern separation (produce distinct responses to similar inputs according to the learned pattern most similar to each input). The learned patterns become attractor states of the network (see figure 8.3).

1a.   Define four matrices of size 17×17 (you can call them `pattern1`, `pattern2`, `pattern3`, and `pattern4`) whose entries are 1s and 0s arranged in a distinct pattern that you can recognize (in the lower panel of figure 8.3, the light entries correspond to active neurons with an entry of 1 in the pattern and the dark entries correspond to inactive neurons with an entry of 0 in the pattern). You may choose your own distinct patterns (with neurons defined on a 17 $\times$ 17 grid) or use the patterns in the lower row of figure 8.3. Do not use patterns with a lot of overlap between any of them.

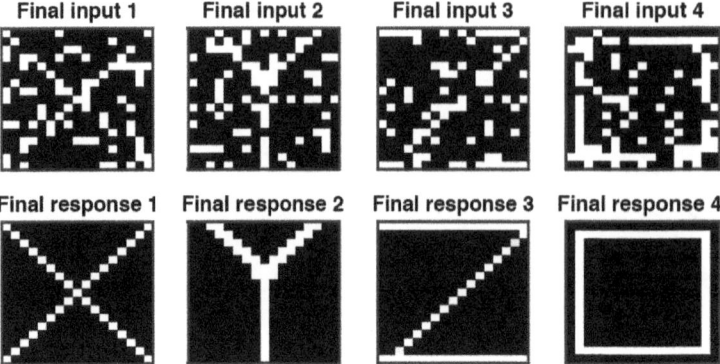

Figure 8.3
Pattern completion and pattern separation of corrupted exemplars following Hebbian learning. Each small square in the 17×17 array corresponds to a unit. The coordinated activity of a set of units can be entirely produced by input (top row) or, after learning, can represent a particular memory stored in the connections between units (bottom row). In this example, inputs generate an activity pattern with approximately 80 percent overlap with one of the "ideal" templates (20 percent of units have their input switched sign, top row). After learning, the structure of feedback connections within the network causes the units to retrieve the activity corresponding to the closest "ideal" template (bottom row). That is, the network has learned an "ideal" template, even though the exact template was never presented during Hebbian learning—the corrupted exemplars were sufficient. See tutorial 8.1. This figure was produced by the online code `Tutorial_8_1.m`.

1b. Set up a loop of 400 trials, and on each trial:

(i)   Select at random one of your four patterns to be the input. (The MATLAB function `randi` is useful here. Otherwise, you can take a random number between 0 and 1, multiply it by 4, add 0.5 to your result and round to the nearest integer.)

(ii) Randomly, with probability of 0.1 (i.e., again use the `rand` function), flip entries from 0 to 1 or 1 to 0, to generate a corrupted pattern, `input_pattern`, with on average 90 percent overlap with the original one.

(iii) Assign each element of `input_pattern` to produce the applied current to a particular firing-rate unit (so you will have $N_{units} = 17^2 = 289$ units). Simulate for 1 s the set of firing-rate units whose activity will be stored in an array of size $N_T \times N_{units}$, where N_T is the number of time points, using:

$$\tau_r \frac{dr_i}{dt} = -r_i + \frac{r^{max}}{1 + \exp[-(I_i - I_{th})/\Delta_I]}$$

$$I_i = \sum_{j=1}^{N_{units}} r_j W_{ji} + I_i^{app}(t)$$

(see equations 6.1–6.3 for a review of the notation).

Use the parameters $\tau_r = 10$ ms, $r^{max} = 50$ Hz, $I_{th} = 10$, $\Delta_I = 1$. Initially, the connection matrix between the units, W_{ji}, provides uniform inhibition with $W_{ji} = -0.3 / N_{units}$ identical across all connections. All firing rates should be initialized as zero. The applied current, $I_i^{app}(t)$, should only be present for the first 0.5 s of the trial. Its value is 50 for those units, i, receiving input (i.e., those units whose entry in input_pattern is 1) and its value is 0 otherwise.

Note that in MATLAB, although the trial-dependent variable input_pattern is a matrix, typing "input_pattern(:)" produces the required vector of length N_{units}.

(iv) At the end of each trial, alter the connection strengths according to rule 4 (figure 8.1D) which can be written as:

$$\Delta W_{ji} = \varepsilon_+ \int_{t=0}^{t=1} \Theta(r_j - r_T) \cdot \Theta(r_i - r_T) dt - \varepsilon_- \int_{t=0}^{t=1} \Theta(r_T - r_j) \cdot \Theta(r_i - r_T) dt$$

where the time integrals are across the trial of 1 s, $r_T = 25$ Hz, $\varepsilon_+ = 0.1 / N_{units}$ and $\varepsilon_- = \varepsilon_+ / 4$. In this equation, the rate of the presynaptic cell is r_j and the rate of the postsynaptic cell is r_i. The Heaviside functions, for example $\Theta(r_i - r_T)$, give 1 if the term in parenthesis is greater than zero (e.g., if $r_i > r_T$ in the example) and otherwise 0. The parameters ε_+ and ε_- determine the rate of plasticity—if too slow, more trials are needed, if too fast, random errors can be imprinted in the circuit.

MATLAB allows these integrals to be evaluated rapidly via matrix multiplication. For example, the first of the two terms can be evaluated as:

```
epsilon_plus*(rate' > r_T)*(rate > r_T)*dt
```

if firing rate is stored as a matrix, rate, with the first index (row number) corresponding to the time point and the second index (column number) corresponding to the neural unit number. dt is the timestep used in the code (so 1/dt is the number of time points), r_T the threshold, r_T, and epsilon_plus is the simulation's term for the parameter ε_+.

(v) Ensure no connection strengths pass beyond a maximum value of $W_{max} = 8 / N_{units}$, or a minimum value of $W_{min} = -8 / N_{units}$, by clamping them to these values if necessary.

1c. Once the 400 trials with plasticity are completed, simulate four new trials, one with each input pattern but now with 20 percent of entries to the input matrix flipped from 0 to 1 or from 1 to 0. Plot, using image the set of firing rates at the end of the simulation (i.e., the last time point), being careful to assign the rate of the unit labeled i (where i runs from 1 to $N_{units} = 289$) to its correct location on the 17×17 grid. For example, you can use the reshape command in MATLAB to convert a vector to a matrix.

Challenge

2. Repeat the learning paradigm using one or more of the other plasticity rules (rules 1–3 of figure 8.1) and/or with altered values for ε_+ and ε_-. Assess whether some rules are better than others at achieving perfect recognition of corrupted patterns by using increasingly corrupted patterns during learning and testing (only allowing 400 trials of learning).

Note: When you use rule 1, in order to prevent all connections from becoming maximally strong, you should (after updating the connections) renormalize them to maintain a fixed mean incoming connection strength to each unit. For example, a line of code like:

```
W = W−ones(Nunits,1)*mean(W)+ W0_mean;
```

where W0_mean is the initial fixed connection strength ($-0.3 / N_{units}$), ensures the mean incoming synaptic inputs to each unit never change.

8.3   Spike-Timing Dependent Plasticity (STDP)

Spike-timing dependent plasticity (STDP)   Changes in connection strength that depend on the relative timing of presynaptic and postsynaptic spikes, not just their rates.

While Hebb's postulate contains an important causal direction, inasmuch as cell A can "take part in" firing cell B without cell B's taking part in firing cell A, the rules for strengthening of connections described so far in this chapter lack such causality. Indeed, they may be summarized by the distorted shortening of Hebb's postulate, "cells that fire together wire together." Yet, shortly before the turn of the millennium, the first experimental data were published demonstrating the importance of the temporal order of neural activation.[4,5]

In particular, in protocols with electrodes implanted in both members of a pair of connected excitatory neurons, a connection increased in strength when the postsynaptic neuron produced an action potential within a small time window (on the order of tens of ms) after the action potential of the presynaptic neuron. Conversely, when the ordering of spikes was reversed the synaptic connection strength decreased. This process means that in a bidirectionally connected pair (a common motif in cortical circuits) whenever one connection strengthens, the reciprocal connection weakens, because the relative order of presynaptic and postsynaptic spikes is reversed for the reciprocal connection.

Some caveats were noted for these data at the time and over later years[20]—a minimum rate of spike pairs was required,[21] combined with a minimum number of pairings to produce significant change; after a few tens of pairings no further change occurred;[22] the variability in the amount of change across different synapses undergoing the same protocol was huge.[5] Yet to produce simple models of spike-timing dependent plasticity (STDP) whose consequences can be easily explored

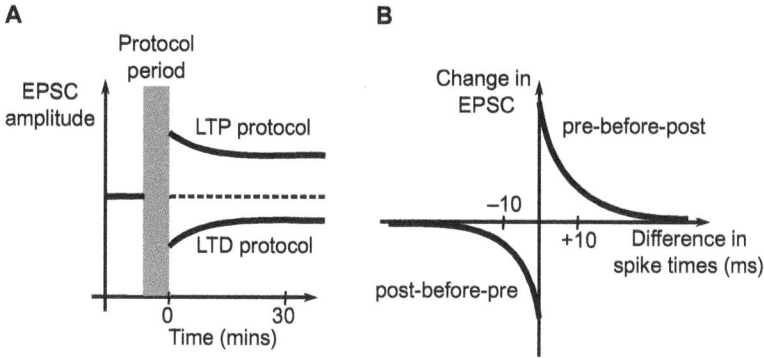

Figure 8.4
Spike-timing dependent plasticity. (A) The amplitude of excitatory postsynaptic current (EPSC) can change following a pairing protocol. The change in amplitude can be sustained for tens of minutes or many hours following the protocol (the experiments are in vitro, so there is a time limit for sustaining healthy cells). The protocols comprise several minutes of combined presynaptic spikes and postsynaptic depolarization or hyperpolarization or spiking, and they last several minutes. Depending on the protocol long-term potentiation (LTP, an increase in evoked EPSC) or long-term depression (LTD, a decrease in evoked EPSC) can be produced. (B) In a protocol using 10 Hz trains of presynaptic spikes (a spike every 100 ms) the change in synaptic strength between two excitatory cells is positive if the postsynaptic cell is caused to spike within 20 ms following each presynaptic spike (pre-before-post), and is negative if it is caused to spike in a similar time window preceding each presynaptic spike (post-before-pre). The dependence on the time difference between the two spikes can be fitted as two exponentials. Such spike-timing dependent plasticity demonstrates a causal requirement for a change in connection, a requirement that was present in Hebb's original postulate.[1]

and understood, we ignore these caveats in this book. We therefore define STDP as a change in synaptic strength that is produced by each pair of presynaptic and postsynaptic spikes, and that depends only on the time difference between the two spikes (figure 8.4B).

Even with this simplification of the process, alternative methods still remain for simulating the total change in synaptic strength arising from the trains of action potentials of two connected cells.

First is the choice of whether to include all spikes from one neuron as the partner to form a spike pair with each and every spike of the other neuron, or to include only "nearest-neighbor" spikes. At low firing rates this choice makes little difference, but once either of the neurons produces a significant number of spikes separated by intervals smaller than the time window of STDP, then the choice to omit many spike pairs can even shift the net result from LTD to LTP, or vice versa (a result derived in appendix A and shown in figure 8.13).

Since the original data used to produce figures such as figure 8.4B were acquired from the sum of all successive synaptic changes at a fixed rate of pairings, the original data do not constrain which rule should be used. That is, so long as two rules can correctly model the original data, that data cannot favor one rule over the other even when they produce opposite results for a different protocol.

Second is the choice of whether to use a batch method (simulate the spike trains for a trial and only update the synaptic strengths after the simulated trial) or a continuous method (update the synaptic strengths at the time of each spike within a simulated trial). The positive feedback that is inherent in standard rules for STDP—a change in connection leads to a change in spike patterns that makes the same change in connection more likely to happen again—means that the continuous method of updating, which more rapidly employs the positive feedback, leads to greater overall changes in synaptic strength than the batch method. Again, since synaptic changes are typically measured after many pairings whose individual effects can accumulate, the original data do not provide a reason to support one method over another.

Batch updating method A protocol with which all synaptic strengths are updated at the end of discrete blocks of simulation, or trials, with the set of synaptic strengths kept constant during each simulated block of time.

Continuous updating method A protocol for updating synaptic strengths within a simulation, typically at the time of each spike, but possibly on every timestep.

8.3.1 Model of STDP

The standard model of STDP requires an update of the connection strength, W_{ij}, from neuron i to neuron j whenever either neuron i or neuron j produces a spike. At the time, t_i, of a presynaptic spike the update, ΔW_{ij}, is negative, because any postsynaptic spikes used to update the synapse would be at earlier times, t_j, so the spike order is the postsynaptic cell's spike before the presynaptic cell's spike (post-before-pre):

$$\Delta W_{ij} = -A_- \exp\left[\frac{-(t_i - t_j)}{\tau_-}\right] \quad (\text{with } t_i > t_j). \tag{8.3}$$

Conversely, at the time, t_j, of every postsynaptic spike the update is positive, because any presynaptic spikes have arrived at earlier times, so the spike order is the presynaptic cell's spike before the postsynaptic cell's spike (pre-before-post):

$$\Delta W_{ij} = A_+ \exp\left[\frac{-(t_j - t_i)}{\tau_+}\right] \quad (\text{with } t_j > t_i). \tag{8.4}$$

By specifying these changes at the time of each spike we are able to implement the continuous update rule. It is computationally straightforward at the time of each spike of each cell to update weights based on only the most recent spike of the connected cell, or to store all prior spike times of connected cells and add together the contributions of all those prior spikes to changes in synaptic strength. In tutorial 8.2, we will just assume the most recent spike is used in each update.

When implementing STDP via the batch method, it is straightforward to simply sum the contributions of all presynaptic spikes of a given presynaptic cell for a given postsynaptic spike, then sum over all presynaptic cells for each postsynaptic spike, and finally sum over all postsynaptic spikes in a set of nested loops.

For the continuous method, it is better to continually update two additional variables for each neuron, i—one of which, $a_i(t)$, represents the impact of all presynaptic (afferent) spikes, the other of which, $e_i(t)$, represents the impact of all postsynaptic (efferent) spikes (figure 8.5). These variables follow:

$$\frac{da_i}{dt} = -\frac{a_i}{\tau_+} \text{ and } \frac{de_i}{dt} = -\frac{e_i}{\tau_-}. \tag{8.5}$$

(If you are modeling neurons that are only either presynaptic or postsynaptic but not both, then only one of these variables per neuron is needed.) At the time of a spike of neuron i, the variables are updated in the same manner, either being incremented by 1 if all spikes are used (e.g., $a_i \to a_i + 1$) or set to 1 if nearest-neighbor spikes are used (e.g., $a_i = 1$).

Then, at the time of each postsynaptic spike of cell j the synaptic strength is incremented by an amount $\Delta W_{ij} = A_+ a_i$ and at the time of each presynaptic spike of cell i the synaptic strength is decremented according to $\Delta W_{ij} = -A_- e_j$.

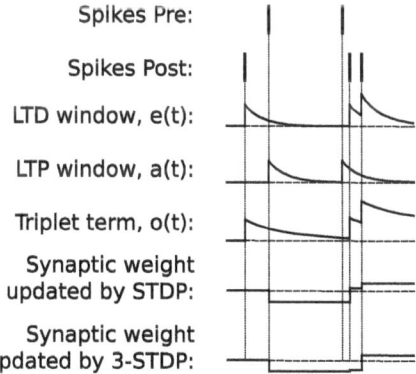

Figure 8.5
Continuous update rule for STDP and triplet-STDP (3-STDP). Synaptic weights can increase (LTP) at the time of a postsynaptic spike (row 2) and decrease (LTD) at the time of a presynaptic spike (row 1). In STDP (row 6) the amount of LTD is given by $e(t)$ (row 3), which depends on the time since the prior postsynaptic spike, whereas the amount of LTP is given by $a(t)$ (row 4), which depends on the time since the prior presynaptic spike. In triplet STDP (bottom row) the amount of LTP is also multiplied by an extra factor, $o(t)$ (row 5), which depends on prior postsynaptic spikes. This leads to minimal LTP at the time of the second postsynaptic spike but enhanced LTP at the time of the third postsynaptic spike that shortly follows.

8.3.2 Synaptic Competition via STDP

Synaptic competition   A situation in which the increase of strength of some synapses causes a decrease in strength of others.

You will notice that each of the update rules has two parameters, an amplitude such as A_+, which determines the change in synaptic strength for near-coincident spikes and a decay time constant such as τ_+, which determines how close together in time the two spikes must be to cause a significant change. The area between each curve and the x-axis is given by the product of amplitude and decay time constant, i.e., $A_+\tau_+$ for LTP and $A_-\tau_-$ for LTD. The relative values of these two areas is important, because if two spike trains are uncorrelated such that the differences in their spike times fall with equal likelihood across the region of the STDP window, then the mean amount of potentiation is proportional to the area under the LTP curve while the mean amount of depression is proportional to the area under the LTD curve. (This result assumes all spike pairs are used to update synaptic strength, not just nearest neighbor pairs, but it also holds if only nearest-neighbor pairs are used so long as the mean interspike intervals of each cell are significantly longer than the decay time constants of the plasticity rule, i.e., at low firing rates—see appendix A.)

Since we only want net strengthening if there is a causal relationship between presynaptic and postsynaptic spike times—and a causal relationship produces correlations—it is important that the mean amount of LTP is less than the mean amount of LTD for uncorrelated spike trains. That is, in models we require $A_-\tau_- > A_+\tau_+$. Such a requirement allows for competition for the following reasons:

1. If a subset of input synapses is strengthened then postsynaptic activity increases.
2. Increased postsynaptic activity leads to more pairs of presynaptic and postsynaptic spikes and shorter time intervals between one neuron's spikes and the other's. Both of these effects lead to enhanced synaptic plasticity.
3. If other presynaptic neurons have spikes that are uncorrelated with the additional spikes in the postsynaptic cell then their input synapses to the postsynaptic cell become more rapidly depressed, because the net effect of uncorrelated presynaptic and postsynaptic spikes is depression.
4. The combined effect is competition—the increase in strength of one subset of input synapses to a neuron causes the decrease in strength of another subset of its inputs.

Therefore, STDP can cause a postsynaptic cell to enhance its response to one set of inputs at the expense of another set of inputs. Such selection is called synaptic competition, which can be useful in two ways that we will investigate in tutorial 8.2.

First, it can lead to a separation of two distinct sets of mixed inputs.[23] For example, if a mixture of inputs, each originating from activity in the retina of either the left eye or the right eye, impinge in equal amounts on a set of cells, one cell may eventually respond to only the left eye's inputs while another cell responds only to the right eye's inputs. Such selectivity, called ocular dominance, is observed in the primary visual cortex of many mammals.[24,25]

Second, if a sensory neuron is receiving a subset of correlated inputs amid a majority of uncorrelated inputs, it is likely that the correlated subset contains the most useful information from the environment. In this case STDP allows the neuron to respond more strongly to the correlated set and to "filter out" the noise produced by the uncorrelated inputs.[23] Such plasticity produces a more faithful neural response to environmental signals.

8.3.3 Sequence Learning via STDP

If a sequence of external stimuli causes neurons to fire in a particular order repeatedly, then STDP strengthens connections from those neurons that fire early in the sequence to those neurons that fire later in the sequence (figure 8.6A). These strengthened connections can be sufficient to cause the postsynaptic neuron to fire following a presynaptic spike, in the absence of the external stimuli.[26] Thus, STDP provides a mechanism for sequence completion.[27] Relatedly, in the phenomenon of prospective coding, neurons that fire at specific locations on a track (called place cells) shift their receptive fields to earlier locations if an animal passes along the track many times.[28] The explanation[29–31] relies upon a temporally asymmetric plasticity rule (like STDP, but suggested before the discovery of STDP) strengthening connections in a directional manner from neurons that respond at one location to neurons that respond at a later location on the route. Following sufficient synaptic strengthening, those neurons that initially responded later are driven to respond earlier.

Perhaps sequence learning is most important in auditory processing—sound is inherently dynamic and words with meaning are formed by sequences of phonemes, whose order is essential. It takes minimal effort for us to continue phrases or verses of songs we know, when provided with the beginning. A similar type of learning is essential for juvenile songbirds to learn a stereotypical tune from an adult songbird, known as the "tutor."[32] The songs are sequences of syllables—shorter musical phrases—which can either be learned as a fixed sequence or with some variability based on transition probabilities between syllables.[33,34] Therefore, the songbird system has proven to be of great help in our understanding of the plasticity mechanisms that can underlie the learning and generation of sequences.[35]

8.3.4 Triplet STDP

The standard method of implementing STDP, in which all spike pairs are included in the update of synaptic strength, lacks the qualitative rate-dependence observed in early experiments.[36] The early experiments showed that synaptic stimulation at high frequency produced potentiation, whereas at low frequency it produced depression. Later STDP protocols suggested a similar

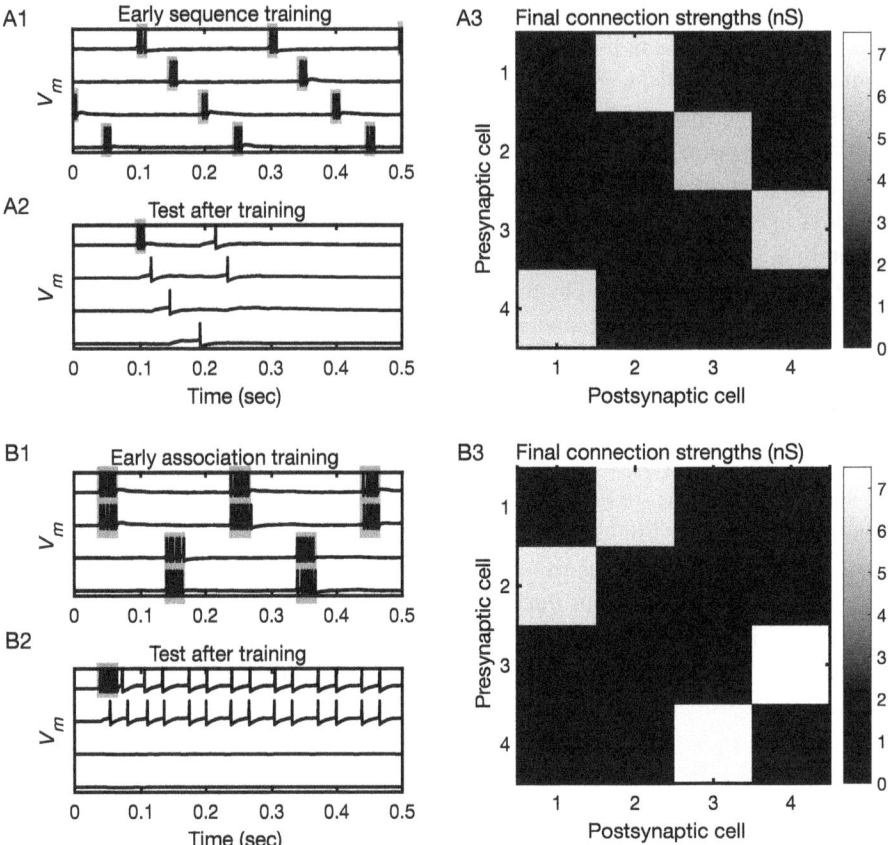

Figure 8.6
Encoding of sequences and paired associations using triplet STDP. Four excitatory neurons are stimulated repeatedly by external input (gray shaded rectangles), either in sequence (A1) or simultaneously as pairs of two neurons (B1). During training the connections between the four neurons undergo synaptic plasticity using a mechanism of triplet STDP (figure 8.5 and equation 8.6). After training, stimulation of a single neuron (single gray shaded rectangle) causes a replay of the sequence (A2) or of activity in the neuron paired with the stimulated neuron (B2) according to the prior training. These responses arise because triplet STDP can encode asymmetric synaptic weights in a unidirectional pattern following sequence training (A3), or autoassociate the paired neurons in a symmetric, bidirectional pattern following paired training (B3). This figure is based on the works of Pfister, Clopath, Busing, Vasilaki, and Gerstener.[39,41] The code used to generate this figure is available online as STDP_recurrent_final.m. Feedback inhibition is used in the code to enhance stability and it is worth noting that following further training in pattern B, the pairs of neurons would maintain activity in the delay between stimulations. Thereafter the sequence of stimulations lacks a silent period of activity between pairings, so the pairs of neurons connect with each other in sequence (as in A) until eventually all neurons are highly connected. Therefore, the simple model used here is ultimately unstable.

frequency-dependence of synaptic plasticity, with potentiation favored as the rate of the postsynaptic cell increased (as in figures 8.1A and 8.2).

The "nearest-neighbor" approach to implementing STDP does include such a qualitative rate-dependence; however, the crossover rate for the postsynaptic cell to switch the plasticity from depression to potentiation increases with the presynaptic firing rate (see appendix A for the derivation). Therefore, in this implementation high presynaptic rates favor depression, unlike the observed data.

To better extract and model the combined rate and timing dependence of STDP, in the years following its initial discovery, various groups assessed the impact of more complex or naturalistic protocols including triplets and quadruplets of presynaptic and postsynaptic spikes on synaptic strength.[37,38] Here we discuss a rule, triplet STDP,[39] that fits these later empirical data extremely well and that produces both firing-rate dependence as well as spike-timing dependence of the direction of synaptic plasticity.

As its name suggests, the triplet rule relies on the relative times of a triplet of three spikes. These can be two presynaptic and one postsynaptic, or one presynaptic and two postsynaptic (figure 8.5). In the formulation used here, we only include the latter pattern and only for the potentiating term of STDP, maintaining the depression term's dependence on only a pair of spikes (one presynaptic and one postsynaptic).

Triplet STDP   Changes in synaptic strength that depend on the relative times of up to three different spikes from the connected neurons, such as one from the presynaptic cell and two from the postsynaptic cell.

In this formulation, the modification of the standard STDP rule is to include an additional multiplicative factor when potentiating synapses. The factor decays exponentially from the time of the previous postsynaptic spike. Therefore, the amplitude of potentiation depends on two exponential factors. The first factor is the one we have seen already that is based on the time difference between the presynaptic spike and postsynaptic spike with a time constant on the order of 20 ms. The second is an extra term, whose decay time constant is slower, on the order of 100 ms. Specifically, we consider all triplets of spike times t_i, t_j, and $t_{j'}$, where t_i is the time of a presynaptic spike, t_j and $t_{j'}$ are the times of postsynaptic spikes with $t_i < t_j$ and $t_{j'} < t_j$. We then increase the synaptic strength by an amount:

$$\Delta W_{ij} = A_3^+ \exp\left(\frac{t_i - t_j}{\tau_+}\right)\exp\left(\frac{t_{j'} - t_j}{\tau_3}\right). \tag{8.6}$$

In this case, the continual update rule can be more efficient, since we can represent the contribution to the extra, final term of equation 8.6 from all the postsynaptic spikes of a neuron with a single variable, $o(t)$, which increases by one at the time of each postsynaptic spike and, thereafter decays with the appropriate time constant (figure 8.5).

The term producing depression is unchanged from that of standard STDP (equation 8.3), so depression is present at low postsynaptic rates. Therefore, when it is rare for two postsynaptic spikes to fall within a window of 100 ms of each other, potentiation is suppressed so depression dominates if spike times are uncorrelated. However, at high postsynaptic firing rates, potentiation is possible and—if its peak amplitude, A_3^+, is high enough—potentiation can dominate, even when spike times are uncorrelated (producing a positive feedback that could drive coactive synapses to their maximum strength).

In appendix B, we solve for the rate dependence of triplet STDP when the presynaptic and postsynaptic spike trains are produced as two uncorrelated Poisson processes at the given rates. The result is:

$$\frac{dW_{ij}}{dt} = r_i r_j \left(A_3^+ \tau_+ \tau_3 r_j - A_- \tau_- \right), \tag{8.7}$$

which increases linearly with the presynaptic rate, r_i, but is quadratic in the postsynaptic rate, r_j, switching from net depression to net potentiation when the postsynaptic rate increases past a threshold of $A_- \tau_- / \left(A_3^+ \tau_+ \tau_3 \right)$. Notice the equivalence of equation 8.7, produced from a model of spike timing, with equation 8.2 and figure 8.2, which were based on firing-rate models.

8.3.5   A Note on Spike-Timing Dependent Plasticity

Any empirical rule for altering the synaptic strength between two neurons on the basis of the times of presynaptic and postsynaptic spikes is an approximation that ignores a great deal of intracellular complexity. The electrophysiological and biomolecular processes set into motion in the postsynaptic cell, particularly those in the neighborhood of one synapse, depend on many additional factors—local densities of ion channels, local concentrations of enzymes, proximity and activity of other synapses, distance of the synapse from the soma, and the size and shape of the cell. Such factors are typically unknown and can be treated as a source of random variability, or simply ignored if the system is already variable enough, or one can attempt to account for as much of the underlying biophysics as possible.

However, if we want to simulate many neurons and their interactions, it can be convenient to use an empirical rule for updating synaptic strengths, rather than attempt to simulate all of the intracellular processes that are rarely measured in a functioning circuit. In many cases the empirical plasticity rule can substitute for a more complex process. For example, in some cases a dendritic calcium spike may be essential to induce synaptic plasticity and without such a spike no plasticity is induced.[40] Since such calcium spikes are highly correlated with a period of high-frequency dendritic input and with postsynaptic activity, we can simply produce a rule for synaptic plasticity that requires a high rate of postsynaptic spikes or presynaptic inputs to produce a change in strength. Such a procedure is more efficient than extending the simulations of each neuron to the level where we can more accurately predict the timings of the underlying calcium spikes.

8.3.6 Mechanisms of Spike-Timing Dependent Synaptic Plasticity

Given the importance of synaptic plasticity for brain development and memory formation, a large wealth of data has been accrued over the last fifty years with the goal of pinning down the signaling pathways involved. While significant differences can arise across brain areas, across species, across developmental stages, and across cell types, for the majority of—if not all—mechanisms, the time dependence of the local calcium concentration is the key initiating factor for synaptic plasticity. While in some cells it is calcium entry through voltage-gated calcium channels that is most important, one of the most understood mechanisms that can relate causality of neural firing to the level of calcium concentration is when considering calcium entry through NMDA receptors.

NMDA receptors require both the binding of glutamate—the neurotransmitter released by an excitatory presynaptic cell in the cortex—and depolarization of the membrane potential in order to open and admit entry to calcium ions. Being situated in the membrane of the postsynaptic cell, they provide the ideal mechanism for detecting a coincidence of presynaptic and postsynaptic spikes, as long as the postsynaptic spike causes significant depolarization of the membrane potential in the vicinity of the synapse. For synapses on the dendrites far from the soma, a dendritic calcium spike may be essential for the necessary depolarization to occur.

NMDA receptor activation is strongest when the presynaptic spike precedes the postsynaptic spike. This is because the glutamate ligand remains bound to the receptor for many tens of millisecond (50 ms is a typical time constant at in vivo temperature), whereas the membrane potential decays within a millisecond of a spike. Therefore, NMDA receptors can implement the causal aspect of "pre-before-post" in Hebb's postulate that is measured in STDP protocols (see figure 8.7).

8.4 More Detailed Empirical Models of Synaptic Plasticity

In nearly all simulations based on spiking neurons, the membrane potential of the neuron is the most significant variable in the model. I say "nearly all" because it is possible to simulate the spikes of a neuron simply by gathering statistics of its inputs and its history of spike times to determine (usually probabilistically) whether a spike should be generated at any instant. However, when the membrane potential itself is simulated—as in all spiking models considered in this book—then there is minimal simulation cost to basing synaptic plasticity on the membrane potential rather than the spike times alone.

In a successful voltage-dependent rule that is compatible with the empirical data,[41] the timing of presynaptic spikes combines with the recent history of postsynaptic membrane potential to determine the direction and magnitude of any synaptic plasticity. Depression arises at the time of a presynaptic spike whenever it follows a period of moderate depolarization. Potentiation arises at the time of high postsynaptic membrane potential (equivalent to the time of a spike) if this follows both a presynaptic spike (as in STDP) and a period of moderate depolarization of the postsynaptic membrane potential. Therefore, the rule is similar to triplet-STDP, the main distinction being that for both depression and potentiation the initial postsynaptic spike is replaced by

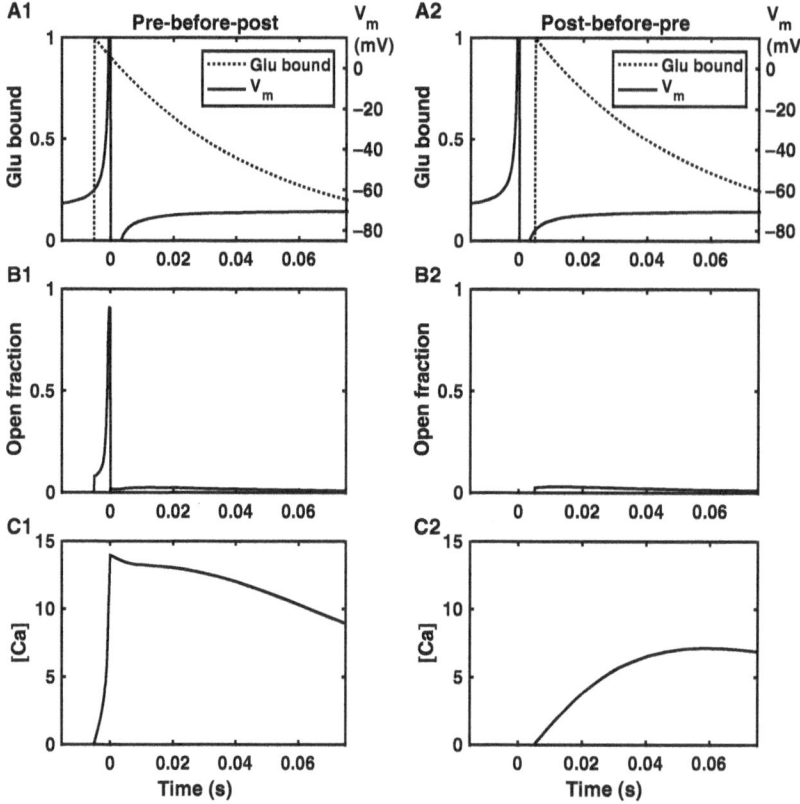

Figure 8.7
NMDA receptor activation as a temporal-order detector. (A1) If the presynaptic spike precedes the postsynaptic spike, the transient postsynaptic depolarization (solid line, scale on right axis) overlaps with the period of glutamate binding to the NMDA receptor (dotted line, scale on left axis). (B1) As a result, NMDA receptors open rapidly to admit calcium. (C1) Calcium concentration then rises rapidly to a high level in a manner that could lead to LTP. (A2) If the postsynaptic spike precedes the presynaptic spike, then the rapid membrane depolarization is over before glutamate binds to the NMDA receptor. (B2) Receptors open only partially. (C2) Calcium concentration rises slowly to a lower peak value in a manner that could lead to LTD. This figure is produced by the online code NMDA_activation.m.

a period of moderate depolarization that need not include a spike. In particular, LTD can occur following presynaptic spikes in the absence of postsynaptic spiking, as observed empirically.

If one additional variable should be simulated to make better contact with realistic plasticity mechanisms, then the calcium concentration in the postsynaptic cell is the clear choice. However, care is needed because the calcium concentration varies tremendously across different domains and organelles of any cell. The calcium concentration that drives independent plasticity at each synapse is necessarily different from synapse to synapse. So, if calcium concentration is to be used then a separate variable is needed for each synapse. In section 8.6 we will see that somatic calcium concentration, which depends on the cell's firing rate, can be used for coordinated, cell-wide homeostatic regulation of all synapses in the cell.

> **Homeostatic regulation** Control of a system via feedback from the measurement of at least one variable, with the goal of ensuring the system remains in a stable and useful operating regime.

8.5 Tutorial 8.2: Competition via STDP

Neuroscience goals: Understand STDP in action; discover how responses can become selective either by spontaneous symmetry breaking, or selection of correlated inputs.

Computational goals: Practice at defining and manipulating distinct subsets of an array differently; practice at updating variables appropriately within nested loops.

In this tutorial, you will see how STDP can produce a selective response in a neuron by preferentially strengthening one subset of its incoming synapses while weakening all other inputs. This tutorial is based on a paper by Sen Song and Larry Abbott (2001).[42]

1. Generation of input spike trains and synaptic conductance.

   a. Define 50 input spike trains, grouped into two subsets (a, b) of 25. Each subset follows a single time-varying rate (r_a or r_b) given by:

   $$r_{a,b}(t) = \frac{r_{max}}{2}[1 + \sin(2\pi v t + \phi_{a,b})],$$

   with maximum rate of $r_{max} = 60$ Hz, frequency of oscillation of $v = 20$ Hz, and phase offsets of $\phi_a = 0$ and $\phi_b = \pi$. The two phase offsets can represent stimuli received via the two eyes.

   b. Produce an array of input spikes, with 50 rows and a number of columns equal to the number of timesteps per trial (see below). Produce spikes according to an inhomogeneous (time-varying) Poisson process with rate $r_a(t)$ for rows 1–25 (the first set of 25 inputs) and with rate $r_b(t)$ for rows 26–50 (the second set of 25 inputs). Each row represents a single trial of inputs, for which you can use a timestep of $\delta t = 0.1$ ms and a duration of 0.5 s.

   c. Produce a vector of initial values of the synaptic strength, G_i, of each input synapse, labeled i (one per distinct incoming spike train) with all entries sampled from a normal distribution with mean 500 pS and standard deviation 25 pS.

   d. Produce a single vector representing the total (summed) synaptic excitatory conductance, $G_{syn}(t)$. Increment $G_{syn}(t)$ by an amount, G_i, at the time of each incoming spike at each input synapse, i. Between spikes $G_{syn}(t)$ decays exponentially to zero, with time constant $\tau_{syn} = 2$ ms.

2. Simulate the spike train of a leaky-integrate and fire (LIF) neuron receiving these inputs.
    a. Simulate the membrane potential of an LIF neuron (see chapter 2) that follows:

$$C_m \frac{dV}{dt} = G_L(E_L - V) + G_{syn}(E_{syn} - V); \text{ where if } V > V_{th} \text{ then } V \rightarrow V_{reset}$$

with parameters $C_m = 100$ pF, $G_L = 5$ nS, $E_L = -70$ mV, $E_{syn} = 0$ mV, $V_{th} = -50$ mV, and $V_{reset} = -80$ mV, initialized with $V(0) = E_L$.
    b. Record all the spike times of the LIF neuron produced during the trial.
    c. Plot the total synaptic conductance, $G_{syn}(t)$, and membrane potential, $V(t)$, as a function of time.

3. Update the input synaptic strengths using the STDP rule after the trial (batch method).
    a. For each input neuron and each input spike compare the time, t_{pre}, of the input (presynaptic) spike to the time, t_{post}, of each spike of the LIF neuron (the postsynaptic spike). Change the synaptic strength, G_i, of that input connection according to the rule:

$$G_i \rightarrow G_i + \Delta G_{LTP} \exp\left(\frac{t_{pre} - t_{post}}{\tau_{LTP}}\right) \text{ if } t_{pre} < t_{post}, \text{ or}$$

$$G_i \rightarrow G_i - \Delta G_{LTD} \exp\left(\frac{t_{post} - t_{pre}}{\tau_{LTD}}\right) \text{ if } t_{pre} > t_{post},$$

with $\Delta G_{LTP} = 20$ pS, $\Delta G_{LTD} = 25$ pS, and $\tau_{LTP} = \tau_{LTD} = 20$ ms.
    b. Ensure synaptic strengths are neither less than zero nor greater than a maximum bound, $G^{max} = 2$ nS.
    c. Record the mean synaptic strength for each set of 25 inputs and plot the set of synaptic strengths as a function of input number.
    d. Explain how the synaptic strengths change across trials and how the changing synaptic strengths impact the time variation of the input conductance and the pattern of the LIF neuron's spike train. Repeat the simulation one or two times and explain any differences you see.

4. "Train" the system.
    a. Repeat 1–3 for 200 trials, but only producing plots every 20 trials.
    b. Plot the mean synaptic strengths of the two sets of 25 inputs as a function of trial number.

5. Alter the phase offsets of two stimuli.

Repeat (1)–(4) using a phase offset that alternates across trials, such that $\phi_a = 0$ on all trials, while $\phi_b = \pi/2$ on odd trials and $\phi_b = -\pi/2$ on even trials. The sign reversals represent stimuli moving or eyes moving in either direction—input from one eye does not consistently precede the other.

6. Correlated versus uncorrelated stimuli.

   a. In this question, the phase offset of each oscillation producing the 50 inputs will vary, so you should produce a set of 50 rate vectors,

   $$r_i(t) = \frac{r_{max}}{2}[1 + \sin(2\pi v t \pm \phi_i)],$$

   where $\phi_i = 0$ for $1 \leq i \leq 10$ and ϕ_i is selected as a random number from 0 to 2π for $i > 10$. On odd trials set the phase offset as $+\phi_i$, and on even trials set it as $-\phi_i$. Use a frequency of $v = 10$ Hz and otherwise repeat questions (1)–(4) with no further alterations.

   b. Comment on any differences you see and plot the mean synaptic strength of the correlated group separately from the mean synaptic strength of uncorrelated inputs, as a function of trial number.

   c. Plot the final synaptic strength of each neuron as a function of its phase offset.

8.6 Homeostasis

Homeostasis (literally "similar standing" or "equilibrium") is the process by which all living things maintain a stable internal environment that is conducive to their well-being. Perhaps the best-known example is temperature regulation—we sweat to cool down if we are too hot and shiver to warm up if we are too cold. Since biophysical processes depend on chemical reactions, whose rates depend on temperature, salt concentrations, and pH, regulation of all these parameters is important. Also, at the cellular and organismal level properties such as size and energy use/supply must be tightly regulated.

Regulation is a term for negative feedback. "Feedback" means that the outputs of a system are monitored and used to alter the system's inner workings, or its inputs. The "negative" part means that the effect of feedback is in the direction that counteracts or ameliorates any change, whereas its opposite, positive feedback, would enhance any change. It is a curiosity of language that positive feedback, which sounds desirable in human interactions, is very bad for a complex system as it pushes processes toward instability and extremes. On the other hand, in biology and engineering, negative feedback is valuable and indeed essential, to ensure those processes remain in their desired operating regime.

Neural circuits have their own specific needs of such regulation or control. Space is at a premium in the brain, so neurons need the right level of connectivity and, therefore, appropriate dendritic ramification (branching). Neural activity is energetically costly, so neurons should not fire excessively, but neither should they be always unresponsive since they take up space and resources. Finally, the role of a neuron is to process information—minimally to pass on information that it receives, but usually to process the information with a useful filter or operation. Therefore, neurons should respond with more spikes at some times and fewer spikes at other times—i.e., they should possess an input-dependent range of firing rates. Both their connectivity and their internal excitability contribute to their ability to produce the necessarily variable, information-rich responses.

When simulating neurons or neural circuits, you will have noticed that their behavior depends quite sensitively on parameters such as the maximum calcium conductance, or the strength of synaptic connections within a circuit. If these parameters were not regulated in vivo, then neural circuits would quickly produce excessive activity (a cause of epilepsy) or become quiescent. These model parameters represent the combined effect of numerous proteins, which must be expressed, translocated, and inserted to the appropriate part of the cell membrane. Since the proteins are not static, but are replaced on a timescale of hours to days in a process called turnover, any maintenance of conductance or synaptic strength is based on a dynamic equilibrium as proteins are manufactured and inserted at the rate at which they are lost. The maintenance of these levels in the range that allows neurons to operate is, therefore, an example of homeostasis.

8.6.1 Firing-Rate Homeostasis

For a neuron to convey or process information, it should be able to modulate its firing rate in response to changes in its inputs. In particular, any neuron that is so unexcitable it remains quiescent, or that is so excitable it fires maximally at all times, would convey no information.

Figure 8.8C and F show the typical response needed of neurons, albeit in the most simplified case of two inputs. In the example, the neuron is required for coincidence detection, so should produce a spike whenever its two inputs arrive at the same time, but not when they arrive alone. If the neuron is not excitable enough (figure 8.8E) or its incoming synapses are too weak (figure 8.8B), it does not respond to the coincident inputs. None of the types of synaptic plasticity we have looked at would rescue this situation—in fact, they may decrease synaptic strength making it even harder for the neuron to respond. Conversely, if the neuron is too excitable (figure 8.8G) or its incoming synapses are too strong (figure 8.8D) then it responds to every input—a situation that would worsen when any of the Hebbian-like plasticity mechanisms we have considered cause strengthening of the incoming synapses.

You will have noticed that when the neuron fails, in one situation it does not respond to any input (figure 8.8B and E) so it would be quiescent, whereas in the other situation (figure 8.8D and G) it responds to every input so it would have excessively high activity. Therefore, if the neuron could monitor its firing rate and adjust the strengths of its afferent (incoming) synapses and/or its intrinsic excitability via a homeostatic firing-rate dependent mechanism, it could arrive at the situation shown in figure 8.8C and F, where it performs a computation.

It is worth noting that the desired behavior (figure 8.8C and F) could be achieved in a number of ways—for example, the neuron still responds as a coincident detector even with a fivefold increase in synaptic strengths, so long as the `A_type` conductance is simultaneously increased threefold. This is a common feature of neural systems—there is a large range of "sweet spots" producing useful function that may be maintained by coordinated or correlated changes in different parameters.[43] However, the systems have an even larger space of possible parameter combinations that lead to loss of function.

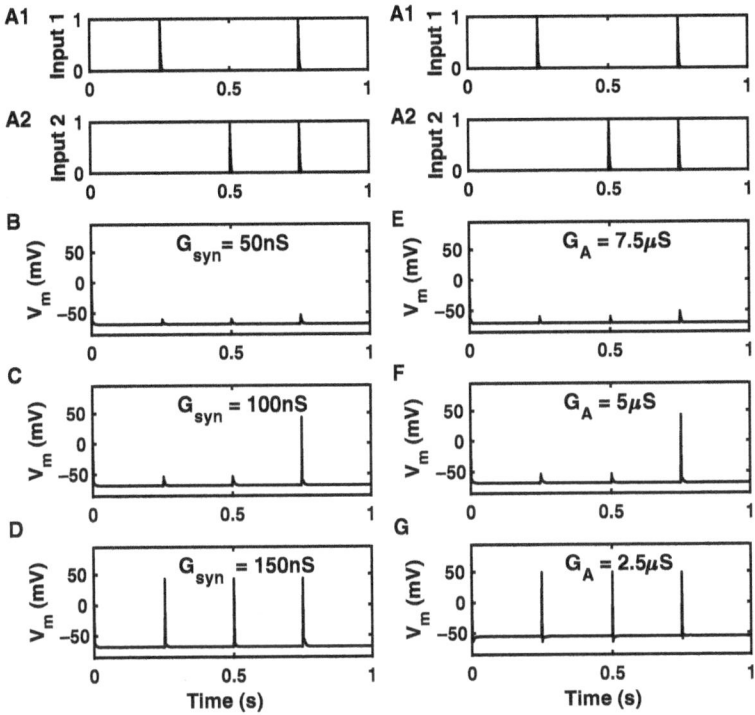

Figure 8.8
Example of the need for homeostasis in a neuron that responds to coincident inputs. (A1–2) Synaptic inputs 1 and 2 to the cell. These should not produce a response individually (first spike in each panel) but only when they arrive together (second spike in each panel). (B, E) If synaptic conductance is too low (B) or A-type potassium conductance too high (E), then the neuron never responds. (C, F) At the standard level of synaptic and A-type potassium conductance, the neuron responds with a spike only to coincident inputs. (D, G) If synaptic conductance is too high (D) or A-type potassium conductance is too low (G), then the neuron responds to every input, even single ones. Hebbian plasticity can rescue neither the unresponsive case (B, E) nor the overly responsive case (D, G). Figure produced by the online code CS_3spikesin.m.

Tetrodotoxin (TTX)   A sodium channel blocker.

Bicuculline   An antagonist of (meaning a chemical that binds without activating) GABA$_A$ chloride channels.

Both intrinsic excitability and excitatory synaptic strengths have been shown to change in a homeostatic manner following strong perturbations from equilibrium.[44,45] For example, if neurons are prevented from producing spikes (by application of tetrodotoxin, TTX, a venom found in species such as pufferfish) for a day or more, they become more excitable and their excitatory inputs become stronger. Conversely, application of bicuculline, which leads to more rapid firing by preventing inhibitory inputs from affecting a neuron, causes a reduction of excitatory synaptic strengths and a reduction of intrinsic excitability.[46]

While the underlying molecular mechanisms are still being elucidated, overwhelming evidence favors a calcium-dependent negative feedback pathway.[47] The negative feedback is such that an increase in a neuron's firing rate causes an increase of somatic calcium, which ultimately causes a reduction in AMPA receptors or sodium channels (or increase in potassium channels) and a decrease in the neuron's firing rate. Here we gain some idea of the versatility of calcium-dependent signaling pathways. While a transient, high-amplitude pulse of calcium in the vicinity of a synapse can lead to local insertion of extra AMPA receptors to strengthen synapses in an LTP protocol, a long-lasting small increase in calcium in the soma can have the opposite effect on all of a neuron's incoming excitatory synapses.

To understand the impact of such homeostasis on circuit activity, we can bypass the internal mechanisms and let either a neuron's excitatory inputs or its intrinsic excitability—for example, via the spiking threshold of a LIF model—depend negatively on its average firing rate:

$$\tau_g \frac{dG_E}{dt} = r_{goal} - r(t), \tag{8.8}$$

or

$$\tau_{th} \frac{dV_{th}}{dt} = r(t) - r_{goal}. \tag{8.9}$$

In equations 8.8 and 8.9, the time constants, τ_g and τ_{th}, should be long, on the order of hours, to mimic the observed slowness of homeostasis. However, when simulating the effect of homeostasis on a network it is sometimes necessary to reduce the time constants, in order to simulate hours rather than days of activity—the latter can take inordinately long when a timestep at the submillisecond level is needed to capture spikes, especially if large networks are being modeled.

8.6.2   Homeostasis of Synaptic Inputs

A particular need for homeostasis arises at the level of synaptic inputs to a cell, since those inputs undergo changes throughout the life of an animal as it develops and learns. Most Hebbian mechanisms—even STDP—lead to instability via positive feedback: Stronger synapses are more likely to produce a postsynaptic response, which for Hebbian mechanisms leads to increased strengthening of those synapses and a greater response in the postsynaptic cell. In a recurrent circuit this effect is exacerbated, since the increased postsynaptic response produces greater input to other cells, leading to the potential for runaway excitation (as in figure 6.4C).

One of the first suggestions for homeostasis of synaptic inputs, the "BCM rule,"[48] was via "metaplasticity," meaning that the parameters of the synaptic plasticity mechanism should change slowly over time. In this case, the boundary between potentiation and depression of synapses that depends on postsynaptic firing rate (figure 8.2) would slowly increase if mean postsynaptic rate were too high (making potentiation harder to achieve) or slowly decrease if the rate were too low (making potentiation easier to achieve).

However, the results of TTX treatment suggest that the synapses strengthen or weaken in a homeostatic manner without a requirement for synaptic input. Moreover, the distribution of synaptic strengths scales multiplicatively, a result that is consistent with the change in synaptic strength being proportional to the initial synaptic strength. These results suggest an actual rule of the form of equation 8.8, but with an extra term on the right-hand-side:

$$\tau_g \frac{dG_E}{dt} = G_E \left[r_{goal} - r(t) \right]. \tag{8.10}$$

A rule like that of equation 8.10 keeps the ratios of all synaptic strengths onto a neuron the same, since they all change in proportion. In this manner, any information stored in the synaptic strengths via prior Hebbian learning is maintained, while the overall input scales up or down to ensure the mean postsynaptic firing rate approaches r_{goal}.

8.6.3   Homeostasis of Intrinsic Properties

We saw in section 2.3.1, that neurons of different sizes can function alike, to a first approximation, if the specific conductance that is, conductance per unit area—of each ion channel is the same. This implies that, for example, if a neuron's activity should not change as it grows, then the density of ion channels should remain constant. If the proteins comprising ion channels have a fixed lifetime, the total rate of manufacture and insertion into the cell's membrane of the proteins should be proportional to the cell's total membrane area. Yet it is unlikely that the cell's combined machinery of transcription and translation would, in the absence of modification, lead to protein expression at a rate that is proportional to the surface area of its membrane. Therefore, a homeostatic feedback process is likely to be necessary to ensure neurons retain their functional role as they change in size.[49]

If a neuron is a pacemaker, its ability to produce bursts at the requisite frequency depends on parameters such as the density of calcium channels (figure 8.9), the relative sizes of regions of the cell with distinct profiles of channel densities, and how well these different regions are electrically connected (tutorial 4.3). It is not always clear how these parameters can be controlled. For example, from figure 8.9, it is not apparent from the membrane potential whether a neuron spiking tonically at a high rate, as in panels A or D, should respond homeostatically by producing more calcium channels (the correct response in A) or fewer calcium channels (the correct response in D). It turns out that monitoring of dendritic calcium would answer this question, suggesting it could serve a role in regulation of bursting, as that quantity (not unexpectedly) increases monotonically with the maximum calcium conductance.

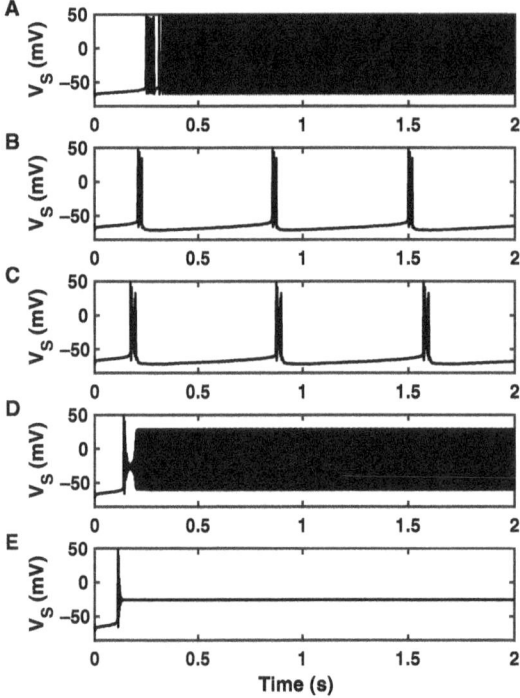

Figure 8.9
Need for homeostasis to produce intrinsic bursts. The somatic membrane potential V_S, indicates the pattern of neural firing in the Pinsky-Rinzel model (section 4.7) when the maximum calcium conductance in the dendrite is: (A) $A_D \times 1$ μS, or (B) $A_D \times 1.5$ μS, or (C) $A_D \times 2$ μS, or (D) $A_D \times 2.5$ μS, or (E) $A_D \times 3$ μS (where A_D is relative dendritic area). Here, mean membrane potential is nonmonotonic in the maximum calcium conductance (with lowest mean membrane potential in panel C) so a feedback mechanism based on the membrane potential or firing rate would not be able to return dendritic calcium channel density to its ideal level (panel C) following all deviations. For example, panels A and D are similar, but require opposite homeostatic responses. This figure was produced by the online code PR_euler_5cellversions.m.

On the other hand, the frequency of bursts—an important quantity in the pacemaker of any motor system, in particular for heart rate or respiration rate—depends in this model on the after-hyperpolarization-activated potassium conductance, G_{KAHP} (figure 8.10). An *increase* in this conductance slows the rhythm of the pacemaker cell, which in turn causes a reduction of net calcium and a reduction in the calcium-activated potassium current, I_{KCa}, yielding an overall *decrease* in flow of potassium ions out of the cell (recall, potassium channels provide an outward current). Regulation of rhythmic frequency by G_{KAHP} would be difficult, therefore, since the system is operating in a regime such that the net effect on potassium concentration is opposite to the direct effect produced by the channel. Moreover, a feedback mechanism based on potassium concentration would be confounded by a change in the other dendritic potassium conductance G_{KCa}.

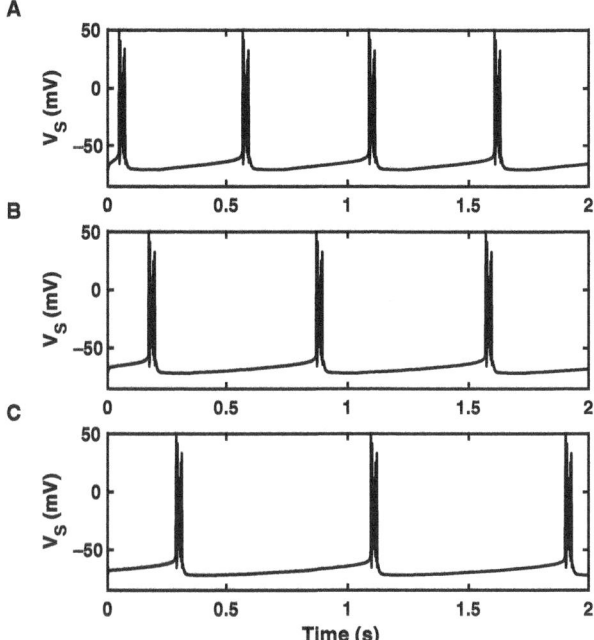

Figure 8.10
Need for homeostasis to control frequency of an intrinsically bursting neuron. The frequency of bursting depends on the maximum after-hyperpolarization conductance, a potassium conductance in the dendrites, with values of: (A) $A_D \times 2$ nS; (B) $A_D \times 4$ nS; and (C) $A_D \times 6$ nS, where A_D is the relative dendritic area. In these examples the mean potassium current across the dendritic membrane decreases (because bursts are less frequent) even as this particular potassium conductance increases. The figure is produced by the online code `PR_euler_3cellversions_freq.m`.

8.7 Supervised Learning

Examples of learning can be grouped into two broad categories, supervised and unsupervised. In the first part of this chapter we focused on Hebbian plasticity, which can underlie much unsupervised learning. The generation of preferential neural responses to the most common patterns of sensory input allows our brains to extract prevalent features in the environment (as in tutorial 8.1 and figure 8.3). This process is an example of unsupervised learning, which enables us to learn to recognize objects we have seen before, but cannot teach us what to do with those objects or how to decide what to look at. The latter requires us to ascribe values and likelihoods to the outcomes that can be achieved if we take a particular action. The values and likelihoods of outcomes depend on our current state and on whatever environmental inputs we receive. The process of learning how to improve our actions in any situation in order to achieve a better outcome requires supervised learning.

Supervised learning   Learning that depends on feedback based on the outcome of a behavior.

Reinforcement learning   Supervised learning, in which the feedback signal is a scalar quantity, either positive or negative depending on the outcome of the behavior.

Supervised learning involves feedback, so that we can adapt our actions based on their consequences, not simply on the correlation of events. If unsupervised learning were the only type present, all we would have would be random, ill-formed habits—the reason for an action would simply be the history of such actions in the same circumstances, much as persistent repetitions of a sequence result in its spontaneous occurrence under STDP (figure 8.6A). Such repetition, like a song replaying in our head, however annoying it is to us, is independent of whether the action provides a beneficial outcome for the animal. The addition of a feedback signal provides the necessary information for us to alter our behavior in a beneficial direction. The feedback signal can be binary ("good" or "bad"), or scalar (graded from "extremely good" through "OK" to "extremely bad"), or multidimensional ("I can feel myself tipping a bit to the left so should shift my weight a bit to the right"). The feedback itself can be mediated through self-sensation (e.g., balance or proprioception), through direct sensory input (e.g., pain from heat, or pleasure from food), or more acquired measures (e.g., advice, social approval, or the promise of future reward).

The form of supervised learning that we will consider here is reinforcement learning.[50] In reinforcement learning the feedback provided after an action is a scalar signal, meaning it can be ascribed a single numerical quantity. A better than expected outcome produces a positive reinforcement signal, while a worse than expected outcome produces a negative reinforcement signal. Unlike more sophisticated forms of supervised learning, reinforcement learning does not contain within the feedback any information as to how the action should be altered in order to improve the outcome. It is therefore ideal for use in situations with discrete alternatives. For example, if there are only two alternatives, positive reinforcement (commonly called "reward") should lead to a greater likelihood of a repetition of a choice, while negative reinforcement (commonly called "punishment") should lead to a greater likelihood of switching to the other alternative.

Since supervised learning is intimately connected with changes in behavior, it can be thought of as shaping our decisions. Indeed, in the task we consider later, (the weather-prediction task, section 8.7.3) the supervised learning is framed in terms of its impact on a decision-making circuit. The quantification and evaluation of human decisions is at the heart of the field of economics, so the study of how neural circuits respond to feedback in order to shape future behavior has been termed neuroeconomics.

> **Neuroeconomics**   The study of the neural basis of decision making in situations where the outcomes of those decisions are quantifiable.

8.7.1   Conditioning

> **Unconditioned stimulus**   A stimulus that causes a response or behavior automatically, or innately, without any need for training to produce the response.

> **Conditioned stimulus**   A stimulus that causes a particular behavioral response only after training (conditioning) in which it is associated with an unconditioned stimulus.

> **Classical conditioning**   Also called Pavlovian conditioning, the altering of an animal's behavior via stimuli that are independent the animal's behavior, by associating a conditioned stimulus with an unconditioned stimulus.

> **Operant conditioning**   Also called instrumental conditioning, the altering of an animal's behavior, often through reinforcement learning, with stimuli that can be rewarding and/or aversive and depend on the animal's action.

The study and shaping of an animal's responses through presentation of stimuli coupled with reinforcement is the subject of conditioning. Two categories of conditioning should be clearly distinguished, classical conditioning and operant conditioning.

In "classical," also called "Pavlovian," conditioning, an initial stimulus (the conditioned stimulus, CS) is paired with a reinforcement signal (the unconditioned stimulus, US) on all trials, whatever action is taken by the animal. Such conditioning causes any innate, reflexive, or previously learned response to the unconditioned stimulus to become a response to the conditioned stimulus.

In "operant," also called "instrumental," conditioning, the reinforcement signal depends upon the animal's action. Therefore, the study of operant conditioning becomes important if we want to understand decision making in terms of why we choose some actions or outcomes over others. The key to operant conditioning is learning to associate the value of a later outcome with the decision(s) that led to that outcome.

Reward prediction error   The difference between the actual value of an outcome, positive if good, negative if bad, following an action compared to the expected value beforehand obtained from the weighted average of all possible outcomes.

8.7.2   Reward Prediction Errors and Reinforcement Learning

A neurotransmitter that plays an important role in conditioning is dopamine. While the complete set of circumstances that lead to dopamine release—including novelty and, in certain contexts, aversion—is still being explored, it has been long established that the activity of many dopamine-releasing neurons signals a reward prediction error.[51–53]

A reward prediction error is positive if a rewarding stimulus, such as food, is received at a time when it was not entirely expected. Conversely, if an expected reward is omitted, the reward prediction error is negative. The greater the expectation of reward at a given time, as based on past experience, the smaller the reward prediction error when the reward is received, and the more negative the reward prediction error if the reward is not received. Importantly, the reward prediction error transfers from the primary reward, such as juice delivery, to the prior cues or stimuli that reliably predict such rewards. The phasic (i.e., transient, time-locked) activity of a large subset of dopaminergic neurons appears to follow these patterns typifying a reward-prediction error signal.

Such dopamine signals can be used as the scalar feedback signal underlying reinforcement learning. In general, reinforcement learning enables an animal to progress from a period of exploration, during which different alternative responses or behaviors are tried out, to a period of exploitation, once the consequences—usually evaluated in terms of the "value" of the state reached following a given action in a given situation—have been learned. Reinforcement learning can also help the animal learn to change established behaviors, so as to overcome intrinsic biases or undo previous learning when it is no longer applicable. For reinforcement learning to be effective, the likelihood of repeating an action should increase with the value of the state reached by that action.

In the following section we will study a simple model of reinforcement learning (figure 8.11) that can be simulated using circuits of model neurons.[54] We will implement reward as a binary signal that is either present or absent on any trial, with a probability that depends on the stimuli that are present and the action that is chosen. The underlying plasticity rule will depend on three terms: presynaptic activity, representing the stimulus; postsynaptic activity, representing the response; and dopamine release, indicating the reinforcement signal (see table 8.1).

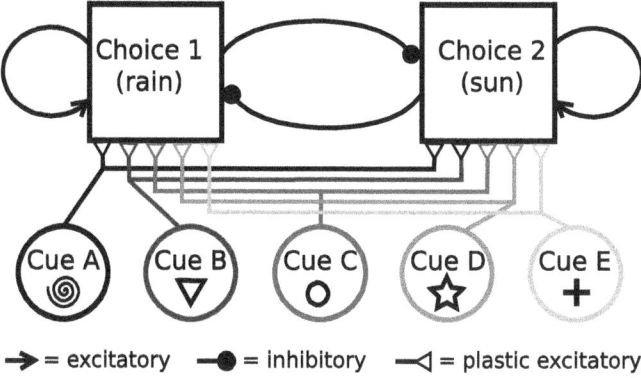

→ = excitatory   ─● = inhibitory   ◁ = plastic excitatory

Figure 8.11
The weather-prediction task: learning to respond in an uncertain environment. The task is a two-alternative forced choice task that could be considered as similar to the choice of clothing to wear based on a prediction of the day's weather—often an uncertain affair. In each trial of the task, one of the responses is considered correct and rewarded. Depending on the correct response, each cue appears with a given probability. For example, cue A appears with 95 percent probability if choice 1 is correct but only with 5 percent probability if choice 2 is correct, while cue E appears with 5 percent probability if choice 1 is correct but 95 percent probability if choice 2 is correct. Cue C is uninformative, appearing with 50 percent probability, whichever choice is correct. Cues B and D are intermediate, with probabilities of 75 percent versus 25 percent and 25 percent versus 75 percent respectively if choice 1 versus choice 2 is correct. In the standard version of the task each choice is correct with 50 percent probability on any trial. The model circuit used to solve the task—i.e., to produce correct choices given the uncertain cues—relies on reward-dependent plasticity at the excitatory synapses between the cue-responsive units and the decision-making units. Initially the noise in the decision-making circuit generates random choices (the exploration stage), but the synaptic plasticity causes biases so that eventually the response is dominated by activity of units responsive to the most predictive cues.

Table 8.1
A three-component plasticity rule that can be used for supervised learning in problems such as the weather-prediction task

Presynaptic rate	Postsynaptic rate	Reward prediction error	Change in synaptic strength
High	High	+	+
High	Low	+	−
Low	High	+	0
Low	Low	+	0
High	High	−	−
High	Low	−	+
Low	High	−	0
Low	Low	−	0

The reward prediction error is positive (+) when reward is delivered during training (before an animal is certain the response is correct and the reward will arrive) and is negative (−) when the potential reward does not arrive. When reward arrives, the synaptic plasticity is Hebbian (as in figure 8.1C) whereas in the absence of reward the response is anti-Hebbian. Other combinations of rule are possible and different versions of Hebbian plasticity are possible (cf. figure 8.1A–D) when there is reward, combined with their opposite in the absence of reward. The impact of these different variants will be tested in tutorial 8.3. A general requirement for all useful rules is that the change in synaptic strength is either zero or equal to the parity of the three prior columns. That is, if we convert the rates of the presynaptic cell and of the postsynaptic cell to values of "+1" for high rate and "−1" for low rate, then multiply together those two values with the sign of the reward prediction error, the product indicates the required direction of any change in synaptic strength (i.e., a result of "+1" means increase strength and "−1" means decrease strength).

8.7.3 The Weather-Prediction Task

The weather-prediction task (figure 8.11) is one that requires the subject to learn how well different stimuli predict the desired response. The task is difficult because stimuli appear together, so that good performance requires the subject to weight the more predictive stimuli more strongly than the less predictive ones, while on any one trial it is not clear which of the presented stimuli is more associated with the reward and so should be reinforced.

The name of the task arises from the need to choose appropriate clothing, or whether to bring an umbrella when heading out, based on a number of pieces of information such as cloud cover, weather forecast, time of year, current temperature, and so on. Ideally, we would always bring an umbrella or raincoat on a day trip when it rains and avoid the extra baggage on a day trip without rain. While in the real world the costs of the two errors are not equal, in laboratory tasks set to mimic this type of decision, a correct choice in either direction is equally rewarded.

It turns out that a three-component plasticity rule is sufficient to produce the correct behavior, at least qualitatively.[54] Moreover, the required rule has some evidence in favor of it from the impact of dopamine on synaptic plasticity.[55,56] The basis of the rule is one of reward-dependent plasticity: If a decision is correct it leads to reward and this reward causes (or enhances) Hebbian plasticity between the neurons that led to the reward. In particular, in the circuit of figure 8.11, if specific stimuli are present and a specific choice is made, then if that choice results in reward the connections from the neurons responsive to those stimuli to the neurons generating that choice should be strengthened. We discuss a few alternative three-component plasticity rules that could lead to successful in the following paragraphs.

Rule 1: In order to prevent chance coincidences from strengthening all connections, a method for reducing synaptic strengths is also necessary. For example, the standard rules for synaptic depression with high presynaptic activity and low postsynaptic activity (rule 3 of figure 8.1 and/ or figure 8.2) can be used, if a requirement of a positive reward signal is added. In this case, when the correct choice is made, the synapses being depressed correspond to those promoting the unchosen, incorrect alternative. While such a method has the benefit of counterbalancing potentiation with depression, it has the disadvantage of only producing synaptic change when a correct response is made. The problem with such a plasticity rule kicking in only when a pulse of dopamine arrives following a correct choice, is that the circuit could get "stuck," always producing incorrect responses and without the ability to alter its behavior. That is, one could not unlearn behaviors that initially, by chance, seemed advantageous but later produced negative outcomes.

Rule 2: Therefore, alternatively, it can be better to ensure that incorrect choices cause a decrease in the likelihood of that stimulus-response combination occurring again. To achieve this, the same synapses that are potentiated when a choice is correct should be depressed in an anti-Hebbian manner, if the same choice made with the same stimuli proves to be incorrect. Table 8.1 indicates such a situation.

Rule 3: Finally, it is worth thinking about the desired plasticity mechanisms from a Bayesian perspective. If a particular stimulus is very rare, but when it appears a particular choice is always correct, then that stimulus is highly predictive of the required choice. Therefore, the synapses connecting neurons responsive to the stimulus with neurons generating the correct choice should remain strong and not be diminished on the many trials when the presynaptic neurons are inactive because the stimulus did not appear. That is, the probability of the response given the stimulus is much higher than the probability of the stimulus given the response when the stimulus is rare. If the strength of the synapse were to alter when a response occurred (postsynaptic activity) without the stimulus (presynaptic activity) then the final state of the synapse would convey, primarily, information about the probability of the stimulus given the response. However, optimal decision making requires the probability of the response given the stimulus. This argument suggests rule 3 of figure 8.1, which produces plasticity only when the presynaptic cell is active, is more appropriate than rules 2 or 4, following a correct response, at least in cases when stimuli are not equally likely—a hypothesis we can test in tutorial 8.3.

In laboratory tasks the reward arrives immediately after the choice, so that one can assume stimulus-related activity and choice-related activity coincide with reward delivery. This is the assumption behind the model of tutorial 8.3. However, in reality a reward may arrive after all the predictive activity has dissipated. This leads to a problem of eligibility: Of all the neurons that have been active in the brain prior to the reward, which ones should have their synapses potentiated? It turns out that with a slowly decaying "eligibility trace,"[57] which tags the synapses eligible for alteration when a reward signal arrives, it is possible to solve this problem over time and even for the reward signal to propagate back to arise at the time of predictive stimuli, as seen in vivo.[58]

8.7.4 Calculations Required in the Weather-Prediction Task

In the weather-prediction task, a subject must predict which of two responses is the rewarded one based on which symbols appear beforehand. Each symbol appears on a given trial with a probability that depends on which of the two responses is correct (rewarded) on that trial. Across many trials, the subject can learn the degree to which a symbol is predictive of one response or the other and choose the response that is the most likely to be correct given the observed set of symbols.

The optimal method—in terms of maximizing the number of rewarded responses made—for combining together the information from the presented symbols is for the subject to multiply together the likelihood ratios (or add the log-likelihood ratios) produced by each symbol's presence or absence.

Likelihood ratio   The ratio of the probabilities of two different outcomes (in this case sunny versus rainy) given the observed data.

Example   Suppose choice 1 and choice 2 are rewarded equally often, and three symbols can appear. Symbol A appears on 3/4 of the trials when choice 1 is rewarded and 1/4 of the trials when choice 2 is rewarded. Symbol B appears on 2/3 of the trials when choice 1 is rewarded and 1/3 of the trials when choice 2 is rewarded. Symbol C appears on 1/3 of the trials when choice 1 is rewarded and 2/3 of the trials when choice 2 is rewarded. If symbols A and C appear, what is the preferred choice?

We use the notation for probability (see section 1.4.3) where $\tilde{B}$ means 'NOT B', so that $P(\tilde{B}) = 1 - P(B)$. We want to compare the probability that Choice 1 is correct:

$$P(1|A,\tilde{B},C) = \frac{P(1)P(A,\tilde{B},C|1)}{P(A)P(\tilde{B})P(C)} = \frac{P(1)P(A|1)P(\tilde{B}|1)P(C|1)}{P(A)P(\tilde{B})P(C)} \tag{8.11}$$

with the probability that choice 2 is correct:

$$P(2|A,\tilde{B},C) = \frac{P(2)P(A,\tilde{B},C|2)}{P(A)P(\tilde{B})P(C)} = \frac{P(2)P(A|2)P(\tilde{B}|2)P(C|2)}{P(A)P(\tilde{B})P(C)}. \tag{8.12}$$

If $P(1|A,\tilde{B},C) > P(2|A,\tilde{B},C)$ the subject should make choice 1, otherwise the subject should make choice 2. We make the comparison by testing if the likelihood ratio is greater than 1, i.e., we make choice 1 if $\left(P(1|A,\tilde{B},C)\right)/\left(P(2|A,\tilde{B},C)\right) > 1$. This likelihood ratio is given by the ratio of the priors, $(P(1))/(P(2)) = 1$, multiplied by the individual, stimulus-specific likelihood ratios:

$$\frac{P(1|A,\tilde{B},C)}{P(2|A,\tilde{B},C)} = \frac{P(1)}{P(2)} \cdot \frac{P(A|1)}{P(A|2)} \cdot \frac{P(\tilde{B}|1)}{P(\tilde{B}|2)} \cdot \frac{P(C|1)}{P(C|2)} = (1)\left(\frac{3}{1}\right)\left(\frac{1}{2}\right)\left(\frac{1}{2}\right) = \frac{3}{4}. \tag{8.13}$$

So, since $\left(P(1|A,\tilde{B},C)\right)/\left(P(2|A,\tilde{B},C)\right) < 1$ the subject should make choice 2. Notice that the absence of symbol B provided information that swung the balance in favor of choice 2 from choice 1, so the absence of a common stimulus also provides information that should impact our decisions.

In tutorial 8.3 we will assess how well a neural circuit can achieve a similar result through reward-dependent plasticity.

8.8   Tutorial 8.3: Learning the Weather-Prediction Task in a Neural Circuit

Neuroscience goal: Learn how a single reinforcement signal can lead to changes in specific connections in a manner conducive to the learning of tasks.

Computational goal: Keep track of connections in different types of connectivity matrix.

In this tutorial, you will produce a firing-rate model network comprising five input units (labeled A–E) that provide feedforward input to two units in a decision-making circuit (labeled choice 1 and 2, figure 8.11). You will simulate 800 trials. On each trial the weather is sunny with probability $P(sun)$, in which case choice 2 is correct (so $P(2) = P(sun)$); otherwise it is rainy, in which case choice 1 is correct (so $P(1) = 1 - P(sun)$). During a trial the input units receive applied current if the corresponding cue is present; otherwise they receive no applied current. If it is sunny the cue corresponding to input units A, B, C, D, and E is present with respective probabilities, $P(A|2) = 0.95$, $P(B|2) = 0.75$, $P(C|2) = 0.5$, $P(D|2) = 0.25$, and $P(E|2) = 0.05$. If it is rainy the cue corresponding to input units A, B, C, D, and E is present with respective probabilities, $P(A|1) = 0.05$, $P(B|1) = 0.25$, $P(C|1) = 0.5$, $P(D|1) = 0.75$, and $P(E|1) = 0.95$.

Simulate the firing rate of each unit, i, as a sigmoid function of its input current, $I_i(t)$, without noise, as:

$$\tau \frac{dr_i}{dt} = -r_i + \frac{r_{max}}{1 + \exp\left[\dfrac{I_{th} - I_i(t)}{I_\sigma}\right]},$$

with parameters $\tau = 20$ ms, $r_{max} = 100$ Hz, $I_{th} = 50$, and $I_\sigma = 5$. Each trial should last 500 ms, with a timestep of $\Delta t = 0.001$ s.

The input current, $I_i(t)$, for the input units whose corresponding cue is present on a trial should be set as $I_i(t) = 50$ for $t > 0.1$ s (otherwise the input current, $I_i(t)$, for the input units is zero).

The input current to the decision-making units depends on the rates of the input units, through a 5×2 matrix of connections, $W_{ji}^{(In)}$, and the rates of the decision-making units through a 2×2 matrix of recurrent connections, $W_{ji}^{(Rec)}$, plus an additional noise term, $I^{(rnd)}(t)$:

$$I_i(t) = W_{Ai}^{(In)} r_A + W_{Bi}^{(In)} r_B + W_{Ci}^{(In)} r_C + W_{Di}^{(In)} r_D + W_{Ei}^{(In)} r_E + W_{1i}^{(Rec)} r_1 + W_{2i}^{(Rec)} r_2 + I^{(rnd)}(t)$$

Initialize all of the connections from input units to decision-making units, $W_{ji}^{(In)}$, to a value of $W_0 = 0.2$. These will be updated by the plasticity rule. The decision-making circuit has fixed connections with self-excitation $W_{11}^{(Rec)} = W_{22}^{(Rec)} = 0.5$ and cross-inhibition, $W_{12}^{(Rec)} = W_{21}^{(Rec)} = -0.5$. The noise current is generated independently for each decision-making unit and independently on each timestep, by repeatedly sampling from a Gaussian distribution of unit standard deviation (randn() in MATLAB) and multiplying by a scaling factor of $\sigma_I / \sqrt{\Delta t}$ where $\sigma_I = 1$. Each separate trial should have independent noise (i.e., regenerate the noise on every trial).

Use a threshold of 40 Hz and determine which of the two decision-making units reaches that rate first (typically only one of the units achieves a high rate)—that unit represents the choice made, (unit 1 = rain, unit 2 = sun), so you should add code to determine if the choice is correct, in which case reward is received.

The synapses between input units and decision units will be updated at the end of each trial according to a three-component rule (table 8.1): (a) low/high rate of input unit; (b) low/high rate of decision-making unit; and (c) correct/incorrect response.

One goal of this tutorial is to compare and contrast the success (in terms of achieving a high rate of correct responses) and stability (in terms of connection strengths remaining within a limited range) of different rules in different conditions. To achieve this, for each simulation calculate the cumulative number of correct responses and the number correct in the final 100 trials. Plot the synaptic strengths as a function of trial number. Compare the behavior of each rule when connection strengths are unbounded versus bounded such that $0 \le W_{ij} \le 2W_0$.

For each rule plot the difference in synaptic strength to the two pools as a function of the log likelihood ratio due to a given input unit's activity. For example, for input unit A, you would plot $W_{A2}^{(In)} - W_{A1}^{(In)}$ against $\log[(P(A|2))/(P(A|1))]$, where $P(A|2)$ and $P(A|1)$ are the probabilities given in the question, while $W_{A2}^{(In)}$ and $W_{A1}^{(In)}$ are the connection strengths produced by learning. In this manner, for each rule you will obtain five data points, one for each input unit. Which rule produces the most linear result? A linear result would indicate that the total difference in input currents to the two decision-making units is proportional to the log-likelihood ratio based on the probability of each unit producing the correct response.

Rule A
Assume the reward prediction error, E, is +0.5 on correct trials and −0.5 on incorrect trials. Update synapses only from input units that were active in the trial (i.e., the corresponding symbol was present).

Update connections from active input units to the active decision-making unit by an amount, which is positive on correct trials and negative on incorrect trials:

$$\Delta W_{ij} = +\varepsilon E$$

and from active input units to the inactive decision-making unit by an amount, which is negative on correct trials and positive on incorrect trials:

$$\Delta W_{ij} = -\varepsilon E,$$

where $\varepsilon = 0.04$ sets the plasticity rate.

Treat a trial without a decision as incorrect, so that ΔW_{ij} is positive from all active input units to both decision-making units.

Rule B
Follow rule A, but calculate the reward prediction error as $E = R - \langle R \rangle$, where $R = 1$ if the trial was rewarded and $R = 0$ otherwise, while $\langle R \rangle$ is the mean reward across the prior ten trials (assume $\langle R \rangle = 0.5$ over the first ten trials).

Rule C
Calculate the reward prediction error as in rule A, but update synapses using soft bounds, so that if the calculated ΔW_{ij} is negative, multiply it by the current value of W_{ij} before reducing the connection strength and if ΔW_{ij} is positive, multiply it by a factor of $W_{max} - W_{ij}$ before increasing the connection strength. Use $W_{max} = 2W_0$. That is, update connections from active input units to the active decision-making unit by an amount:

$$\Delta W_{ij} = +\varepsilon E (W_{max} - W_{ij}) \quad \text{if } E > 0$$

or

$$\Delta W_{ij} = +\varepsilon E W_{ij} \qquad\qquad \text{if } E < 0.$$

Similarly, update connections from active input units to the inactive decision-making unit by an amount:

$$\Delta W_{ij} = -\varepsilon E (W_{max} - W_{ij}) \quad \text{if } E < 0$$

or

$$\Delta W_{ij} = -\varepsilon E W_{ij} \qquad\qquad \text{if } E > 0.$$

Rule D
Follow rule B, but only update synapses to the active decision-making unit from active input pools (as in figure 8.1A).

Rule E
Follow rule D, but also update synapses to the active decision-making unit from inactive input units (as in figure 8.1D) by an amount:

$$\Delta W_{ij} = -\varepsilon E.$$

Part B: Optional

Alternative Training Protocols   In the standard training protocol, $P(sun) = 0.5$ and 800 trials are used.

1. Try $P(sun) = 0.2$. How do the final connection strengths change and does performance change? Which rules cope better with this change? If performance deteriorates try to explain the reason.

2. Increase the number of trials to 4,000. For which plasticity rules do the connection strengths reach a steady state? Explain your findings.

Challenge   Generate a rule such that connection strengths reach a stable steady state and the network performs well with both alternative training protocols.

8.9   Eyeblink Conditioning

Eyeblink conditioning is a form of classical conditioning in which an animal (often a rabbit) learns to associate a tone (the conditioned stimulus, CS) with an aggravating puff of air aimed at its eyes (the unconditioned stimulus, US). The puff of air reliably follows the tone after a predetermined time interval that can range from 150 ms to 1.5 s. The puff of air produces an unconditioned response, the eyeblink, which, after a sufficient number of trials (more than 100 for a rabbit!), is produced just before the puff of air arrives. While this task is known as classical conditioning because the final response, the eyeblink, is the same as the innate response to the US, it could arguably be thought of as instrumental conditioning: the animal must learn to make the eyeblink at the correct time, and when it does so it avoids the "punishment" of an aggravating puff of air in the eye.

The protocol can be one of delay conditioning, in which the tone remains on, or trace conditioning in which the tone switches off to leave a silent pause before the puff of air. The distinction between trace and delay may not seem important, but trace conditioning requires processing in the hippocampus, whereas delay conditioning may be entirely a function of the cerebellum.

In tutorial 8.4 we will simulate a simplified cerebellar model (figure 8.12) for delay conditioning, based on models by Buonomano, Medina, and Mauk,[59,60] which provide an example of the Marr-Albus-Ito theory of cerebellar function.[61-63] To solve the delay-conditioning task, two requirements must be satisfied. First, some function of the neural activity following the CS must reliably predict the US at the correct time. Therefore, the neural activity must contain information about time. Second, a plasticity rule must allow the circuit to extract that time information

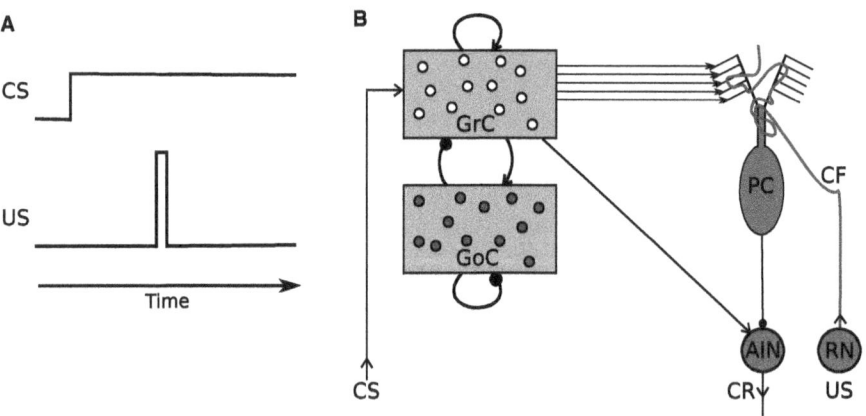

Figure 8.12
Simplified model of cerebellar eyeblink conditioning. (A) The conditioned stimulus (CS) is typically a tone, which stays on during a trial. The unconditioned stimulus (US), a puff of air into the eye, arrives a fixed time interval later. (B) In the simplified model circuit, the CS provides input to excitatory Granule Cells (GrC), which are in a recurrent circuit with inhibitory Golgi cells (GoC), and which provide input to the inhibitory Purkinje cell (PC) via parallel fibers. The US, as well as producing the automatic response of eyelid closure, also initiates the reinforcement signal for learning. To this end, it generates transient activity in the excitatory cells of the red nucleus (RN), whose activity is transmitted via the climbing fiber (CF), which winds around the dendrites of the Purkinje cell. The CF provides a reinforcement signal to the synapses from the Granule cells to the Purkinje cell. The output of the anterior interpositus nucleus (AIN) causes the conditioned response, when a temporary pause in the Purkinje cell's activity causes disinhibition, while it is receiving activity from the granule cells (model connections from GrC to AIN are not present empirically, but in this simplified model, they are necessary to ensure the AIN can only be active when the CS is on). The model is based on the work of Medina et al.[60]

in order to produce the response at the correct time. The plasticity rule—at the synapses onto the cerebellar Purkinje cell—achieves the latter by association of activity in a particular subset of granule cells with a reinforcement signal arriving via the climbing fiber (see figure 8.12).

While the subset of active units in our circuit is strongly dependent on time, so can be used as an indication of time since CS onset, we use a chaotic circuit to produce such a time signal—and as seen in section 7.8, activity in a chaotic circuit is not robust to changes in initial conditions or to noise fluctuations. Therefore, in our simulations the initial conditions must be identical on each trial and the neural responses should be noise-free, to ensure identical dynamics in every trial. Such a circuit is valuable here as a demonstration of cerebellar learning arising from highly irregular granule cell activity as occurs in the cerebellum. However, we still do not understand the means by which neural circuits robustly represent time, so this important topic is a matter of ongoing theoretical and experimental investigation.[64,65]

8.10   Tutorial 8.4: A Model of Eyeblink Conditioning

Neuroscience goals: Gain understanding of how correct timing of behavior can be learned with a reinforcement signal; learn some elements of cerebellar circuitry and its function.

Computational goals: More practice at coupling multiple types of array and variable; build on prior code; use of the scalar product of two vectors to calculate an integral numerically.

In this tutorial, we will use a model network found online as `highD_chaos.m`, comprising a large number of strongly coupled excitatory and inhibitory units that produced the "high-dimensional chaos" shown in figure 7.13. We will make the false assumption that the complex trajectory of neural activity is reproducible from a given starting point, in order to train the activity pattern of a particular point in time to produce an eyeblink. A model of the circuit we will produce—highly simplified from the actual circuitry in the cerebellum—and a schematic of the task are shown in figure 8.12.

All neurons will be treated as firing-rate model units with sigmoidal f-I curves following (see equations 6.1–6.3 for notation):

$$\tau_i \frac{dr_i}{dt} = -r_i + f(I_i)$$

where

$$f(I_i) = \frac{r_i^{max}}{1 + \exp\left[-(I_i - I_i^{th})/I_i^{\sigma}\right]},$$

and I_i is the input current to the cell group. Other parameters of the f-I curve depend on the cell-type in the circuit as described in table 8.2.

Table 8.2
Properties of different cell types in the cerebellar eye-conditioning model of tutorial 8.4

Cell	Symbol	τ_i (s)	r_i^{max} (Hz)	I_i^{th}	I_i^{σ}
Granule cell (E)	$r_i = r_{GrC}$	0.010	100	10	1
Golgi cell (I)	$r_i = r_{GoC}$	0.005	100	10	1
Purkinje cell	$r_i = r_{PC}$	0.010	30	40	10
Climbing fiber	$r_i = r_{CF}$	0.002	30	15	1
Anterior interpositus nucleus	$r_i = r_{AIN}$	0.010	100	25	4

The granule and Golgi cells will be simulated in the chaotic network available online (see next paragraph). For other cells in this model, their input currents are:

$$I_{PC}(t) = \sum_n W_n^{GP} r_n^{GrC}$$

$$I_{CF}(t) = I_{US}(t)$$

$$I_{AIN}(t) = W^{GN} \sum_n r_n^{GrC} - W^{PN} r_{PC},$$

where r_n^{GrC} is the firing rate of the granule cell labeled n, which has a unique connection strength, W_n^{GP}, to the Purkinje cell. Initially set $W_n^{GP} = 10/N_{GrC} = 0.05$ for all such connections, though these strengths will change during learning. Also set $W^{GN} = 5/N_{GrC} = 0.025$ and $W^{PN} = 30$. $N_{GrC} = 200$ is the number of granule cells.

You can use the code highD_chaos.m to implement the balanced network of 200 excitatory granule cells coupled to 200 inhibitory Golgi cells. In this combined inhibitory-excitatory network, each connection is present with a fixed probability, $p^{(conn)} = 0.05$. Connection strengths are drawn randomly from a uniform distribution then inputs are scaled, so the sum of excitatory connection strengths to each granule cell is 90, as is the sum of inhibitory connection strengths to each granule cell, whereas the sum of excitatory connection strengths to each Golgi cells is 80, as is the sum of inhibitory connection strengths to each Golgi cell. In addition to the input currents to these cells already generated in the network, you should include an additional input of $I_{CS}(t)$ to each granule cell, which will initiate activity in the network at the time of the CS.

In each trial, of duration 2 s, you should initialize the firing rates of all cells at zero. At a time of 200 ms, the conditioned stimulus will switch on, to a value of $I_{CS}(t) = 2$ for $t > 200$ ms. At a later time, t_{US}, the unconditioned stimulus will be switched on for 20 ms, to a value of $I_{US}(t) = 100$, for $t_{US} \leq t \leq t_{US} + 20$ ms. At all other times these inputs are fixed at zero.

Learning will occur via plasticity of the connection strengths, W_n^{GP}, between granule cells and the Purkinje cell. The plasticity depends on the overlap between the synaptic gating variable of each synapse and the activity of the climbing fiber. Each synaptic gating variable follows the firing rate of the corresponding granule cell, according to:

$$\tau_s \frac{dS_n}{dt} = -S_n + \frac{r_n^{GrC}}{r_{GrC}^{max}},$$

where $\tau_s = 50$ ms (and the factor $r_{GrC}^{max} = 100$ Hz ensures $0 \le S_n \le 1$).

The learning rule is then

$$W_n^{GP} \to W_n^{GP} \left\{ 1 - \Delta \left[\int S_n(t) r_{CF}(t) dt - \Theta \right] \right\},$$

where you can evaluate the integral by summing the product of $S_n(t)$ and $r_{CF}(t)$ across all time bins then multiplying by the timestep in the simulation. If this integral is larger than the threshold, $\Theta = 0.1$, it indicates the climbing fiber fires when there is synaptic input from that particular granule cell and the corresponding synapse depresses. Otherwise the synapse increases in strength (as is necessary to prevent all synapses from eventually depressing to zero). The factor $\Delta = 0.1$ sets the rate of change of synaptic strength.

Update the synaptic strengths at the end of each two-second trial using the above method, then constrain the strengths to be no less than zero and no greater than a maximum value of $50/N_{GrC} = 0.25$.

1. Train the system for 200 trials with the unconditioned stimulus arriving 800 ms after the conditioned stimulus (i.e., $t_{US} = 1$ s, when $t_{CS} = 200$ ms). Plot the total input current to the Purkinje cell, the firing rate of the Purkinje cell, and the firing rate of the interpositus nucleus before and after training. (It can be useful to plot these variables every 20 trials, overwriting the figure each time, to assess progress through the simulation.)

Note: Since there is no added noise in this simulation, you can speed it up by increasing the value of Δ and correspondingly reducing the number of trials if the time taken is too long. A timestep of 0.2 ms should be sufficiently fine for this simulation of firing-rate model neurons.

2. Repeat, but with $t_{US} = 500$ ms.

3. Repeat, but with $t_{US} = 1500$ ms. Is there a limit to the range of values of t_{US} that will produce a correctly timed conditioned response?

4. How do you expect your results to change (if at all) if the synaptic time constant for granule cells to the Purkinje cell is reduced from 50 ms to 2 ms? Explain any changes both in terms of the function of the circuit and how it would impact the animal's behavior.

5. The granule cells of the cerebellum are the most numerous cells of a given class in the brain. Do you expect the results to be more reliable or less reliable if the circuit contained more granule cells and Golgi cells? Test your expectation if time allows. Care is needed when reducing the number of cells to ensure each cell receives some inhibitory input and some excitatory input—either increase connection probability, or sample networks until all cells receive connections, or use another method for ensuring each neuron receives both excitatory and inhibitory connections.

Final notes: In this tutorial, we have simulated a circuit that can achieve a conditioned response. The circuit has many similarities with the circuitry of the cerebellum that is responsible for eye-blink conditioning, but it is important to note some differences.

1. In reality, the Purkinje cell is active (due to other inputs) prior to the conditioned stimulus.
2. There are no direct connections from granule cells to the AIN, though AIN does receive input when the conditioned stimulus is present.
3. We have omitted a parallel path, which may be dominant, in which other Purkinje cells increase their firing rate at the time of the unconditioned stimulus over the course of learning. The timed output of these cells inhibits a parallel pathway that, when active, prevents eyelid closure. That is, eyelid closure can arise via disinhibition of such an indirect pathway.
4. The tiniest bit of noise or change in initial conditions would destroy the reliable alignment of activity in a specific subset of granule cells with the unconditioned stimulus. Such easy disruption of information means the network we have produced would be ineffective in practice. How networks can have chaotic properties but respond reliably and robustly to certain stimuli (such as the conditioned stimulus) is a question of ongoing, active research.[64]

Questions for Chapter 8

1. How does standard STDP (equations 8.3 and 8.4) fully implement Hebb's postulate, and what does STDP include that is not mentioned in Hebb's postulate?
2. A type of synapse is found that increases in strength when the postsynaptic cell fires before the presynaptic cell and decreases in strength when the cell fires in reverse order. Exponential fits of $-0.02e^{-\Delta t/10}$ and $+0.015e^{-\Delta t/15}$ are made respectively for the fractional amount of depression and potentiation when any pair of presynaptic and postsynaptic spikes are separated by an absolute time difference of Δt measured in ms.
   a. Will the synapse strengthen or depress when the two neurons spike in an uncorrelated manner? (Assume all possible spike pairs contribute to plasticity.)
   b. How will the change in synaptic strength impact the correlation between spikes from the two neurons?
   c. Do you expect the rule to lead to a stable nonzero solution of synaptic strength?
3. State which factors in the rule for triplet-STDP impact the postsynaptic rate-dependent crossover between depression and potentiation, and explain qualitatively how changes in those factors cause an alteration in the crossover rate.
4. Explain the benefit of a signal for positive or negative reinforcement in supervised learning being proportional to the reward prediction error, rather than being just proportional to the reward itself.

8.11   Appendix A: Rate-Dependent Plasticity via STDP between Uncorrelated Poisson Spike Trains

Here we will consider successively the two distinct types of standard rule for STDP, first using all spike pairs and second only using the most recent spike of the other neuron during each update (i.e., nearest-neighbor pairs).

We assume both the presynaptic and the postsynaptic neuron produce spikes as a Poisson process with fixed rates (r_i for the presynaptic cell and r_j for the postsynaptic cell). All spike times are uncorrelated with each other. The probability of a neuron producing a spike in a time interval δt is constant for each neuron and proportional to its firing rate ($r_i \delta t$ for the presynaptic cell and $r_j \delta t$ for the postsynaptic cell). We will use the result (section 3.7) that for a neuron emitting spikes as a Poisson process, the probability, $P(T_D)\delta T_D$ that the time since the most recent spike is in a given time interval, T_D to $T_D + \delta T_D$, is exponential: $PT_D \delta T_D = r \exp(-rT_D)\delta T_D$.

1.   Plasticity Due to All Pairs of Spikes

If all pairs of spikes contribute to plasticity, then for each postsynaptic spike the average amount of LTP is calculated as the sum over all time-differences, T_D, of the probability of a presynaptic spike occurring that amount earlier ($r_i \delta T_D$) multiplied by the amount of LTP generated by a presynaptic spike at that amount earlier ($A_+ \exp[-T_D / \tau_+]$). Mathematically the sum over all time-differences becomes an integral over Δt, so the expected (mean) amount of LTP per postsynaptic spike is:

$$\int_0^\infty A_+ \exp[-T_D / \tau_+] r_i dT_D = A_+ \tau_+ r_i. \tag{8.14}$$

Since the rate of postsynaptic spikes is r_j the mean rate of change of synaptic strength due to the LTP window is $A_+ \tau_+ r_i r_j$.

A similar argument based on the amount of LTD per presynaptic spike leads to the mean rate of synaptic change due to the LTD window as $-A_- \tau_- r_i r_j$.

Thus, the net rate of plasticity is proportional to the rate of production of spike-pairs, which is the product of the two firing rates, $r_i r_j$, and the mean level of potentiation or depression per spike pair, which is equal to the total area of the STDP curve, $A_+ \tau_+ - A_- \tau_-$.

2.   Plasticity Due to Only the Most Recent Spike

In the second formulation, whenever the postsynaptic neuron emits a spike, the amount of LTP is $A_+ \exp[-T_D / \tau_+]$, where T_D is the time since the most recent presynaptic spike. Now, to obtain the expected (mean) amount of LTP per postsynaptic spike we multiply the amount of LTP for each possible time interval by the probability of that time interval—which is $r_i \exp(-r_i T_D)dT_D$ for a Poisson process—then sum over all possible intervals. Or, using calculus to convert the summation to an integral, we have:

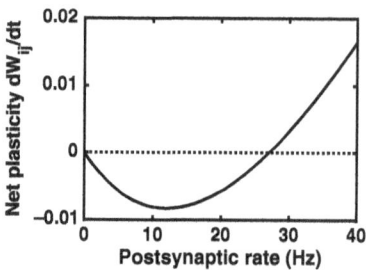

Figure 8.13
Rate-dependence of STDP with nearest-neighbor rule. A plot of dW_{ij}/dt using equation 8.16, using parameters $A_+ = 0.01$, $A_- = 0.011$, $\tau_+ = \tau_- = 20$ ms, $r_i = 20$ Hz, and variable postsynaptic rate, r_j. Notice the similarity in shape with figure 8.2. This figure was produced by the online code `STDP_Poisson_nn.m`.

$$\int_0^{\infty} A_+ \exp[-T_D / \tau_+] r_i \exp(-rT_D) dT_D = \frac{A_+ \tau_+ r_i}{1 + \tau_+ r_i}. \tag{8.15}$$

Since the rate of postsynaptic spikes is r_j the mean rate of change of synaptic strength due to the LTP window is $A_+ \tau_+ r_i r_j / (1 + \tau_+ r_i)$. Notice that this is like the result when all spikes are included, but includes the denominator of $(1 + \tau_+ r_i)$, which becomes important if the presynaptic rate is not much less than $1 / \tau_+$.

Similarly, the mean rate of change of synaptic strength due to the LTD window is $-A_- \tau_- r_i r_j / (1 + \tau_- r_j)$ so that the net expected rate of change of synaptic strength is:

$$\frac{dW_{ij}}{dt} = r_i r_j \left(\frac{A_+ \tau_+}{1 + \tau_+ r_i} - \frac{A_- \tau_-}{1 + \tau_- r_j} \right). \tag{8.16}$$

Interestingly, if we keep the presynaptic rate, r_i, fixed and vary the postsynaptic rate, r_j, the resulting dependence is biphasic, with increasing net depression at low postsynaptic rates switching to potentiation at high postsynaptic rates (figure 8.13). However, unlike the observed data, equation 8.16 indicates that with this rule depression is favored at high presynaptic rates and potentiation is favored at low presynaptic rates.

8.12 Appendix B: Rate-Dependence of Triplet STDP between Uncorrelated Poisson Spike Trains

Here we assume plasticity is due to all pairs of spikes. Since the model of depression is unchanged, the rate of change of synaptic strength due to depression is identical to that of appendix A, part 1, and is $-A_- \tau_- r_i r_j$ (for symbol meanings see appendix A).

For the potentiation term, which we add whenever there is a postsynaptic spike (at time, t_j), we must include all possible times for a preceding postsynaptic spike (at time, $t_{j'}$) as well as all possible times for a preceding presynaptic spike (at time, t_i). The summation over all possible times of preceding presynaptic and postsynaptic spikes requires evaluation of a double integral of the update rule for each triplet of spikes (equation 8.6):

$$\Delta W_{ij} = A_3^+ \int\limits_{-\infty}^{t_j} \exp\left(\frac{t_i - t_j}{\tau_+}\right) r_i dt_i \int\limits_{-\infty}^{t_j} \exp\left(\frac{t_{j'} - t_j}{\tau_3}\right) r_j dt_{j'}. \tag{8.17}$$

In equation 8.17, we have substituted $r_i dt_i$ for the fixed probability of a presynaptic spike in the time interval t_i to $t_i + dt_i$, and $r_j dt_{j'}$ for the fixed probability of a postsynaptic spike in the time interval $t_{j'}$ to $t_{j'} + dt_{j'}$. The integrals can be evaluated to give $\Delta W_{ij} = A_3^+ r_i r_j \tau_+ \tau_3$. Since this is the average increment in synaptic strength per postsynaptic spike, the rate of change of synaptic strength due to potentiation gains an extra factor of the postsynaptic firing rate, r_j, so is $A_3^+ r_i r_j^2 \tau_+ \tau_3$.

Combining the rates of potentiation and depression we see that the net rate of change of synaptic strength is (on average):

$$\frac{dW_{ij}}{dt} = r_i r_j \left(A_3^+ \tau_+ \tau_3 r_j - A_- \tau_- \right). \tag{8.18}$$

It can be seen that this term is quadratic in the postsynaptic rate, r_j, and switches from net depression at low postsynaptic rates ($r_j < A_- \tau_- / A_3^+ \tau_+ \tau_3$) to potentiation at high rates ($r_j > A_- \tau_- / A_3^+ \tau_+ \tau_3$). Therefore, equation 8.18 produces a curve that is identical to the one proposed for firing-rate models, shown in figure 8.2.

9

Analysis of Population Data

Given the vast numbers of neurons involved in cognition and information processing, it is perhaps surprising that observation of a single neuron's activity has ever yielded much insight into brain function. The focus through much of the history of systems neuroscience has been on the analysis of those neurons whose firing rates change strongly and reliably in a particular direction, either in response to a stimulus or as an indication of a forthcoming motor response. This focus arose in part because of limitations of the experimental apparatus available—when recordings of individual neurons are hard to come by, it is important to find and select those neurons with the strongest task-dependent response. Moreover, the observation of a distinct change in a neuron's firing rate is both easier to explain and more dramatic—therefore, more newsworthy and publishable—than any correlation with behavior extracted more painstakingly from changes that are subtle if observed at the level of the individual neuron. Therefore, the desire to find individual neurons with strong, task-specific responses continues today.

Models of neural circuits—such as the majority of those presented in this book—are similarly focused on explaining the strong, task-dependent responses of certain neurons. In part, such models are easier to formulate and understand than those in which the encoding of information is almost impossible to pinpoint. In this book, the one network we have considered with near impenetrable encoding is the network of granule cells in the cerebellum, which in our model (tutorial 8.4) encoded time. While the network activity can produce a temporally precise change in Purkinje cell activity, we would be hard-pressed to find a "time-tuned" neuron in the network. It is quite likely that much neural activity beyond the primary sensory and primary motor areas of any animal's brain is of this ilk.[1–9] We spend less time on such circuits in this introductory book, simply because they are far from being understood and they are more difficult to simulate (see refs. 5, 10–16 for examples and further reading)—not because they are less important for brain function.

The first step toward understanding the mechanisms, or even the general principles and algorithms involved, when they are not revealed by straightforward changes in the firing rates of individual neurons, is to employ appropriate methods to combine activity from many cells.[17–21] In this chapter, we consider three such methods of analysis, each of which can be considered the simplest prototype of a large set of related methods.

9.1   Principal Component Analysis (PCA)

Principal component analysis (PCA)   A method for reducing high-dimensional data to fewer dimensions in a manner that aims to minimize loss of information.

Dimensionality of data   The number of distinct values, or coordinates, used to define one data point. The dimensionality can be the number of variables in the system, or the number of variables multiplied by the number of time points measured.

Before describing PCA, it is important to understand the general concept of "high-dimensional data" and why "dimensionality reduction" can be useful. The ideas are of general applicability and value to multiple fields of data analysis, ranging from the sorting of spikes (described in the next subsection), to the unbiased scoring of elected officials by analysis of their voting records.

The high dimensionality refers to the number of values used to characterize a data point. For example, a particular curve, such as a voltage trace, $V(t)$, can be characterized by a large number of sampled points along the curve. In this manner, a curve characterized by 100 points along the trajectory can be represented by a single point in 100-dimensional space. A dataset with many such curves would then be represented by one or more clusters of such points in the 100-dimensional space.

Similarly, the complete set of votes on a resolution can be characterized by a set of values, one value for each legislator voting. For example, votes in favor of, or against the resolution, or abstentions, can be ascribed values of "+1" or "−1" or "0" respectively. In this manner, the votes of 100 senators would be represented by a single point in 100-dimensional space. A dataset containing many such voting outcomes would then be represented by one or more clusters of such points in the 100-dimensional space, just as in the previous example.

PCA is one of several methods for extracting useful information from such clusters of data points in high-dimensional space. At the heart of PCA is the idea of choosing a basis in geometry—which means choosing a new set of coordinates. The new basis is chosen based on the variance of the data—and the new coordinates are ordered by the amount of variance they account for. This ends up being useful for several reasons, one of which being that in high-dimensional data, random, uncorrelated fluctuations do not produce such large variance in the data as does correlated variation. In many situations, the correlations among the data points are of prime interest, either because they reveal an underlying mechanism or coordination, or because different underlying processes, which produce distinct sets of correlations, can be separated and revealed.

9.1.1 PCA for Sorting of Spikes

In order to isolate the spikes of individual neurons from multielectrode data, it is common to run PCA on the voltage traces of the electrodes (figure 9.1). In this context, the separate dimensions correspond to the values of voltage in different small time bins around the peak of a spike, as measured on separate electrodes. If, for example, 40 time bins are used to quantify the voltage trace at each spike, then the number of dimensions would be 40 multiplied by the number of voltage traces. With such characterization, it is possible to extract multiple distinct "spike signatures" and identify more neurons than electrodes used. For example, the data shown as overlapping clusters of traces for a single electrode in figure 9.1B appear separable as two distinct clusters of points in figure 9.1C.

As noted, PCA is helpful when there are correlations between the dimensions. For example, in multielectrode recordings of electrical activity, any source of electric field, such as a single neuron, whose spike affects the voltage in more than one electrode, produces a correlation between the values recorded in the different electrodes. That is, whenever the neuron spikes, a certain shape of deflection of the voltage in one electrode will occur at the same time as a different shape of deflection of the voltage in another electrode. A different neuron would produce a different pair of simultaneous deflections in the two electrodes. Therefore, observation of a shape of deflection in one electrode correlates with the shape of deflection caused by the same cell in another electrode. Moreover, the values of a voltage trace at successive time points are highly correlated—if an electrode is at a point of high potential then a fraction of a millisecond later, the potential is likely to remain higher than average.

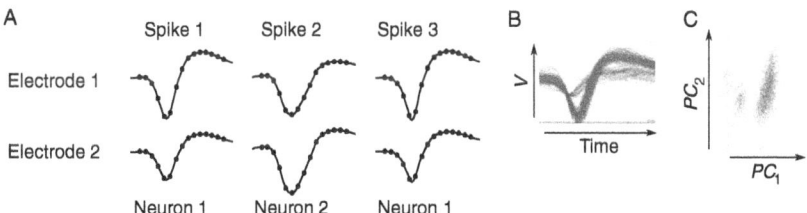

Figure 9.1
Separation of neural spikes by PCA. (A) When a spike is detected, the waveform is characterized by a number of values of the voltage (small circles), which together describe the characteristic shape produced on each electrode. Three spikes are depicted, each spike characterized by fifteen values on each of two electrodes, which would produce a single point in thirty dimensions for each spike. In this illustration, spikes 1 and 3 are similar because they arise from the same neuron, whereas spike 2 is significantly different because it arises from a different neuron. (B) Example traces of the multiple waveforms produced by several hundred spikes on a single electrode implanted in rat gustatory cortex. These traces group around two distinct forms, so most likely arise from two distinct neurons. Notice that the voltage trace, dependent on electric fields outside the neurons, can have a very different shape from the membrane potential of the neuron producing a spike. (C) Each spike appears as a dot in high-dimensional space. The spikes separate into two clusters, representing the two putative neurons. Following PCA, the separation of the clusters is visible here when just the first two principal components of the data are plotted against each other. Forty data points per spike from a single electrode were used for this analysis. (Data provided by the laboratory of Don Katz, Brandeis University.)

While one could imagine designing an algorithm to pick out the important features that differentiate the voltage traces of figure 9.1A or 9.1B, these distinguishing features could vary from electrode to electrode and from cell to cell. The value of PCA in this context is that, once all traces are represented as individual points, it can allow us to see visually whether multiple clusters exist and use simple methods to separate the clusters.

9.1.2 PCA for Analysis of Firing Rates

Once the activities of individual neurons are separated, correlations abound in the dynamics of extracted firing rates. In this case, correlations between the activities of different cells can be caused by connections between the cells within a network, or by common "upstream" input. The interesting signal causing the correlations across dimensions can be more easily observed or extracted by choosing an appropriate linear combination of the measured data. For example, in a binary decision-making task (section 6.5) one may want to add together the firing rates of cells responsive to one choice then subtract all the firing rates of cells responsive to the opposite choice, while ignoring the firing rates of cells with no choice preference. PCA can provide a systematic method for determining how much (and of what sign) each cell's firing rate should contribute to a particular "readout."

When analyzing neural activity, the initial, high number of dimensions is typically the number of neurons, with each dimension indicating the firing rate of a particular neuron. The variation of activity as a function of time would be a curve, or trajectory in the high-dimensional space as neural firing rates covary. If the covariation of activity lies predominantly along a 1D line, or a 2D plane, PCA would reveal this. For example, if the rates of all neurons deviated from their mean by different scaled amounts of the same function of time—i.e., their rate changes were all in proportion to each other—the trajectory would be a straight line in firing-rate space. Even with noise added to each neuron's firing rate, PCA could reveal the straight line and produce a much less noisy estimate of the function of time coordinating the dynamics than that available from analysis of any neuron alone. In so doing, PCA may reveal the most important correlate of a stimulus or behavior to be found in the noisy neural activity. In tutorial 9.1 we will practice such techniques.

9.1.3 PCA in Practice

The first step of PCA is the extraction of a new set of perpendicular axes sorted by the amount of variance of the data in the direction of each new axis. Figure 9.2 indicates how PCA would extract new axes from a pair of neurons whose firing rates are anticorrelated, meaning a higher rate of one neuron coincides with a lower rate of the other neuron and vice versa. In this example, the axis containing the most variance in the data, now labeled PC_1, is at $45°$ to the original axes and corresponds to the direction $r_A + r_B = constant$ (it could as easily have been in the direction direction, $-r_B - r_A = constant$). The second axis extracted, labeled PC_2, must be perpendicular to the first, so in this case is constrained to be in one of the two directions along the line $r_A = r_B$.

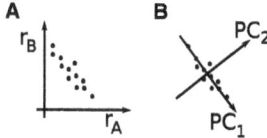

Figure 9.2
Selecting new axes with principal component analysis. (A) Each data point corresponds to a specific measurement and comprises a pair of firing rates, r_A and r_B. (B) The axis corresponding to the first principal component, PC_1, contains the most variance in the dataset. The second axis, PC_2, is perpendicular to the first.

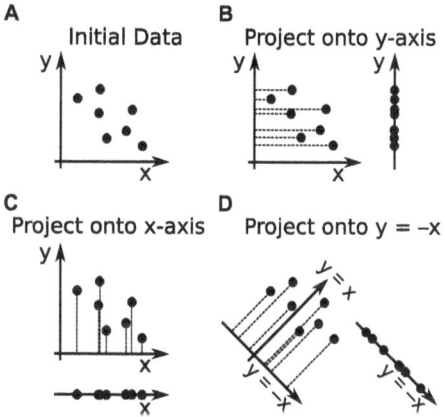

Figure 9.3
Projection of data. (A) A set of data points in two dimensions can be projected to a single dimension in multiple ways. (B–D) Projection of each data point (x_i, y_i) onto an axis is shown as a dashed line. The resulting set of projected data falls on a single line, so each data point is represented by a single number, its y-value (y_i) in B, its x-value (x_i) in C, or the scaled difference between the two $\left((x_i - y_i)/\sqrt{2}\right)$ in D.

Projection of data   Projection onto an axis (or plane) reduces the dimensionality of data in the same way that a 3D image can be projected onto a 2D plane as a photograph—all depth information is lost. For example, if a set of (x, y) points are projected onto the x-axis then only the x-values remain and the y-values are lost.

The variance in the data accounted for by PC_1 is so much greater than that accounted for by PC_2 in this example, that a description of each data point by its first principal component alone—its value of PC_1—would not cause much loss of information. That is, if the data points in figure 9.2B were moved to the nearest point on the PC_1 axis (such a movement is perpendicular to the axis and is called a projection onto the axis—see figure 9.3) then little change in their position would ensue.

If the original data were three-dimensional (so had labels x, y, and z, or equivalently, r_1, r_2, and r_3), it is possible that all the data points fall within a shape that looks like a tilted dinner plate. Such data would be closer to 2D as a plate is approximately a flat surface and a flat surface is a 2D plane. PCA would pick out two directions at right angles to each other within this plane, where the greatest spread in the data is found. If one plotted the data points on coordinates produced by these first two principal components one could see the data sitting essentially in an ellipse. The third principal component (in the direction of the thickness of the dinner plate, perpendicular to its plane) could most likely be ignored with little loss of information about the coordinates of any data point. Reducing the number of dimensions (or coordinates) used to describe data with minimal information loss is a form of data compression that can be used in some forms of data transfer.

Perhaps a key step in the use of PCA, or any other method for analysis of high-dimensional data, is the choice of what components of the data are used to produce new axes and extra dimensions for each data point, versus what components are used to produce new data points. For example, when firing rates of many neurons are measured as a function of time, the dimensionality could be the number of neurons. In this case, each time point would provide a single data point. The set of data points for a single trial can then be joined together as a single curve, indicating the trajectory of neural activity as a function of time. PCA allows us to visualize such trajectories.[5,22]

Alternatively, each time point could add a new set of dimensions, such that the total dimensionality of the system becomes the number of neurons multiplied by the number of time bins. In this case, each trial would correspond to a single high-dimensional point (like each time window around a spike in figure 9.1). Multiple trials are likely to produce a cluster of points, with trials from a different stimulus producing an alternative cluster (as in figure 9.1C). PCA is likely to extract the most significant distinctions between stimuli. These distinguishing features could then be mapped back to distinct signatures in the firing rate trajectories of the neurons.

Covariance   The degree to which two series of data vary in the same direction as each other (positive covariance) versus in the opposite direction to each other (negative covariance). Zero covariance can arise if the two series vary independently of each other, or if either one or both of the series has no variation (zero variance).

Eigenvectors   Eigenvectors are a set of orthonormal vectors—meaning perpendicular vectors of unit length—associated with a particular matrix. The eigenvectors are special, in that when multiplied by the matrix they are simply scaled rather than altered in direction.

> **Eigenvalues** An eigenvalue corresponds to a particular eigenvector and is the amount by which that eigenvector is scaled when multiplied by the matrix.

9.1.4 The Procedure of PCA

In this section, we describe the algorithmic steps that are required for PCA. In this description, each data point will correspond to the values of firing rates of each cell in a single time bin, so the dimensionality is the number of neurons.

1. Center the original data by subtracting the mean in each dimension—i.e., for every data point, subtract the mean (time-averaged) firing rate of the corresponding neuron, so we are analyzing deviations from the mean. The mean values should be stored if firing rates are to be recalculated. Subtraction of the mean facilitates rapid calculation of variances and covariances (recall that variance is calculated using deviations from the mean).

2. Calculate the covariance matrix of the centered data (this can also be done without step 1, since the covariance is mean subtracted). The covariance matrix will be a symmetric $P \times P$ matrix if there are P neurons. Each entry in the covariance matrix is calculated for two neurons, by taking the product of their firing rates together in each time bin and then calculating the mean of these values across time bins. The diagonal of the matrix is simply the variance in firing rate of the corresponding neuron.

3. Find the eigenvectors of the covariance matrix with their eigenvalues. (An eigenvector is a particular direction or axis, such that a vector in this direction, when multiplied by the matrix, is simply scaled in the same direction. The amount of scaling is the eigenvalue—see figure 9.4. Many coding languages can produce eigenvectors and eigenvalues of matrices using a single command—it is `eig` in MATLAB.)

4. Sort, then order, the eigenvectors from the one with the largest eigenvalue to the one with the smallest eigenvalue (all eigenvalues are nonnegative real numbers, since the covariance matrix is real and symmetric). The sum of the eigenvalues is the sum of the variances of firing rates of all neurons. When each eigenvalue is divided by this sum, the result yields the fraction of the total variance of neural activity due to coherent variation of the system in the direction of the corresponding eigenvector.

The eigenvectors, each of which defines a direction providing a new axis, are perpendicular when created, so we have generated a new set of perpendicular axes for the data—i.e., a new basis. The axes are ordered according to the directions with highest to lowest variance in the original data, which is the order of the principal components.

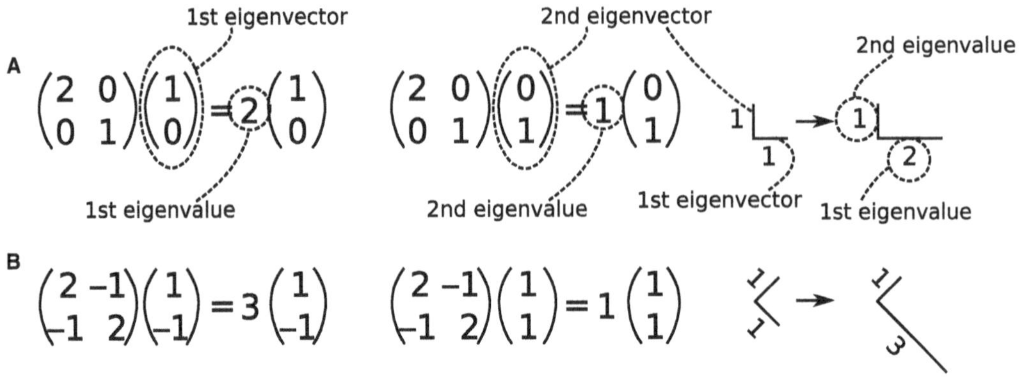

1st eigenvector   2nd eigenvector

2nd eigenvalue

A
$$\begin{pmatrix} 2 & 0 \\ 0 & 1 \end{pmatrix}\begin{pmatrix} 1 \\ 0 \end{pmatrix} = 2\begin{pmatrix} 1 \\ 0 \end{pmatrix} \qquad \begin{pmatrix} 2 & 0 \\ 0 & 1 \end{pmatrix}\begin{pmatrix} 0 \\ 1 \end{pmatrix} = 1\begin{pmatrix} 0 \\ 1 \end{pmatrix}$$

1st eigenvalue   2nd eigenvalue   1st eigenvector   1st eigenvalue

B
$$\begin{pmatrix} 2 & -1 \\ -1 & 2 \end{pmatrix}\begin{pmatrix} 1 \\ -1 \end{pmatrix} = 3\begin{pmatrix} 1 \\ -1 \end{pmatrix} \qquad \begin{pmatrix} 2 & -1 \\ -1 & 2 \end{pmatrix}\begin{pmatrix} 1 \\ 1 \end{pmatrix} = 1\begin{pmatrix} 1 \\ 1 \end{pmatrix}$$

Figure 9.4
Eigenvectors and eigenvalues. (A) The matrix $\begin{pmatrix} 2 & 0 \\ 0 & 1 \end{pmatrix}$ has an eigenvector $\begin{pmatrix} 1 \\ 0 \end{pmatrix}$ directed along the x-axis with an eigenvalue of 2, because multiplication of that eigenvector by the matrix simply scales the vector by a factor of 2. The second eigenvector, $\begin{pmatrix} 0 \\ 1 \end{pmatrix}$, is perpendicular to the first, directed along the y-axis, and has an eigenvalue of 1. Notice that the eigenvectors could equally have been chosen to be in the opposite direction, e.g., $\begin{pmatrix} -1 \\ 0 \end{pmatrix}$ is an eigenvector, and they would still be perpendicular, each with an unchanged eigenvalue. (B) The matrix $\begin{pmatrix} 2 & -1 \\ -1 & 2 \end{pmatrix}$ has eigenvectors rotated by 45°. Since eigenvectors should be of unit length, the actual eigenvectors are $\frac{1}{\sqrt{2}}\begin{pmatrix} 1 \\ -1 \end{pmatrix}$ and $\frac{1}{\sqrt{2}}\begin{pmatrix} 1 \\ 1 \end{pmatrix}$, with eigenvalues of 3 and 1 respectively. Notice that when the matrix multiplies a vector that is not an eigenvector, the result is not a scalar multiple. e.g., $\begin{pmatrix} 2 & -1 \\ -1 & 2 \end{pmatrix}\begin{pmatrix} 1 \\ 0 \end{pmatrix} = \begin{pmatrix} 2 \\ -1 \end{pmatrix}$ means a vector along the x-axis is rotated clockwise as well as scaled by the matrix.

9.2   Tutorial 9.1: Principal Component Analysis of Firing-Rate Trajectories

Neuoscience goal: Use PCA to reduce the noise in a data set.

Computational goal: Rebuild arrays after projection of data onto a subset of axes.

In this tutorial, you will first generate firing rate trajectories of fifty neurons. Each neuron's firing rate is produced by a weighted combination of two input signals, mixed in with a lot of noise. You will then use PCA to produce less noisy firing rates for each cell (denoising) and to extract the input signals from the noisy set of firing rates.

1a. Define two vectors to represent time-dependent oscillating inputs, I_A and I_B, such that $I_A = A\sin(2\pi ft)$ and $I_B = B\cos(2\pi ft)$, with frequency $f = 0.5$ Hz, and amplitudes $A = 20$ and $B = 10$. You can use timesteps of 1 ms and a duration of 10 s.

1b. Set up a matrix to contain the firing rates of fifty neurons across this time span. Generate the firing rate of each neuron as a column in the matrix, with each row representing a separate time point. The time-dependent firing rate of a neuron, labeled i, is given by:

$$r_i(t) = 100 + W_i^{(0)} I_0 + W_i^{(A)} I_A + W_i^{(B)} I_B + \sigma \cdot \eta_i(t),$$

where $I_0 = 50$, the static input weights, $W_i^{(0)}$, $W_i^{(A)}$, and $W_i^{(B)}$, are each independent numbers selected separately, once for each neuron, from the Normal distribution with unit standard deviation and zero mean (defined as $N(0,1)$ via `randn()` in MATLAB). $\sigma = 10$ scales the noise, and $\eta_i(t)$ is a series of normally distributed random variables, $N(0,1)$, selected independently at each time point for each neuron.

1c. Carry out principal component analysis on the rate matrix, using (in MATLAB) the command:

```
[COEFF,SCORE,LATENT,TSQUARED,EXPLAINED,MU] = pca(rate);
```

Note that the `pca` function in MATLAB requires the data in each dimension to be stored as a set of column vectors, with each column being a different dimension. In MATLAB you can type `help pca` to find the meanings of the outputs. The important ones for this tutorial are:

COEFF, which defines the new basis. Each column vector in COEFF is a principal component, which is an eigenvector of the covariance matrix. The vector has unit length with elements given by the relative contribution of each neuron (in this example) to the principal component.

SCORE, which contains the representation of the centered data in the new basis. Each row of SCORE corresponds to a single data point, with each entry its value in the new coordinate system, that is, its projection on each successive principal component.

EXPLAINED, which is a vector containing the percentage of variance explained by each principal component.

MU, which is a vector containing the mean firing rates of each neuron, as needed to recreate the original rates.

1d. Plot the first column of coefficients, COEFF, (corresponding to the contribution of each neuron to the first principle component) against the vector of input weights, $W_i^{(A)}$, and, on a separate subplot, the second column of coefficients (corresponding to contribution of each neuron to the second principal component) against the vector of input weights $W_i^{(B)}$. Does the sign of the slope of any observed trends matter?

1e. Plot the variable, EXPLAINED, to ensure the bulk of the variability in the data is contained in the first two principal components and calculate that fraction explained by summing its first two values.

1f. To denoise the data, you will produce a new matrix of firing rates by multiplying the first two columns of SCORE by the first two rows of COEFF′ (the transpose of COEFF) and adding to all entries of each column produced in this manner the value of the corresponding entry of MU (the mean rate of that neuron).

In this step, we are assuming that once in the coordinates of the principle components, any changes in rates on principal component axes 3 or higher correspond to noise fluctuations that can be ignored. Therefore, we are recreating the firing rates of the original neurons just using the projection of each neuron's firing rate on the axes of the first two principal components.

1g. Compare the behavior of the denoised data with the original data by plotting (on separate subplots) the original rates of two neurons as a function of time, then the corresponding columns of the new matrix to reveal the denoised rates as a function of time.

1h. Select two neurons and plot the rate of one against the other, both using the original noisy rates, and, in a separate subplot, using the denoised rates.

1i. Plot separately the first and second columns of the matrix, SCORE, to show the time-dependence of the system's first and second principal components.

Be sure to comment on all of your observations and relate them to the initial set of inputs.

2. Repeat question 1, but in part (b) define $I_A = A\sin(2\pi f_A t)$ and $I_B = B\cos(2\pi f_B t)$, where $f_A = 1$ Hz and $f_B = 0.5$ Hz, keeping all other parameters the same.

3. Repeat question 1, but define one input as a slowly ramping current, $I_A = At$ (from $t = 0$ to 10 s), and the other as a transient signal during the ramp, $I_B = B\sin(2\pi ft)$ for 4 s $< t <$ 5 s, otherwise $I_B = 0$.

9.3 Single-Trial versus Trial-Averaged Analyses

Most analyses of population data based on the firing rates of neurons initially average over many trials using the methods of chapter 3. Such across-trial averaging can be necessary to produce a smoothly varying firing rate from the series of instantaneous spikes. However, the averaging procedure implicitly assumes that the multiple trials contain fundamentally identical data, and when this is not the case, the point of alignment matters. For example, in the model of a decision-making network (tutorial 6.2) the average activity depended on whether trials were aligned to stimulus onset or to response time.

In the absence of reliable, reproducible circuit dynamics following alignment of activity to an externally identifiable time point, it is preferable to analyze the data with single-trial methods. These are methods that either take into account across-trial variability or do not assume multiple trials. In this chapter, we will consider two such methods, hidden Markov modeling (HMM, section 9.5) and Bayesian filtering (section 9.6).

Single-trial methods are particularly beneficial when multiple neurons are recorded simultaneously, or more generally, when many simultaneously acquired data streams can be combined. Modern methods of electrophysiology, such as the use of high-density electrode arrays,[23] allow for the extracellular recording of spikes from many neurons, even hundreds, simultaneously. Optical imaging, most commonly using calcium-responsive (and increasingly voltage-sensitive) dyes,[24] allow for the activity of an even greater number of cells, perhaps thousands,[25,26] to be observed almost simultaneously (only "almost," because there is typically a need for repeated scanning of the image line by line). Furthermore, noninvasive recordings of neural activity in humans typically acquire multiple data streams simultaneously. For example, using electroencephalography (EEG) and functional magnetic resonance imaging (fMRI), neuroscientists record from tens of electrodes and millions of voxels (volume elements) respectively. All of these methods produce data that have the potential to contain rich information within temporal correlations that may be obscured if the data are averaged across multiple trials.

Electroencephalography (EEG)   The analysis of data acquired from multiple electrodes placed onto the scalp. Each electrode, typically a small flat circle, detects the electric fields produced by neural activity of groups of aligned cells within the brain. The electrodes are usually connected together in a "hood" so as to surround the head.

Functional magnetic resonance imaging (fMRI)   A measurement of blood flow via its oxygenation level in many thousands of tiny cuboids called voxels throughout the brain, using the behavior of hydrogen nuclei in very strong magnetic fields. Since increased neural activity causes increased blood flow to the active neurons, fMRI indicates which voxels contain responsive neurons and how those voxels are correlated as a subject carries out different cognitive tasks, while lying within a huge magnet.

9.4   Change-Point Detection

As a prelude to HMM, we consider the problem of a single, noisy spike train whose underlying firing rate changes discretely at a single point in time. The goal is, from observations of the spikes alone, to detect if and when the firing rate changes. If the change is drastic, then the change point may be visible by eye alone. If multiple trials are carried out and the change point is at an identical time in each trial, then the PSTH (section 3.1.2) could reveal the change point, even if the rates differ by a small amount. However, with just a single trial and a small change in firing rate, the change point can be estimated only probabilistically. Here we will consider how to find the probability of a change point at any time point in the trial and use the maximum of this probability to estimate the change point in trials when we predict there is such a change. To proceed, we assume spikes are produced with Poisson statistics at unknown underlying rates (see section 3.3.3).

If we split a spike train with a number of spikes, N, and with a duration, T, into two intervals, the first interval containing N_1 spikes and with a duration of T_1, then the second interval contains

$N_2 = N - N_1$ spikes and has a duration of $T_2 = T - T_1$. We can first ask the probability of such an apportioning of spikes by chance.

Although the rates of the two processes are unknown, we can show (appendix B) that the optimal rates are $r_1 = N_1 / T_1$ and $r_2 = N_2 / T_2$. These rates are what one might expect—if we count a given number of spikes in a given time interval our best guess at the rate of any underlying process is the mean rate observed, calculated as number of spikes divided by the time interval.

The probability of the observed sequence of spikes arising from two processes with a transition at a certain time point is equal to the product of the two individual sequences on either side of that time point arising, each with its own rate. Some analysis (appendix B) shows that this product is proportional to:

$$P(r_1, N_1) P(r_2, N_2) \propto (r_1)^{N_1} \exp(-N_1)(r_2)^{N_2} \exp(-N_2).$$ (9.1)

Computationally, since the product of terms $\exp(-N_1)\exp(-N_2) = \exp[-(N_1 + N_2)]$ depends only on the total number of spikes, not on the change-point, we just need to look for the maximum of:

$$(r_1)^{N_1} (r_2)^{N_2} = (N_1 / T_1)^{N_1} (N_2 / T_2)^{N_2}$$ (9.2)

as we vary T_1 (with $T_2 = T - T_1$), to find the most likely change-point.

It should be noted that we have used Bayes' theorem (section 1.4.3, equation 1.32) to obtain this result, assuming a uniform prior on firing rates (i.e., all rates are equally likely prior to the observation) and a uniform prior on the time of the change point (the change in rates is equally likely to occur at any point in the time window prior to observation). Given those assumptions of equal priors (which defines a maximum likelihood estimate), we have been able to assume the probability of a particular firing rate and time interval given the spike train is proportional to the probability of that particular spike train given the firing rate and time interval. That is, Bayes' theorem simplifies:

$$P(r_i, T_i | \{s\}) = \frac{P(\{s\} | r_i, T_i) P(r_i, T_i)}{P(\{s\})} \propto P(\{s\} | r_i, T_i),$$ (9.3)

where the final proportionality is valid if the prior, $P(r_i, T_i)$, is independent of r_i and T_i. In equation 9.3, r_i is the rate and T_i is the duration of any interval of fixed rate and $\{s\}$ indicates the set of spike times within the interval.

We can take the calculation of equation 9.2 one step further and compare the likelihood of the change point at a given time, T_1, to the likelihood of the spikes arising from a process with the same fixed rate at all times. This requires a division of equation 9.2 by $(N / T)^N$ and produces a likelihood ratio (figure 9.5). Simulations suggest that a likelihood ratio that exceeds 50 for some values of T_1 is an indication of a change in rates within the entire time interval (see tutorial 9.2). Again, the value of T_1 that produces the greatest likelihood ratio is the value best chosen as the change point. Using this method, the subintervals so obtained can be further investigated to assess whether there is evidence for an additional transition in firing rates within each subinterval.

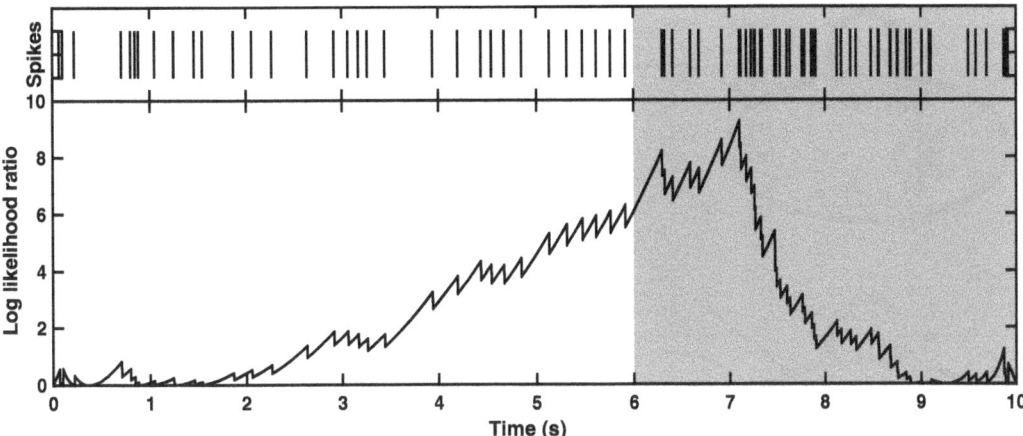

Figure 9.5
Detecting a change in firing rate. (Top) Each vertical line indicates the time of a spike generated by a Poisson process whose rate changes from 5 Hz (white region) to 10 Hz (gray region) at a time of 6 s (the change point). (Bottom) The log likelihood ratio is the log of the probability of the spikes being produced by a Poisson process whose rate changes at that given point of time divided by the probability of those spikes being produced by a process whose rate does not change. In practice, it is rare that the log likelihood ratio exceeds 4 in the absence of a rate change. This figure was produced by the online code `single_jumprate_detect.m`.

9.4.1   Computational Note

When calculating the probabilities of spike trains, the product of many quantities yields terms that grow exponentially with the number of spikes, N. These terms can produce "overflow"—a condition where the corresponding numbers are greater than those able to be stored given the computer's standard memory allocation per number—or, when the numbers are combined to calculate the probability, "underflow"—a condition where the result can be so small that the computer equates the very small probability to zero. Such problems can be avoided by calculating the logarithm of the probabilities (or the log likelihood ratio in this case). The product of individual probabilities then converts to the sum of their logarithms, producing numbers on the order of N, which can be dealt with easily. The point of maximum probability is identical to the point at which its logarithm is a maximum, so the latter can be used to estimate a change point.

9.5   Hidden Markov Modeling (HMM)

Markov process   A dynamical process based on states, with the probability of a state transition per unit time being based only on the current state, not on the prior history of states, except in so far as such history led to the current state. In this sense, Markov processes are considered memoryless.

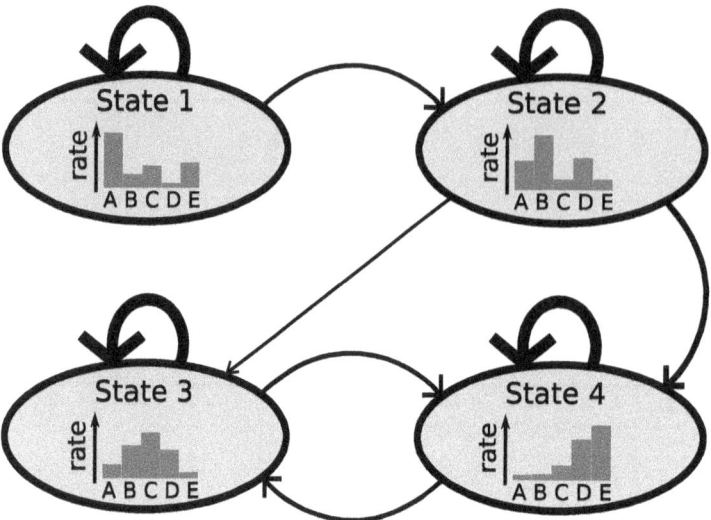

Figure 9.6
Example of a hidden Markov model. A model with four states (ovals), as shown, would be defined by a 4 × 4 transition matrix, indicating the probability per time bin of a transition from one state to another (represented as arrows between ovals). In models of neural activity, the strongest transition is found to be from a state to itself, indicating the persistence of a state across many time bins. Each state is defined by the probability of spike emissions, defined by the firing rate of each cell. In this example, rates of five neurons (labeled A–E) define the states and are depicted as histograms. The set of rates is unique to each state.

A hidden Markov model (HMM) describes a system in terms of discrete states with transitions between them (figure 9.6). The parameters of the states are not directly observable, but can be estimated from the times of emitted events, which in our formulation will be the spike trains. In this sense, hidden Markov modeling resembles the change point detection method of the previous section, where the spike trains are observed, but the underlying firing rate (the hidden parameter) is unknown. Also like the change point detection method, a hidden Markov model can produce the probability of a change in state as a function of time, when initially the number of state changes and their timings (if there are any) are unknown.

Hidden Markov modeling works well in the analysis of neural activity patterns, when multiple neurons change their rates coherently. For example, if one neuron were to change its firing rate from 5 Hz to 10 Hz, it would take a few hundred milliseconds around the transition point to accumulate enough spikes to signify a rate change. However, if ten such neurons all changed their rate at the same transition point, both the duration of trial needed to detect the change and the uncertainty in the time of the transition would reduce by tenfold.

Transition matrix A matrix denoting the probability per timestep of the system making a transition from the current state (the row index) to another state (the column index). Since the system will be in one state or another on the next timestep, the elements on each row of a transition matrix sum to one. The diagonal elements correspond to the probability of remaining in that state on successive time points.

Emission matrix A matrix denoting the state-dependent probability per timestep of emitting a particular event, which in neural recordings could be a spike from a particular neuron, a set of multiple spikes, or an absence of spikes. Since the set of events must cover all possibilities, the entries in each row of an emission matrix, which corresponds to the probabilities of each event in a given state, must sum to one.

An HMM possesses the Markov property in two manners. The Markov property means that the probability of a transition or an emission depends only on the current state of the system, not on the past sequence of states. The first manner in which the Markov property arises in HMMs is in the transition probability, which can be represented as a matrix denoting the probability per unit time of a transition from one state (which must be the current state) to another. Given the current state, the probability per unit time of transitioning to any new state is constant, independent of how long the system has spent in the current state. The second manner in which the Markov property arises in HMMs is in the emission probability, which again can be represented as a matrix denoting the constant probability per unit time of any event given the current state. In practice, for neural spike trains, this means that spikes are emitted as a Poisson process (section 3.3.3).

Hidden Markov models are more complicated than the example of change point detection we saw in the prior subsection, for several reasons:

1. Spike trains from multiple neurons contribute to the analysis, so an individual "event" in a small time bin is not just the emission or absence of a spike, but can be a spike from any one of the neurons, or a combination of spikes from different neurons. Related to this, a state of the system would be described by the probabilities of all of the possible "events," requiring as a minimum the firing rates of all neurons.

2. The number of possible states is usually greater than two, so several sets of firing rates coordinated across the measured neurons is common.

3. Transitions can be possible in a nonsequential order, with reverse, or back-and-forth transitions also possible (for example, commencing from state 1 in figure 9.6, one can follow arrows to produce a sequence of states such as 1–2–4–3–4–3).

4. The output of HMM produces the probability per unit time of being in each state and so provides an indication of the abruptness of any state changes.

The model is then defined by two matrices, the matrix of emission probabilities and the matrix of transition probabilities, plus an initial state, all of which are optimized by an iterative algorithm.[27,28]

Iterative algorithm   A computational procedure that repeatedly loops through a set of calculations, using the result of the previous loop as an input to the next loop. An initial input must be provided, which often constitutes an initial "guess" at the solution, which gets improved via an optimization process on successive loops through the code.

In the example shown in figure 9.6, the emission probabilities for the five neurons should be contained in a matrix with a size of at least 4×6, like:

$$
\begin{pmatrix}
0.20 & 0.03 & 0.05 & 0.01 & 0.06 & 0.65 \\
0.10 & 0.20 & 0.03 & 0.10 & 0.03 & 0.56 \\
0.05 & 0.12 & 0.15 & 0.10 & 0.02 & 0.56 \\
0.01 & 0.01 & 0.04 & 0.14 & 0.20 & 0.60
\end{pmatrix}.
$$

The matrix must have four rows, one for each state (1–4), and at least six columns, given the possibility in a time bin of a spike from any of the five neurons (A–E) plus the possibility of an absence of spikes in the time bin. Each entry in such an emission matrix is the probability per time bin of a spike being produced by a given neuron (or no neuron) in the corresponding state. In this formulation, the possibility of multiple spikes per time bin is excluded and the first five values in each row are proportional to firing rates of the neurons (the histograms of figure 9.6) while the last column ensures that the rows sum to one. When more than one spike occurs in a time bin in the data, one would have to shift the randomly chosen lost spikes to neighboring time bins, or lose them (in which case rates would be slightly underestimated in the model). More accurately, combinations of multiple spikes can be included in a more extensive emission matrix, with each possible spike combination producing an extra column in the emission matrix.

In the same example, the transition probabilities would be contained in a 4×4 matrix like:

$$
\begin{pmatrix}
0.98 & 0.02 & 0.00 & 0.00 \\
0.00 & 0.97 & 0.01 & 0.02 \\
0.00 & 0.00 & 0.98 & 0.02 \\
0.00 & 0.00 & 0.02 & 0.98
\end{pmatrix}
$$

where the large entries on the diagonal indicate that across consecutive time bins the system is most likely to remain in the same state—a common result of such analysis of neural data[29-31]—while the off-diagonal entries indicate the probability of transitioning from one state (given by a row) to another state (given by the column).

Alongside the model, which is always optimized according to a particular dataset, is the analysis of the data in terms of the model. Such analysis yields the probability, in each time bin, of the system being in one state or another. Rapid changes in these probabilities are indicative of sudden changes in the firing rates underlying the emission of the observed spikes. A single model can be trained from multiple trials of a task, with the final analysis yielding different timings of transitions and even different sequences that best match the data on the different trials.[30-32]

In summary, HMM assumes all trials follow the same probabilistic framework for the progression of neural activity through distinct states, but allows for trial-to-trial differences in the activity patterns that arise from the framework. For example, during perceptual bistability (section 7.5) the same perceptual states are present on different trials, but both the transition times between percepts, and the identity of the initial percept, can vary. Therefore, the trial-averaged data would merge the activity arising from each of the two percepts and obscure the most important features of the neural dynamics that correspond to the perceptual switches.[33, 34] In this and similar examples, the underlying neural activity is best analyzed via HMM (figure 9.7).

9.6   Tutorial 9.2: Change-Point Detection for a Poisson Process

Neuroscience goal: Learn how to extract state changes from noisy spike trains and learn some limitations of the process.

Computational goals: Calculate log-likelihood ratios to combine probabilities; produce scatter plots and a correlation coefficient.

In this tutorial, you will produce a Poisson process for a single neuron, whose rate changes abruptly at a randomly determined time point within a trial. You will produce multiple such trials and assess how well a change point detection algorithm extracts the correct change point. Explain what you observe.

1. Define a set of fifty different change points, each randomly chosen in the interval from 0 to 10 s.

2. Produce a set of fifty spike trains, one spike train for each value of the change point, with Poisson emission of spikes at a rate $r_1 = 5$ Hz before the change-point and at a rate $r_2 = 10$ Hz after the change-point.

3. For the first spike-train, plot as a function of T_1, which can vary in 1 ms increments from 0.001 s to 9.999 s, the log likelihood ratio indicating the probability of the spike train being produced by an inhomogeneous Poisson process with a rate jump at T_1 divided by the

A Data

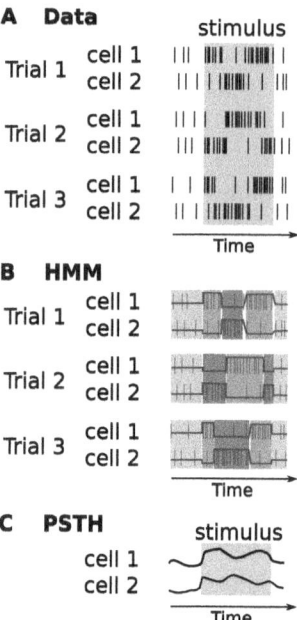

B HMM

C PSTH

Figure 9.7
HMM extracts trial-specific state sequences. (A) Sketch of spike trains from two cells, which respond preferentially to one of the two percepts that arise when a stimulus induces the bistable state. On separate trials the duration of percepts and the initial percept can vary. (B) A hidden Markov model of the system would, most likely, extract three states (indicated by the shade of gray), one prestimulus and poststimulus, the other two corresponding to the distinct activity states present during the distinct percepts when the stimulus is present. The firing rates, shown as continuous bold lines, transition between distinct values, one value for each cell within each state (see figure 9.6). State transitions are evaluated probabilistically, with short periods without shading between states representing periods when the model does not predict a single state's presence with high probability. (C) A peristimulus time histogram (PSTH), like other standard analyses, which begin by averaging data across trials aligned to stimulus onset, would obscure evidence of these state transitions.

probability of it being produced by a single homogeneous Poisson process. That is, you can use the formula:

$$P(T_1) = \left(\frac{N_1}{T_1}\right)^{N_1} \left(\frac{N_2}{T_2}\right)^{N_2} \Big/ \left(\frac{N}{T}\right)^{N}$$

which yields

$$\ln[P(T_1)] = N_1 \ln\left[\frac{N_1}{T_1}\right] + N_2 \ln\left[\frac{N_2}{T_2}\right] - N \ln\left[\frac{N}{T}\right],$$

where $T = 10$ s, $T_2 = T - T_1$, N is the total number of spikes in time T, N_1 is the number of spikes before T_1, and $N_2 = N - N_1$ is the remaining number of spikes after T_1 (in the subinterval of length T_2).

4. Estimate the change-point as the value of T_1 with maximum log-likelihood ratio.

5. Plot the estimated firing rate as a function of time, assuming estimated rates of N_1/T_1 and N_2/T_2 respectively before and after the estimated change-point. On the same graph plot the firing rate used to generate the spike train.

6. Calculate the square root of the mean-squared error in your estimate of the firing rate, and the square root of the mean-squared error if the firing rate were assumed to be fixed (at N/T) for the entire trial. Plot these two values as a point on a scatter graph.

7. Repeat the calculation in (3), but not the plot of the log likelihood ratio, and repeat (4)–(6), for all fifty trials.

8. Calculate the correlation between the estimated change point and the change point used to generate each spike train. Plot these values on a scatter plot and comment on the results.

Hippocampus   Named after its shape as "seahorse," a region of the brain underneath the cerebral cortex, known to be essential for formation of memories in context and of sequences of events (episodic memory). The hippocampus is also highly involved in navigation, since its neural activity is location-dependent.

Place field   The receptive field of a neuron whose firing rate is greatest at specific locations in the environment. The place field is the set of locations producing high firing rate in such a cell, which is called a place cell.

9.7   Decoding Position from Multiple Place Fields

Neurons in the hippocampus have firing rates that depend on the spatial location of an animal. Within a particular environment, the place fields—meaning the set of locations at which a neuron fires—are reliable for periods ranging from a few hours or days, possibly up to several weeks,[26] and are compact, peaking at a particular location in space and decreasing monotonically with distance away from the peak. Cells with such responses are called place cells.

In this section, we will consider how information from the spike trains of multiple place cells can be combined to provide an estimate of an animal's position that is more precise than one might expect given the number of neurons and size of their place fields. The method requires the multiplication together of probability distributions, where each information source can provide one such distribution. For a review of combining probabilities, see section 1.4.3. Two main principles are required in these methods:

1. When combining information from multiple sources, the probability of an occurrence is calculated by multiplying together the contributions from the separate information sources.

2. The sum of probabilities over all possible occurrences yields one. When decoding an animal's position, an occurrence is equivalent to the animal being at a particular location, so the probability distribution is the probability of the animal being at any point in space. This second principle means that the sum over all locations of the probability distribution is one—that is, the animal cannot be in two places at once and it must be somewhere.

An important aspect of the approach used here is the incorporation of the most recent estimate of an animal's position as a probability distribution in the calculation of the estimate of its position at the next time point. Specifically, the latest estimate of position produces the prior probability at each location for the next estimate, as shown in figure 9.8B. The prior probabilities need not be identical to the previous position estimate, because the rat is known to move and its typical movements—how much it shifts position between time points—can be taken into account in the update (see below). Therefore, three distinct sources of information are combined together to update the estimate of the animal's position:

1. The probability distribution produced for the previous estimate of position.

2. A model of the animal's typical movements to update the previous estimate of position into a prior for the next estimate of position.

3. A probability distribution for each cell, proportional to the place field of each cell that spikes, or with a dip at the place field of each cell that does not spike. These probability distributions are multiplied together with the prior to produce a new position estimate.

These methods, called Bayesian filtering,[35] were developed by Emery Brown, Uri Eden, and colleagues[36,37] and are similar to a method called Kalman filtering (see, e.g., Deneve et al.[38] for an example and see Faragher[39] for an introduction).

When decoding position from neural activity in this manner, two stages are required (figure 9.8):

Stage 1: The position must be observed while spikes from each cell are counted. The firing rate of a cell in each position can then be calculated as the number of spikes produced while in a position divided by the total time spent in that position. For such a calculation, one can split the continuous environment into a square grid. Alternatively, as the resulting set of spike counts will be quite noisy, a smooth function can be fitted to the data. The most common function is a two-dimensional Gaussian, which has a maximum at a particular location and a spread whose standard deviation can vary as a function of direction to define an ellipse, whose long axis can have any direction. For the code used to produce figure 9.9, we just assume circular two-dimensional Gaussians, such that the standard deviation is the same in all directions. These two-dimensional Gaussians define the place-fields of each neuron.

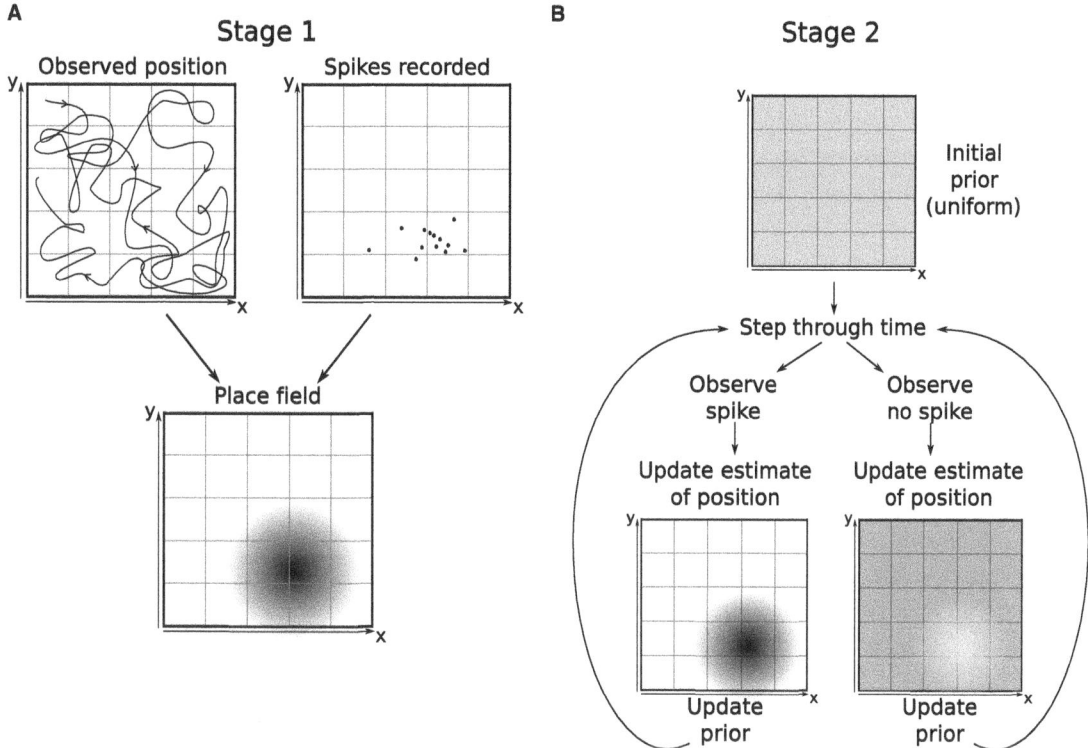

Figure 9.8

Algorithm for decoding position from place fields. (A) In stage 1, a place field (bottom) is calculated for each neuron, by combining the observed position of the animal as a function of time (trajectory, top left) with the spikes emitted by a neuron (dots, top right). (B) In stage 2, the position at each time point is estimated as a probability distribution, which is updated every timestep. An initial prior distribution is assumed to be flat, with no evidence for a particular location of the animal (top). If a neuron emits a spike (left), the neuron's place field is multiplied by the prior, to obtain an updated estimate of the animal's position (bottom left). The updated estimate becomes the new prior in the next timestep. When a neuron does not emit a spike (right), the estimate of position is also impacted, and a new prior is produced (bottom right). In this figure, the darker shading of a region indicates the greater the firing rate of the neuron when the animal is at that location in A, and the greater the probability of the animal being at that location in B.

Stage 2: The estimated place-fields are used to decode the position of the animal using the subsequent spike trains. The firing rates can be converted to a probability of a spike in a small time bin. These probabilities can be subtracted from one to obtain the probability of no spike in a small time bin. Bayes' theorem must then be used to obtain the probability of the animal being in a particular location given a spike (or no spike) in a time bin using the probability of a spike (or no spike) when in that location that was calculated in stage 1. For example,

$$P(position|spike) = \frac{P(position)}{P(spike)} P(spike|position) \qquad (9.4)$$

where

$$P(spike) = \sum_{all\ positions} P(spike|position). \tag{9.5}$$

Use of Bayes' theorem (equation 9.4) for each neuron's place-field is then akin to normalization, so that each probability distribution for location given a spike (or no spike) from a cell sums to one.

The procedure then is to step through time. First, take the previous probability distribution for the animal's location. Alter that distribution as necessary to indicate how the animal could change its location from one time bin to the next, to produce a prior. Then, for each neuron, successively multiply the prior either by its probability distribution for location given a spike (when it spikes, figure 9.8B bottom left) or by its probability distribution for location given no spike (when it does not spike, figure 9.8B bottom right). Finally normalize the resultant distribution so that it sums to one.

The whole procedure can be called Bayesian filtering, because the method of combining the prior probability distribution with the incoming information is an implementation of Bayes' rule. That is, we can expand equation 9.4 for the whole process and write it as Bayes' theorem:

$$P(position|set\ of\ spikes) = \frac{P(position)\,P(set\ of\ spikes|position)}{P(set\ of\ spikes)} \tag{9.6}$$

In equation 9.6, the term $P(position)$ is the prior, based on the probability of a position at the previous time-point and on the model of the animal's movement. The terms $P(set\ of\ spikes|position)$ and $P(set\ of\ spikes)$ are both produced by multiplying together the corresponding probabilities for each neuron.

In the online code (`place_decoding.m`) used to produce figure 9.9, we simulated a random walk for the animal's movement in stage 1. We can then use the random-walk assumption in the model needed to update the prior on the animal's position across timesteps. For example, if we knew the animal's location precisely at a previous time point, then, without any information from neural activity, we would expect the animal to be within a small distance of that prior position, with equal likelihood in any direction. Therefore, without information from spikes, the estimate of position gets less certain—the peak of high probability spreads out—as time goes on. In stage 2 of the code, therefore, with possible positions given as grid-squares, we take the probability distribution of the previous time point and allocate a fraction of the probability on each grid-square equally to neighboring grid-squares. In this manner, we update the position estimate in terms of the animal's movement.

Alternatively, in a more sophisticated method, one could select specific directions based on the prior movement of the animal and the observed tendency to keep going in the same direction. In general, any observed regularity in motion can be incorporated by modifying the update from the previous probability distribution when producing the prior. Similarly, any preference for certain positions in the environment can be incorporated by multiplying by an additional prior

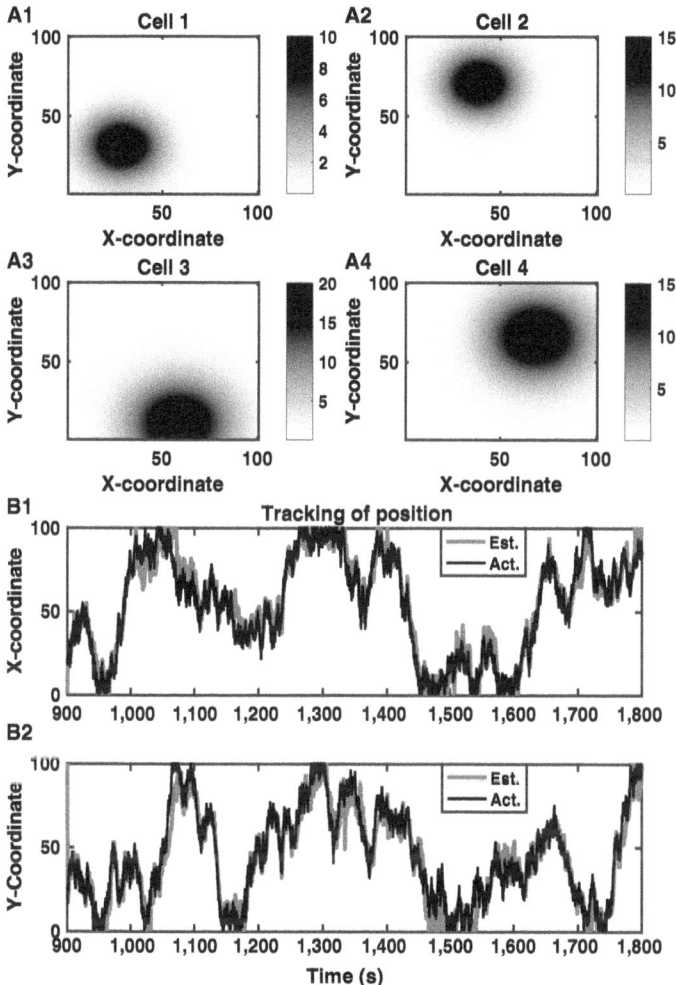

Figure 9.9
Decoding of position using the spikes produces by place-cells. (A1–A4) Circular place fields of simulated cells with the neurons' firing rates peaked at a particular location (higher firing rate is indicated by the darker shading—see color bars, with rate indicated in Hz). (B1–B2) An adaptive decoder allows spikes from the four cells represented in (a) to update *estimates* of the position (gray curves, Est.), tracking the *actual* position (black curves, Act.) with a high degree of accuracy. Indeed, the error between estimates and actual position is generally a lot lower than the standard deviations of the mostly nonoverlapping place fields shown in A. (B1) Estimate and actual value of the x-coordinate as a function of time. (B2) Estimate and actual value of the y-coordinate as a function of time. The second half (shown) of an 1800-s trial was used for decoding (figure 9.8B), while the first half (not shown) was used for estimation of the place fields (figure 9.8A). This figure was produced by the online code `place_decoding.m`. If you run the code, you will see the two-dimensional probability distribution for the estimate of position evolve over time.

proportional to the relative preferences for each position. In general, any prior information can be translated into a probability distribution, which multiplies any current estimate of position. By incorporating more information, such priors tighten the estimate of position, but they could lead to errors, in particular if a probability of zero is given to any location. For example, one might be tempted to provide a probability of zero to any location where the animal has not been observed, but this would be an error—just because the animal has not done something in the past does not make it impossible in the future, yet a probability of zero implies impossibility.

Questions for Chapter 9

1. In a group of ten neurons recorded during a task, the firing rates of six of them vary in proportion with each other, while the firing rates of the other four decrease from 50 Hz in proportion with each other. The mean rate of the group of four is 50 minus 1.5 times the mean rate of the group of six. How many principal components are needed to explain the full variance of the set of ten neurons?

2. When two neurons are recorded during a binary decision-making task, the first principal component is their mean firing rate.
    a. What is the second principal component?
    b. What would it mean about their tuning if the first principal component of activity is most correlated with the decision?
    c. What would it mean about their tuning if the second principal component of activity is most correlated with the decision?

3a. You observe a 1-second spike train produced by an inhomogeneous Poisson process with rate of 4 Hz in the first half-second, then 8 Hz in the second half-second. What is the probability you see more spikes in the first half-second than the second half-second?

3b. What is the probability that a homogeneous 4 Hz Poisson process, observed for a one-second interval, produces at least twice as many spikes in one half of the interval compared to the other half?

4. Two neurons with place fields of identical shape but with different centers, one with a peak rate of 20 Hz, the other with a peak rate of 10 Hz, each emit two spikes in a small, 50 ms, time window. Explain as quantitatively as you can, where the decoder of section 9.7 would place the most likely location of the animal.

9.8   Appendix A: How PCA Works: Choosing a Direction to Maximize the Variance of the Projected Data

The first "principal component" is the direction maximizing the variance of the data when the data are projected onto that direction. To understand this section, a bit of linear algebra is needed, along with the geometric idea of choosing a basis—choosing a new set of axes in terms of the original axes.

Here we consider the original "basis" or set of directions to be given by x, a vector in P dimensions. The original set contains N data points $\{x(1), x(2), x(3), \ldots x(N)\}$. In our examples, the N data points correspond to the N values of time at which firing rates are calculated. Then at each time point the vector, $x(t)$ has P components corresponding to P cells. For example, $x(1) = [x_1(1), x_2(1), x_3(1), \ldots x_P(1)]$, where the (1) indicates the first time-point and the subscripts indicate the cell labels.

We will generate a new basis, u, which will also have P components, where each component in the new basis is a linear combination of components of the old basis,

$$u_i = \sum_{j=1}^{P} M_{ij} x_j. \tag{9.7}$$

In equation 9.7, the matrix, M, simply tells us how each of the new directions in u depends on the old directions in x. In particular, the first principal component (which we desire to be the direction with most variance) is given by:

$$u_1 = \sum_{j=1}^{P} M_{1j} x_j.$$

The matrix, M, is an orthonormal matrix, meaning each row is a vector of unit length at right-angles to all others. The sum over the index j means each entry in a row of the matrix, M, is multiplied by a corresponding entry in the column vector x, then all products are summed together—i.e., the notation describes matrix multiplication (see section 1.4.2).

To proceed (and to make things easier) we will assume that we have already subtracted the mean of each component to produce our data, so each data point has the mean rate of each neuron subtracted. Thus, the mean of all the data points is now zero. Our new basis will be a new set of orthogonal axes going through the same origin, so the mean of the data points in the new basis will also be zero. The variance in one direction such as the principal component is given simply by its mean-square value now. Therefore, we wish to choose the first principal component, $u_1 = \sum_j M_{1j} x_j$ to maximize the sum of the squares of the data points projected onto its axis:

$$\left[\sum_{n=1}^{N} u_1(n) \right]^2 = \left[\sum_{n=1}^{N} \sum_{j=1}^{P} M_{1j} x_j(n) \right]^2$$

$$= \left[\sum_{m=1}^{N} \sum_{k=1}^{P} M_{1k} x_k(m) \right] \left[\sum_{n=1}^{N} \sum_{j=1}^{P} M_{1j} x_j(n) \right]$$

$$= \sum_{j=1}^{P} \sum_{k=1}^{P} M_{1k} M_{1j} \sum_{n=1}^{N} \sum_{m=1}^{N} x_k(m) x_j(n) \tag{9.8}$$

$$= \sum_{j=1}^{P} \sum_{k=1}^{P} M_{1k} M_{1j} C_{kj}$$

where C is the covariance matrix of the original data—a symmetric matrix ($C_{kj} = C_{jk}$) which would be a diagonal matrix (with individual variances on the diagonal) if all neurons had zero correlation with each other.

In order to show that the direction of maximum variance is an eigenvector, we can use the method of Lagrange multipliers outlined in the following. Since eigenvectors must be normalized, there is a constraint:

$$\sum_{k=1}^{P}(M_{1k})^2 = 1. \tag{9.9}$$

To find a maximum of the variance (which we achieve by altering any/all of the coefficients, M_{1k}) we want the derivative of the following to be zero:

$$\sum_{j=1}^{P}\sum_{k=1}^{P} M_{1k}M_{1j}C_{kj} + \lambda\left[1 - \sum_{k=1}^{P}(M_{1k})^2\right] \tag{9.10}$$

with respect to variation in a coefficient. If you have not seen Lagrange multipliers, you can think of the last term, with the added Lagrange multiplied, λ, as a trick, which adds "zero" to the first term. Rather, the Lagrange term incorporates the constraint of equation 9.9 in a manner that is satisfied when equation 9.10 is maximized. Taking the derivative to be zero with respect to M_{1l} (to find the value of the l-th coefficient) we have:

$$\delta_{jl}\sum_{k=1}^{P} M_{1k}C_{kj} + \delta_{kl}\sum_{j=1}^{P} M_{1j}C_{kj} - 2\lambda M_{1k}\delta_{kl} = 0 \tag{9.11}$$

or

$$2\lambda M_{1l} = \sum_{k=1}^{P} M_{1k}C_{kl} + \sum_{j=1}^{P} M_{1j}C_{lj} = 2\sum_{j=1}^{P} C_{lj}M_{1j}, \tag{9.12}$$

where we have used the symmetry of the covariance matrix ($C_{lj} = C_{jl}$). Thus, given the definitions of eigenvectors and eigenvalues (figure 9.4), equation 9.12 shows that the vector of components, M_{1l}, is an eigenvector of the covariance matrix, C, with eigenvalue λ. A similar rearrangement shows that the variance of the projected data (equation 9.8) is simply proportional to the eigenvalue, λ. Therefore, the direction with highest variance is the eigenvector of the covariance matrix with highest eigenvalue.

9.8.1   Carrying out PCA without a Built-in Function

The preceding description of principal component analysis indicates how principal components can be extracted using standard mathematical functions, if your software lacks a built-in function for PCA.

Descriptively, the process is as follows:

1. Center the original data by subtracting the mean in each dimension (i.e., for each cell sub-tract its mean firing rate from every data point so we are recording deviations from the mean). Record those mean values to later recreate neural firing rates. These values comprise the vector MU in tutorial 9.1.

2. Calculate the covariance matrix (this can be done without step 1, since the covariance is mean subtracted). This will be a $P \times P$ matrix if there are P neurons, and it is matrix C in the previous section.

3. Find the eigenvectors of the covariance matrix with their eigenvalues (figure 9.4)—nearly all software packages have standard functions for this.

4. Sort, then order the eigenvectors from the one with the largest eigenvalue to the smallest (all eigenvalues are nonnegative real numbers since the covariance matrix is real and symmet-ric). The vector of sorted eigenvalues is the same as LATENT in tutorial 9.1, and if divided by their sum and multiplied by 100 it is the same as EXPLAINED in tutorial 9.1. The matrix of eigenvectors, after ordering, is equivalent to the matrix COEFF in tutorial 9.1.

5. To obtain the projection of the mean-subtracted original data on the principal components, multiply the matrix of centered original data from (1) by the eigenvectors obtained in (4). The resulting matrix of projections is equivalent to the matrix SCORE in tutorial 9.1.

The eigenvectors are now ordered in terms of the principal components. The eigenvectors are orthonormal when created (perpendicular and of unit length) so you will have generated a new set of axes for the data. The axes are ordered according to the directions with highest to lowest variance in the original data.

Each principal component corresponds to a new axis in the high-dimensional space, but whether one direction along that axis is positive or negative is arbitrary. Therefore, the results of different methods for evaluating principal components can give rise to eigenvectors (columns of COEFF) and the corresponding projections on those eigenvectors (rows of SCORE), which differ together by a sign-flip.

A code that can produce these results follows:

```
% First store the mean of the firing rates
mean_rates = mean(rate); % same as MU
% Use dev_rate as the deviation from mean rate for each cell
dev_rate = rate—ones(Nt,1)*mean_rates;
% Next calculate the covariance matrix of firing rates
C_matrix = cov(rate);
% Finds eigenvectors as columns then eigenvalues in a diagonal matrix
[eig_vectors Diag_evals] = eig(C_matrix);
% Form a column of eigenvalues
eig_vals = diag(Diag_evals);
```

```
% Sort the eigenvalues into descending order
[ordered_eig_vals, new_indices] = sort(eig_vals,'descend');
% Divide by sum of eigenvalues for fraction of variance explained
var_explained = ordered_eig_vals/sum(ordered_eig_vals);
% EXPLAINED
% Sort the eigenvectors according to variance explained
new_basis = eig_vectors(:,new_indices); % new_basis is like COEFF
% Find the projection of the mean-subtracted data on PCs.
PC_rates = dev_rate*new_basis; % PC_rates is like SCORE
```

9.9 Appendix B: Determining the Probability of Change Points for a Poisson Process

9.9.1 Optimal Rate

Our first goal is to show that the best estimate for the unknown firing rate, $r^{(opt)}$, of a Poisson process producing N spikes in time T is $r^{(opt)} = N/T$, if we assume a priori that all firing rates are equally likely (i.e., a uniform prior). The assumption of uniform prior means our estimate is the maximum likelihood estimate (MLE).

With a uniform prior, Bayes' theorem (equation 1.32) tells us that:

$$P(r|N,T) \propto P(N|r,T), \tag{9.13}$$

so we just need to find which firing rate is most likely to yield the observed number of spikes. For this calculation, we use the Poisson formula (section 3.3.3),

$$P(N|r,T) = \frac{(rT)^N e^{-rT}}{N!}. \tag{9.14}$$

The probability approaches zero as r approaches either zero or infinity and is always greater than zero between these values, so has a maximum. Calculus tells us the maximum is the point at which the rate of change of this probability with respect to r is zero, that is $r^{(opt)}$ is the value of r at which

$$\frac{dP(N)}{dr} = 0. \tag{9.15}$$

If, after taking the derivative of equation 9.14 with respect to r, we cancel out the denominator, $N!$ (which is independent of r), the maximum requires:

$$N\left(r^{(opt)}T\right)^{N-1} e^{-rT} - \left(r^{(opt)}T\right)^N e^{-rT}/T = 0 \tag{9.16}$$

which simplifies to

$$r^{(opt)} = N/T. \tag{9.17}$$

9.9.2 Evaluating the Change Point, Method 1

We divide an interval of length T into two subintervals, one of length T_1, the other of length $T_2 = T - T_1$. To test whether the time-point, T_1, is the most likely time-point for a change in rate, we treat the rates before and after T_1 as the optimal values, $r_1 = N_1 / T_1$ and $r_2 = N_2 / T_2$ respectively, where N_1 is the number of spikes in the interval $0 < t \leq T_1$ and N_2 is the number of spikes in the interval $T_1 < t \leq T$.

We can then divide each subinterval into time-bins of width δt. For each time-bin, find the probability of the observed spike or lack of spike given the rate of the process, then multiply together these individual probabilities to obtain the probability of the entire sequence of spikes. There are $N_{T_1} = T_1 / \delta t$ such bins in the first subinterval and $N_{T_2} = T_2 / \delta t$ such bins in the second interval. The probability of the entire sequence, assuming optimal rates, becomes then:

$$P(T_1, T_2 | N_1, N_2) = (r_1 \delta t)^{N_1} (1 - r_1 \delta t)^{N_{T_1} - N_1} (r_2 \delta t)^{N_2} (1 - r_2 \delta t)^{N_{T_2} - N_2} . \tag{9.18}$$

To evaluate equation 9.18 in the limit $\delta t \to 0$, it is easier to take the logarithm before expanding in small δt, so:

$$\begin{aligned}
\ln [P(T_1, T_2 | N_1, N_2)] &= N_1 \ln (r_1 \delta t) + (N_{T_1} - N_1) \ln (1 - r_1 \delta t) \\
&\quad + N_2 \ln (r_2 \delta t) + (N_{T_2} - N_2) \ln (1 - r_2 \delta t) \\
&\cong N_1 \ln (r_1 \delta t) - (N_{T_1} - N_1) r_1 \delta t + N_2 \ln (r_2 \delta t) - (N_{T_2} - N_2) r_2 \delta t \\
&\cong N_1 \ln (r_1) + N_2 \ln (r_2) + N \ln (\delta t) - T_1 r_1 - T_2 r_2 \\
&= N_1 \ln (r_1) + N_2 \ln (r_2) + N \ln (\delta t) - N.
\end{aligned} \tag{9.19}$$

equation 9.19 leads to

$$P(T_1, T_2 | N_1, N_2) \cong r_1^{N_1} r_2^{N_2} \delta t^N e^{-N} . \tag{9.20}$$

Removing the last two terms that do not depend on the separation into subintervals, we find

$$P(T_1, T_2 | N_1, N_2) \propto r_1^{N_1} r_2^{N_2} = \left(\frac{N_1}{T_1} \right)^{N_1} \left(\frac{N_2}{T_2} \right)^{N_2} . \tag{9.21}$$

Computationally we can vary T_1, count the spikes in the two subintervals and evaluate the probability of this being the change point according to the preceding formula. The value of T_1 that maximizes $P(T_1, T_2 | N_1, N_2)$ is selected as the change point.

9.9.3 Evaluating the Change Point, Method 2

In the second method, we will evaluate the ratio of two probabilities: the probability of the total numbers of spikes on either side of a proposed change point being produced by two distinct Poisson processes of different rates, divided by the probability of those spike counts being produced by a single Poisson process of constant rate.

In this case, we use the Poisson formula (section 3.7.1)

$$P(N|rT) = \frac{(rT)^N e^{-rT}}{N!},$$ (9.22)

for the probability of N spikes in an interval of length T, with fixed rate, r.

For the two intervals of length T_1 and $T_2 = T - T_1$, this formula yields for the probability of N_1 spikes in a period of T_1 and the probability of N_2 spikes in a period of T_2 as:

$$P(N_1, N_2|r_1, T_1, r_2, T_2) = \frac{(r_1 T_1)^{N_1} e^{-r_1 T_1}}{N_1!} \cdot \frac{(r_2 T_2)^{N_2} e^{-r_2 T_2}}{N_2!}.$$ (9.23)

We again use Bayes' theorem with a uniform prior to obtain the maximum likelihood estimate of rates as $r_1 = N_1 / T_1$ and $r_2 = N_2 / T_2$ from equation 9.17.

We compare with the case where the entire time interval has no change in firing rate, r, in which case we use equation 9.23 with $r_1 = r$ and $r_2 = r$:

$$P(N_1, N_2|r, T_1, r, T_2) = \frac{(rT_1)^{N_1} e^{-rT_1}}{N_1!} \cdot \frac{(rT_2)^{N_2} e^{-rT_2}}{N_2!}.$$ (9.24)

with $r = N / T$, where $N = N_1 + N_2$.

The ratio of equation 9.23 to equation 9.24 gives us the probability of a change point at T_1 as a likelihood ratio in comparison to the probability of no rate-change in the entire period:

$$\frac{P(N_1, N_2|r_1, T_1, r_2, T_2)}{P(N_1, N_2|r, T_1, r, T_2)} = \left(\frac{N_1}{T_1}\right)^{N_1} \left(\frac{N_2}{T_2}\right)^{N_2} \Big/ \left(\frac{N}{T}\right)^{N}.$$ (9.25)

The advantage of equation 9.25 is that it can allow us to decide whether to introduce a change point or not based on a criterion for the value of the likelihood ratio. Similarly, the likelihood can be recalculated for the subintervals once a change point is found, to assess whether they should be further subdivided with additional change points. In this manner, as with hidden Markov modeling (section 9.5), the total number of states and transitions between them, can be produced as a result of the analysis, rather than prescribed beforehand.

References

Chapter 2

1. Tuckwell HC. 1988. *Introduction to Theoretical Neurobiology.* Cambridge: Cambridge University Press.

2. Dayan P, Abbott LF. 2001. *Theoretical Neuroscience.* Cambridge, MA: MIT Press.

3. Bond CT, Maylie J, Adelman JP. 1999. Small-conductance calcium-activated potassium channels. *Ann N Y Acad Sci* 868: 370–378.

4. Fourcaud-Trocme N, Hansel D, van Vreeswijk C, Brunel N. 2003. How spike generation mechanisms determine the neuronal response to fluctuating inputs. *J Neurosci* 23(37): 11628–11640.

5. Brette R, Gerstner W. 2005. Adaptive exponential integrate-and-fire model as an effective description of neuronal activity. *J Neurophysiol* 94(5): 3637–3642.

Chapter 3

1. Sherrington CS. 1906. *The Integrative Action of the Nervous System.* New York: Scribner & Sons.

2. Spillmann L. 2014. Receptive fields of visual neurons: The early years. *Perception* 43(11): 1145–1176.

3. Nobili R, Mammano F, Ashmore J. 1998. How well do we understand the cochlea? *Trends Neurosci* 21(4): 159–167.

4. Urban NN. 2002. Lateral inhibition in the olfactory bulb and in olfaction. *Physiol Behav* 77(4–5): 607–612.

5. Windhorst U. 1996. On the role of recurrent inhibitory feedback in motor control. *Prog Neurobiol* 49(6): 517–587.

6. Maler L. 2007. Neural strategies for optimal processing of sensory signals. *Prog Brain Res* 165: 135–154.

7. Helmstaedter M, Sakmann B, Feldmeyer D. 2009. Neuronal correlates of local, lateral, and translaminar inhibition with reference to cortical columns. *Cereb Cortex* 19(4): 926–937.

8. Hirsch JA, Gilbert CD. 1991. Synaptic physiology of horizontal connections in the cat's visual cortex. *J Neurosci* 11(6): 1800–1809.

9. Theunissen FE, Sen K, Doupe AJ. 2000. Spectral-temporal receptive fields of nonlinear auditory neurons obtained using natural sounds. *J Neurosci* 20(6): 2315–2331.

10. Hubel DH, Wiesel TN. 1959. Receptive fields of single neurones in the cat's striate cortex. *J Physiol* 148: 574–591.

11. Hubel DH, Wiesel TN. 2009. Republication of The Journal of Physiology (1959) 148, 574–591: Receptive fields of single neurones in the cat's striate cortex. 1959. *J Physiol* 587(Pt 12): 2721–2732.

12. Ringach DL. 2004. Mapping receptive fields in primary visual cortex. *J Physiol* 558(Pt 3): 717–728.

13. Rodieck RW, Stone J. 1965. Analysis of receptive fields of cat retinal ganglion cells. *J Neurophysiol* 28(5): 832–849.

14. Rodieck RW, Stone J. 1965. Response of cat retinal ganglion cells to moving visual patterns. *J Neurophysiol* 28(5): 819–832.

15. Sharpee TO. 2013. Computational identification of receptive fields. *Annu Rev Neurosci* 36: 103–120.

16. Arieli A, Shoham D, Hildesheim R, Grinvald A. 1995. Coherent spatiotemporal patterns of ongoing activity revealed by real-time optical imaging coupled with single-unit recording in the cat visual cortex. *J Neurophysiol* 73(5): 2072–2093.

17. Buno W, Jr, Bustamante J, Fuentes J. 1984. White noise analysis of pace-maker-response interactions and non-linearities in slowly adapting crayfish stretch receptor. *J Physiol* 350: 55–80.

18. Bryant HL, Segundo JP. 1976. Spike initiation by transmembrane current: A white-noise analysis. *J Physiol* 260(2): 279–314.

19. Cai D, DeAngelis GC, Freeman RD. 1997. Spatiotemporal receptive field organization in the lateral geniculate nucleus of cats and kittens. *J Neurophysiol* 78(2): 1045–1061.

20. Holt GR, Softky WR, Koch C, Douglas RJ. 1996. Comparison of discharge variability in vitro and in vivo in cat visual cortex neurons. *J Neurophysiol* 75: 1806–1814.

21. Greiner M, Pfeiffer D, Smith RD. 2000. Principles and practical application of the receiver-operating characteristic analysis for diagnostic tests. *Prev Vet Med* 45(1–2): 23–41.

22. Ratcliff R, Sheu CF, Gronlund SD. 1992. Testing global memory models using ROC curves. *Psychol Rev* 99(3): 518–535.

23. Cowan N, Stadler MA. 1996. Estimating unconscious processes: Implications of a general class of models. *J Exp Psychol Gen* 125(2): 195–200.

24. Sauvage MM. 2010. ROC in animals: Uncovering the neural substrates of recollection and familiarity in episodic recognition memory. *Conscious Cogn* 19(3): 816–828.

25. Ratcliff R, Van Zandt T, McKoon G. 1995. Process dissociation, single-process theories, and recognition memory. *J Exp Psychol Gen* 124(4): 352–374.

26. Fortin NJ, Wright SP, Eichenbaum H. 2004. Recollection-like memory retrieval in rats is dependent on the hippocampus. *Nature* 431(7005): 188–191.

27. Ratcliff R, McKoon G, Tindall M. 1994. Empirical generality of data from recognition memory receiver-operating characteristic functions and implications for the global memory models. *J Exp Psychol Learn Mem Cogn* 20(4): 763–785.

Chapter 4

1. Hodgkin AL, Huxley AF. 1952. A quantitative description of membrane current and its application to conduction and excitation in nerve. *J Physiol* 117(4): 500–544.

2. Hodgkin AL, Huxley AF. 1952. The dual effect of membrane potential on sodium conductance in the giant axon of Loligo. *J Physiol* 116(4): 497–506.

3. Hodgkin AL, Huxley AF. 1952. The components of membrane conductance in the giant axon of Loligo. *J Physiol* 116(4): 473–496.

4. Hodgkin AL, Huxley AF. 1952. Currents carried by sodium and potassium ions through the membrane of the giant axon of Loligo. *J Physiol* 116(4): 449–472.

5. Hodgkin AL, Huxley AF, Katz B. 1952. Measurement of current-voltage relations in the membrane of the giant axon of Loligo. *J Physiol* 116(4): 424–448.

6. Hutcheon B, Yarom Y. 2000. Resonance, oscillation and the intrinsic frequency preferences of neurons. *Trends Neurosci* 23(5): 216–222.

7. Connor JA, Stevens CF. 1971. Prediction of repetitive firing behaviour from voltage clamp data on an isolated neurone soma. *J Physiol* 213(1): 31–53.

8. Whitaker M. 2006. Calcium at fertilization and in early development. *Physiol Rev* 86(1): 25–88.

9. Skelding KA, Rostas JA, Verrills NM. 2011. Controlling the cell cycle: The role of calcium/calmodulin-stimulated protein kinases I and II. *Cell Cycle* 10(4): 631–639.

10. Wiegert JS, Bading H. 2011. Activity-dependent calcium signaling and ERK-MAP kinases in neurons: A link to structural plasticity of the nucleus and gene transcription regulation. *Cell Calcium* 49(5): 296–305.

11. Greer PL, Greenberg ME. 2008. From synapse to nucleus: Calcium-dependent gene transcription in the control of synapse development and function. *Neuron* 59(6): 846–860.

12. Jiang H, Stephens NL. 1994. Calcium and smooth muscle contraction. *Mol Cell Biochem* 135(1): 1–9.

13. Jarvis SE, Zamponi GW. 2005. Masters or slaves? Vesicle release machinery and the regulation of presynaptic calcium channels. *Cell Calcium* 37(5): 483–488.

14. Sjöström PJ, Nelson SB. 2002. Spike timing, calcium signals and synaptic plasticity. *Curr Opin Neurobiol* 12(3): 305–314.

15. Zucker RS. 1999. Calcium- and activity-dependent synaptic plasticity. *Curr Opin Neurobiol* 9(3): 305–313.

16. Verkhratsky AJ, Petersen OH. 1998. Neuronal calcium stores. *Cell Calcium* 24(5–6): 333–343.

17. Ross WN. 2012. Understanding calcium waves and sparks in central neurons. *Nat Rev Neurosci* 13(3): 157–168.

18. Stanley EF. 2016. The nanophysiology of fast transmitter release. *Trends Neurosci* 39(3): 183–197.

19. Takahashi T, Momiyama A. 1993. Different types of calcium channels mediate central synaptic transmission. *Nature* 366(6451): 156–158.

20. Mangoni ME, Couette B, Marger L, Bourinet E, Striessnig J, Nargeot J. 2006. Voltage-dependent calcium channels and cardiac pacemaker activity: From ionic currents to genes. *Prog Biophys Mol Biol* 90(1–3): 38–63.

21. Destexhe A, Sejnowski TJ. 2002. The initiation of bursts in thalamic neurons and the cortical control of thalamic sensitivity. *Philos Trans R Soc Lond B Biol Sci* 357(1428): 1649–1657.

22. Contreras D. 2006. The role of T-channels in the generation of thalamocortical rhythms. *CNS Neurol Disord Drug Targets* 5(6): 571–585.

23. Bower JM, Beeman D. 1998, 2003. *The Book of GENESIS: Exploring Realistic Neural Models with the General NEural SImulation System* (2nd ed.). Free internet ed. New York: Springer-Verlag.

24. Pinsky PF, Rinzel J. 1994. Intrinsic and network rhythmogenesis in a reduced Traub model for CA3 neurons. *J Comput Neurosci* 1(1–2): 39–60.

25. Traub RD, Wong RK, Miles R, Michelson H. 1991. A model of a CA3 hippocampal pyramidal neuron incorporating voltage-clamp data on intrinsic conductances. *J Neurophysiol* 66(2): 635–650.

26. Pape HC. 1996. Queer current and pacemaker: The hyperpolarization-activated cation current in neurons. *Annu Rev Physiol* 58: 299–327.

27. Luthi A, McCormick DA. 1998. Periodicity of thalamic synchronized oscillations: The role of Ca2+-mediated upregulation of Ih. *Neuron* 20(3): 553–563.

28. DeBello WM, McBride TJ, Nichols GS, Pannoni KE, Sanculi D, Totten DJ. 2014. Input clustering and the microscale structure of local circuits. *Front Neural Circuits* 8: 112.

29. Wilson DE, Whitney DE, Scholl B, Fitzpatrick D. 2016. Orientation selectivity and the functional clustering of synaptic inputs in primary visual cortex. *Nat Neurosci* 19(8): 1003–1009.

30. London M, Hausser M. 2005. Dendritic computation. *Annu Rev Neurosci* 28: 503–532.

31. Poirazi P, Brannon T, Mel BW. 2003. Pyramidal neuron as two-layer neural network. *Neuron* 37(6): 989–999.

32. Jadi MP, Behabadi BF, Poleg-Polsky A, Schiller J, Mel BW. An Augmented Two-Layer Model Captures Nonlinear Analog Spatial Integration Effects in Pyramidal Neuron Dendrites. *Proc IEEE Inst Electr Electron Eng.* 2014;102(5).

33. Liu Z, Golowasch J, Marder E, Abbott LF. 1998. A model neuron with activity-dependent conductances regulated by multiple calcium sensors. *J Neurosci* 18(7): 2309–2320.

Chapter 5

1. Stanley EF. 2016. The nanophysiology of fast transmitter release. *Trends Neurosci* 39(3): 183–197.

2. Surmeier DJ, Ding J, Day M, Wang Z, Shen W. 2007. D1 and D2 dopamine-receptor modulation of striatal glutamatergic signaling in striatal medium spiny neurons. *Trends Neurosci* 30(5): 228–235.

3. Seamans JK, Yang CR. 2004. The principal features and mechanisms of dopamine modulation in the prefrontal cortex. *Prog Neurobiol* 74(1): 1–58.

4. Prinz AA, Bucher D, Marder E. 2004. Similar network activity from disparate circuit parameters. *Nat Neurosci* 7(12): 1345–1352.

5. Sharp AA, O'Neil MB, Abbott LF, Marder E. 1993. Dynamic clamp: Computer-generated conductances in real neurons. *J Neurophysiol* 69(3): 992–995.

6. Abbott LF, Regehr WG. 2004. Synaptic computation. *Nature* 431(7010): 796–803.

7. Tsodyks M, Wu S. 2013. Short-term synaptic plasticity. *Scholarpedia.* 8(10): 3153.

8. Hennig MH. 2013. Theoretical models of synaptic short term plasticity. *Front Comput Neurosci* 7: 154.

9. Zucker RS, Regher WG. 2002. Short-term synaptic plasticity. *Annu Rev Physiol* 64: 355–405.

10. Abbott LF, Varela JA, Sen K, Nelson SB. 1997. Synaptic depression and cortical gain control. *Science* 275(5297): 220–224.

11. Mongillo G, Barak O, Tsodyks M. 2008. Synaptic theory of working memory. *Science* 319(5869): 1543–1546.

12. Dayan P, Abbott LF. 2001. *Theoretical Neuroscience.* Cambridge, MA: MIT Press.

13. Ikeda K, Bekkers JM. 2006. Autapses. *Curr Biol* 16(9): R308.

14. Song S, Sjöström PJ, Reigl M, Nelson S, Chklovskii DB. 2005. Highly nonrandom features of synaptic connectivity in local cortical circuits. *PLoS Biol* 3(3): e68.

15. Perin R, Berger TK, Markram H. 2011. A synaptic organizing principle for cortical neuronal groups. *Proc Natl Acad Sci USA* 108(13): 5419–5424.

16. Rubin E. 1915. *Synsoplevede Figurer: Studier i psykologisk Analyse. Forste Del* [Visually experienced figures: Studies in psychological analysis. Part one]. Copenhagen and Christiania: Gyldendalske Boghandel, Nordisk Forlag, University of Copenhagen.

17. Jastrow J. 1899. The mind's eye. *Popular Science Monthly.* 54: 299–312.

18. Wang XJ, Rinzel J. 1992. Alternating and synchronous rhythms in reciprocally inhibitory model neurons. *Neural Comput* 4: 84–97.

19. Marder E, Bucher D, Schulz DJ, Taylor AL. 2005. Invertebrate central pattern generation moves along. *Curr Biol* 15(17): R685–R699.

20. Calabrese RL, Nadim F, Olsen OH. 1995. Heartbeat control in the medicinal leech: A model system for understanding the origin, coordination, and modulation of rhythmic motor patterns. *J Neurobiol* 27(3): 390–402.

21. Calabrese RL, De Schutter E. 1992. Motor-pattern-generating networks in invertebrates: Modeling our way toward understanding. *Trends Neurosci* 15(11): 439–445.

22. Marder E, Haddad SA, Goeritz ML, Rosenbaum P, Kispersky T. 2015. How can motor systems retain performance over a wide temperature range? Lessons from the crustacean stomatogastric nervous system. *J Comp Physiol A Neuroethol Sens Neural Behav Physiol* 201(9): 851–856.

23. Buchanan JT. 1996. Lamprey spinal interneurons and their roles in swimming activity. *Brain Behav Evol* 48(5): 287–296.

24. Zhang C, Guy RD, Mulloney B, Zhang Q, Lewis TJ. 2014. Neural mechanism of optimal limb coordination in crustacean swimming. *Proc Natl Acad Sci USA* 111(38): 13840–13845.

25. Stiefel KM, Ermentrout GB. 2016. Neurons as oscillators. *J Neurophysiol* 116(6): 2950–2960.

Chapter 6

1. Goldman-Rakic PS. 1995. Cellular basis of working memory. *Neuron* 14: 477–485.

2. Wang M, Yang Y, Wang CJ, et al. 2013. NMDA receptors subserve persistent neuronal firing during working memory in dorsolateral prefrontal cortex. *Neuron* 77(4): 736–749.

3. Wang XJ. 1999. Synaptic basis of cortical persistent activity: The importance of NMDA receptors to working memory. *J Neurosci* 19(21): 9587–9603.

4. Seung HS. 1996. How the brain keeps the eyes still. *Proc Natl Acad Sci USA* 93(23): 13339–13344.

5. Aksay E, Gamkrelidze G, Seung HS, Baker R, Tank DW. 2001. In vivo intracellular recording and perturbation of persistent activity in a neural integrator. *Nat Neurosci* 4(2): 184–193.

6. Mazurek ME, Roitman JD, Ditterich J, Shadlen MN. 2003. A role for neural integrators in perceptual decision making. *Cereb Cortex* 13(11): 1257–1269.

7. Wong KF, Wang XJ. 2006. A recurrent network mechanism of time integration in perceptual decisions. *J Neurosci* 26(4): 1314–1328.

8. Miller P, Brody CD, Romo R, Wang XJ. 2003. A recurrent network model of somatosensory parametric working memory in the prefrontal cortex. *Cereb Cortex* 13(11): 1208–1218.

9. Romo R, Brody CD, Hernandez A, Lemus L. 1999. Neuronal correlates of parametric working memory in the prefrontal cortex. *Nature* 399(6735): 470–473.

10. Ratcliff R, McKoon G. 2008. The diffusion decision model: Theory and data for two-choice decision tasks. *Neural Comput* 20(4): 873–922.

11. Wang XJ. 2002. Probabilistic decision making by slow reverberation in cortical circuits. *Neuron* 36: 955–968.

12. Shadlen MN, Newsome WT. 2001. Neural basis of a perceptual decision in the parietal cortex (area LIP) of the rhesus monkey. *J Neurophysiol* 86(4): 1916–1936.

13. Bogacz R. 2007. Optimal decision-making theories: Linking neurobiology with behaviour. *Trends Cogn Sci* 11(3): 118–125.

14. Drugowitsch J, Pouget A. 2012. Probabilistic vs. non-probabilistic approaches to the neurobiology of perceptual decision-making. *Curr Opin Neurobiol* 22(6): 963–969.

15. Miller P, Katz DB. 2013. Accuracy and response-time distributions for decision-making: Linear perfect integrators versus nonlinear attractor-based neural circuits. *J Comput Neurosci* 35(3): 261–294.

16. Shadlen MN, Newsome WT. 1996. Motion perception: Seeing and deciding. *Proc Natl Acad Sci USA* 93(2): 628–633.

17. Ratcliff R, Smith PL, Brown SD, McKoon G. 2016. Diffusion decision model: Current issues and history. *Trends Cogn Sci* 20(4): 260–281.

18. Usher M, McClelland JL. 2001. The time course of perceptual choice: The leaky, competing accumulator model. *Psychol Rev* 108(3): 550–592.

19. Seung HS, Lee DD, Reis BY, Tank DW. 2000. The autapse: A simple illustration of short-term analog memory storage by tuned synaptic feedback. *J Comput Neurosci* 9(2): 171–185.

20. Eckhoff P, Wong-Lin KF, Holmes P. 2009. Optimality and robustness of a biophysical decision-making model under norepinephrine modulation. *J Neurosci* 29(13): 4301–4311.

21. Shea-Brown E, Gilzenrat MS, Cohen JD. 2008. Optimization of decision making in multilayer networks: The role of locus coeruleus. *Neural Comput* 20(12): 2863–2894.

22. Gillespie DT. 1992. *Markov Processes*. Academic Press.

23. Jensen O, Kaiser J, Lachaux JP. 2007. Human gamma-frequency oscillations associated with attention and memory. *Trends Neurosci* 30(7): 317–324.

24. Fries P, Reynolds JH, Rorie AE, Desimone R. 2001. Modulation of oscillatory neuronal synchronization by selective visual attention. *Science* 291(5508): 1560–1563.

25. Sokolov A, Pavlova M, Lutzenberger W, Birbaumer N. 2004. Reciprocal modulation of neuromagnetic induced gamma activity by attention in the human visual and auditory cortex. *Neuroimage* 22(2): 521–529.

26. Whittington MA, Stanford IM, Colling SB, Jefferys JG, Traub RD. 1997. Spatiotemporal patterns of gamma frequency oscillations tetanically induced in the rat hippocampal slice. *J Physiol* 502(Pt 3): 591–607.

27. Whittington MA, Traub RD, Kopell N, Ermentrout B, Buhl EH. 2000. Inhibition-based rhythms: Experimental and mathematical observations on network dynamics. *Int J Psychophysiol* 38(3): 315–336.

28. Geisler C, Brunel N, Wang XJ. 2005. Contributions of intrinsic membrane dynamics to fast network oscillations with irregular neuronal discharges. *J Neurophysiol* 94(6): 4344–4361.

29. Abbott LF, Chance FS. 2005. Drivers and modulators from push-pull and balanced synaptic input. *Prog Brain Res* 149: 147–155.

30. Priebe NJ, Ferster D. 2008. Inhibition, spike threshold, and stimulus selectivity in primary visual cortex. *Neuron* 57(4): 482–497.

31. Priebe NJ. 2016. Mechanisms of orientation selectivity in the primary visual cortex. *Annu Rev Vis Sci* 2: 85–107.

32. Peters A, Payne BR. 1993. Numerical relationships between geniculocortical afferents and pyramidal cell modules in cat primary visual cortex. *Cereb Cortex* 3(1): 69–78.

33. Koch E, Jin J, Alonso JM, Zaidi Q. 2016. Functional implications of orientation maps in primary visual cortex. *Nat Commun* 7: 13529.

34. Wilson DE, Whitney DE, Scholl B, Fitzpatrick D. 2016. Orientation selectivity and the functional clustering of synaptic inputs in primary visual cortex. *Nat Neurosci* 19(8): 1003–1009.

35. Ben-Yishai R, Bar-Or RL, Sompolinsky H. 1995. Theory of orientation tuning in visual cortex. *Proc Natl Acad Sci USA* 92: 3844–3848.

36. Camperi M, Wang XJ. 1998. A model of visuospatial short-term memory in prefrontal cortex: Recurrent network and cellular bistability. *J Comput Neurosci* 5: 383–405.

37. Compte A, Brunel N, Goldman-Rakic PS, Wang XJ. 2000. Synaptic mechanisms and network dynamics underlying spatial working memory in a cortical network model. *Cereb Cortex* 10: 910–923.

38. Funahashi S, Bruce CJ, Goldman-Rakic PS. 1989. Mnemonic coding of visual space in the monkey's dorsolateral prefrontal cortex. *J Neurophysiol* 61(2): 331–349.

39. Song P, Wang XJ. 2005. Angular path integration by moving "hill of activity": A spiking neuron model without recurrent excitation of the head-direction system. *J Neurosci* 25: 1002–1014.

40. Zhang K. 1996. Representation of spatial orientation by the intrinsic dynamics of the head-direction cell ensembles: A theory. *J Neurosci* 16: 2112–2126.

41. Turner-Evans D, Wegener S, Rouault H, et al. 2017. Angular velocity integration in a fly heading circuit. *eLife* 6: e23496.

Chapter 7

1. Izhikevich EM. 2007. *Dynamical Systems in Neuroscience: The Geometry of Excitability and Bursting.* Cambridge, MA: MIT Press.

2. Strogatz SH. 2015. *Nonlinear Dynamics and Chaos.* 2nd ed. Boulder, CO: Westview Press.

3. Shenoy KV, Sahani M, Churchland MM. 2013. Cortical control of arm movements: A dynamical systems perspective. *Annu Rev Neurosci* 36: 337–359.

4. Nagumo J, Arimoto S, Yoshizawa S. An active pulse transmission lines simulating nerve axon. *Proc I R E.* 1962;50:2061–2070.

5. FitzHugh R. 1961. Impulses and physiological states in models of nerve and membrane. *Biophys J* 1: 445–466.

6. Izhikevich EM, Fitzhugh R. 2006. FitzHugh-Nagumo model. *Scholarpedia.* 1(9): 1349.

7. Tsodyks MV, Skaggs WE, Sejnowski TJ, McNaughton BL. 1997. Paradoxical effects of external modulation of inhibitory interneurons. *J Neurosci* 17(11): 4382–4388.

8. Ozeki H, Finn IM, Schaffer ES, Miller KD, Ferster D. 2009. Inhibitory stabilization of the cortical network underlies visual surround suppression. *Neuron* 62(4): 578–592.

9. Rubin DB, Van Hooser SD, Miller KD. 2015. The stabilized supralinear network: A unifying circuit motif underlying multi-input integration in sensory cortex. *Neuron* 85(2): 402–417.

10. Shadlen MN, Newsome WT. 1998. The variable discharge of cortical neurons: Implications for connectivity, computation, and information coding. *J Neurosci* 18(10): 3870–3896.

11. Miller P. 2016. Itinerancy between attractor states in neural systems. *Curr Opin Neurobiol* 40: 14–22.

12. Rubin E. 1915. *Synsoplevede Figurer: Studier i psykologisk Analyse. Forste Del* [Visually experienced figures: Studies in psychological analysis. Part one]. Copenhagen and Christiania: Gyldendalske Boghandel, Nordisk Forlag, University of Copenhagen.

13. Levelt WJ. 1967. Note on the distribution of dominance times in binocular rivalry. *Br J Psychol* 58(1): 143–145.

14. Brascamp JW, Klink PC, Levelt WJ. The "laws" of binocular rivalry: 50 years of Levelt's propositions. *Vision Res.* 2015;109(Pt A):20–37.

15. Levelt WJM. 1965. *On Binocular Rivalry.* Soesterberg, The Netherlands: Institute for Perception.

16. Hupe JM, Rubin N. 2003. The dynamics of bi-stable alternation in ambiguous motion displays: A fresh look at plaids. *Vision Res* 43(5): 531–548.

17. Hupe JM, Rubin N. 2004. The oblique plaid effect. *Vision Res* 44(5): 489–500.

18. Pressnitzer D, Hupe JM. 2006. Temporal dynamics of auditory and visual bistability reveal common principles of perceptual organization. *Curr Biol* 16(13): 1351–1357.

19. Winkler I, Denham S, Mill R, Bohm TM, Bendixen A. Multistability in auditory stream segregation: A predictive coding view. *Philos Trans R Soc Lond B Biol Sci* . 2012;367(1591): 1001–1012.

20. Moreno-Bote R, Rinzel J, Rubin N. 2007. Noise-induced alternations in an attractor network model of perceptual bistability. *J Neurophysiol* 98(3): 1125–1139.

21. Shpiro A, Moreno-Bote R, Rubin N, Rinzel J. 2009. Balance between noise and adaptation in competition models of perceptual bistability. *J Comput Neurosci* 27(1): 37–54.

22. Afraimovich VS, Zhigulin VP, Rabinovich MI. 2004. On the origin of reproducible sequential activity in neural circuits. *Chaos* 14(4): 1123–1129.

23. Huerta R, Rabinovich M. 2004. Reproducible sequence generation in random neural ensembles. *Phys Rev Lett* 93(23): 238104.

24. Rabinovich MI, Huerta R, Varona P, Afraimovich VS. 2006. Generation and reshaping of sequences in neural systems. *Biol Cybern* 95(6): 519–536.

25. Churchland MM, Yu BM, Cunningham JP, et al. 2010. Stimulus onset quenches neural variability: A widespread cortical phenomenon. *Nat Neurosci* 13(3): 369–378.

26. Rajan K, Abbott LF, Sompolinsky H. 2010. Stimulus-dependent suppression of chaos in recurrent neural networks. *Phys Rev E Stat Nonlin Soft Matter Phys* 82(1 Pt 1): 011903.

27. Sussillo D, Abbott LF. 2009. Generating coherent patterns of activity from chaotic neural networks. *Neuron* 63(4): 544–557.

28. Laje R, Buonomano DV. 2013. Robust timing and motor patterns by taming chaos in recurrent neural networks. *Nat Neurosci* 16(7): 925–933.

29. Bertschinger N, Natschlager T. 2004. Real-time computation at the edge of chaos in recurrent neural networks. *Neural Comput* 16(7): 1413–1436.

30. Rabinovich MI, Abarbanel HD. 1998. The role of chaos in neural systems. *Neuroscience* 87(1): 5–14.

31. Nowak MA, Sigmund K. 2004. Evolutionary dynamics of biological games. *Science* 303(5659): 793–799.

32. van Vreeswijk C, Sompolinsky H. 1996. Chaos in neuronal networks with balanced excitatory and inhibitory activity. *Science* 274(5293): 1724–1726.

33. Sompolinsky H, Crisanti A, Sommers HJ. 1988. Chaos in random neural networks. *Phys Rev Lett* 61(3): 259–262.

34. Bak P, Paczuski M. 1995. Complexity, contingency, and criticality. *Proc Natl Acad Sci USA* 92(15): 6689–6696.

35. Bak P, Tang C, Wiesenfeld K. 1987. Self-organized criticality: An explanation of the 1/f noise. *Phys Rev Lett* 59(4): 381–384.

36. Tkacik G, Mora T, Marre O, et al. 2015. Thermodynamics and signatures of criticality in a network of neurons. *Proc Natl Acad Sci USA* 112(37): 11508–11513.

37. Beggs JM, Plenz D. 2003. Neuronal avalanches in neocortical circuits. *J Neurosci* 23(35): 11167–11177.

38. Beggs JM. 1864. The criticality hypothesis: How local cortical networks might optimize information processing. *Philos Trans A Math Phys. Eng Sci* 2008(366): 329–343.

39. Beggs JM, Timme N. 2012. Being critical of criticality in the brain. *Front Physiol* 3: 163.

Chapter 8

1. Hebb DO. 1949. *Organization of Behavior.* New York: Wiley.

2. Lømo T. 1966. Frequency potentiation of excitatory synaptic activity in the dentate area of the hippocampal formation. *Acta Physiol Scand* 68(suppl. 277): 128.

3. Bliss TV, Lømo T. 1970. Plasticity in a monosynaptic cortical pathway. *J Physiol* 207(2): 61P.

4. Markram H, Lubke J, Frotscher M, Sakmann B. 1997. Regulation of synaptic efficacy by coincidence of postsynaptic APs and EPSPs. *Science* 275(5297): 213–215.

5. Bi GQ, Poo MM. 1998. Synaptic modifications in cultured hippocampal neurons: Dependence on spike timing, synaptic strength, and postsynaptic cell type. *J Neurosci* 18(24): 10464–10472.

6. Teyler TJ, Cavus I, Coussens C, et al. 1994. Multideterminant role of calcium in hippocampal synaptic plasticity. *Hippocampus* 4(6): 623–634.

7. Kirkwood A, Bear MF. 1995. Elementary forms of synaptic plasticity in the visual cortex. *Biol Res* 28(1): 73–80.

8. Finch EA, Tanaka K, Augustine GJ. 2012. Calcium as a trigger for cerebellar long-term synaptic depression. *Cerebellum* 11(3): 706–717.

9. Cavazzini M, Bliss T, Emptage N. 2005. Ca2+ and synaptic plasticity. *Cell Calcium* 38(3–4): 355–367.

10. Blair HT, Schafe GE, Bauer EP, Rodrigues SM, LeDoux JE. 2001. Synaptic plasticity in the lateral amygdala: A cellular hypothesis of fear conditioning. *Learn Mem* 8(5): 229–242.

11. Debanne D. 1996. Associative synaptic plasticity in hippocampus and visual cortex: Cellular mechanisms and functional implications. *Rev Neurosci* 7(1): 29–46.

12. Baker KD, Edwards TM, Rickard NS. 2013. The role of intracellular calcium stores in synaptic plasticity and memory consolidation. *Neurosci Biobehav Rev* 37(7): 1211–1239.

13. Brown TH, Chapman PF, Kairiss EW, Keenan CL. 1988. Long-term synaptic potentiation. *Science* 242(4879): 724–728.

14. Artola A, Brocher S, Singer W. 1990. Different voltage-dependent thresholds for inducing long-term depression and long-term potentiation in slices of rat visual cortex. *Nature* 347: 69–72.

15. Collingridge GL, Peineau S, Howland JG, Wang YT. 2010. Long-term depression in the CNS. *Nat Rev Neurosci* 11(7): 459–473.

16. Artola A, Singer W. 1993. Long-term depression of excitatory synaptic transmission and its relationship to long-term potentiation. *Trends Neurosci* 16: 480.

17. Kirkwood A, Bear MF. 1994. Homosynaptic long-term depression in the visual cortex. *J Neurosci* 14(5 Pt 2): 3404–3412.

18. Willshaw D, Buneman OP, Longuet-Higgins H. 1969. Non-holographic associative memory. *Nature* 222: 960–962.

19. Hopfield JJ. 1982. Neural networks and physical systems with emergent collective computational abilities. *Proc Natl Acad Sci USA* 79: 2554–2558.

20. Feldman DE. 2012. The spike-timing dependence of plasticity. *Neuron* 75(4): 556–571.

21. Nelson SB, Sjostrom PJ, Turrigiano GG. 2002. Rate and timing in cortical synaptic plasticity. *Philos Trans R Soc Lond B Biol Sci* 357(1428): 1851–1857.

22. Tsodyks M. 2002. Spike-timing-dependent synaptic plasticity—the long road towards understanding neuronal mechanisms of learning and memory. *Trends Neurosci* 25(12): 599–600.

23. Song S, Miller KD, Abbott LF. 2000. Competitive Hebbian learning through spike-time-dependent synaptic plasticity. *Nat Neurosci* 3: 919–926.

24. Goodhill GJ, Lowel S. 1995. Theory meets experiment: Correlated neural activity helps determine ocular dominance column periodicity. *Trends Neurosci* 18(10): 437–439.

25. Huberman AD. 2007. Mechanisms of eye-specific visual circuit development. *Curr Opin Neurobiol* 17(1): 73–80.

26. Drew PJ, Abbott LF. 2006. Extending the effects of spike-timing-dependent plasticity to behavioral timescales. *Proc Natl Acad Sci USA* 103(23): 8876–8881.

27. Nowotny T, Rabinovich MI, Abarbanel HD. 2003. Spatial representation of temporal information through spike-timing-dependent plasticity. *Phys Rev E Stat Nonlin Soft Matter Phys* 68(1 Pt 1): 011908.

28. Mehta MR, Barnes CA, McNaughton BL. 1997. Experience-dependent, asymmetric expansion of hippocampal place fields. *Proc Natl Acad Sci USA* 94(16): 8918–8921.

29. Blum KI, Abbott LF. 1996. A model of spatial map formation in the hippocampus of the rat. *Neural Comput* 8(1): 85–93.

30. Gerstner W, Abbott LF. 1997. Learning navigational maps through potentiation and modulation of hippocampal place cells. *J Comput Neurosci* 4(1): 79–94.

31. Levy WB. 1996. A sequence predicting CA3 is a flexible associator that learns and uses context to solve hippocampal-like tasks. *Hippocampus* 6: 579–590.

32. Doupe AJ, Solis MM, Kimpo R, Boettiger CA. 2004. Cellular, circuit, and synaptic mechanisms in song learning. *Ann N Y Acad Sci* 1016: 495–523.

33. Troyer TW, Doupe AJ. 2000. An associational model of birdsong sensorimotor learning II: Temporal hierarchies and the learning of song sequence. *J Neurophysiol* 84(3): 1224–1239.

34. Warren TL, Charlesworth JD, Tumer EC, Brainard MS. 2012. Variable sequencing is actively maintained in a well learned motor skill. *J Neurosci* 32(44): 15414–15425.

35. Brainard MS, Doupe AJ. 2002. What songbirds teach us about learning. *Nature* 417(6886): 351–358.

36. Sjöström PJ, Nelson SB. 2002. Spike timing, calcium signals and synaptic plasticity. *Curr Opin Neurobiol* 12(3): 305–314.

37. Froemke RC, Dan Y. 2002. Spike-timing-dependent synaptic modification induced by natural spike trains. *Nature* 416(6879): 433–438.

38. Sjostrom PJ, Turrigiano GG, Nelson SB. 2001. Rate, timing, and cooperativity jointly determine cortical synaptic plasticity. *Neuron* 32(6): 1149–1164.

39. Pfister JP, Gerstner W. 2006. Triplets of spikes in a model of spike timing-dependent plasticity. *J Neurosci* 26(38): 9673–9682.

40. Brandalise F, Carta S, Helmchen F, Lisman J, Gerber U. 2016. Dendritic NMDA spikes are necessary for timing-dependent associative LTP in CA3 pyramidal cells. *Nat Commun* 7: 13480.

41. Clopath C, Busing L, Vasilaki E, Gerstner W. 2010. Connectivity reflects coding: A model of voltage-based STDP with homeostasis. *Nat Neurosci* 13(3): 344–352.

42. Song S, Abbott LF. 2001. Cortical development and remapping through spike timing-dependent plasticity. *Neuron* 32(2): 339–350.

43. O'Leary T, Williams AH, Caplan JS, Marder E. 2013. Correlations in ion channel expression emerge from homeostatic tuning rules. *Proc Natl Acad Sci USA* 110(28): E2645–E2654.

44. Turrigiano GG, Nelson SB. 2000. Hebb and homeostasis in neuronal plasticity. *Curr Opin Neurobiol* 10(3): 358–364.

45. Turrigiano GG, Nelson SB. 2004. Homeostatic plasticity in the developing nervous system. *Nat Rev Neurosci* 5(2): 97–107.

46. Turrigiano GG. 1999. Homeostatic plasticity in neuronal networks: The more things change, the more they stay the same. *Trends Neurosci* 22(5): 221–227.

47. Turrigiano G. 2012. Homeostatic synaptic plasticity: Local and global mechanisms for stabilizing neuronal function. *Cold Spring Harb Perspect Biol* 4(1): a005736.

48. Bienenstock EL, Cooper LN, Munro PW. 1982. Theory for the development of neuron selectivity: Orientation specificity and binocular interaction in visual cortex. *J Neurosci* 2(1): 32–48.

49. Bucher D, Prinz AA, Marder E. 2005. Animal-to-animal variability in motor pattern production in adults and during growth. *J Neurosci* 25(7): 1611–1619.

50. Barto AG. 1994. Reinforcement learning control. *Curr Opin Neurobiol* 4(6): 888–893.

51. Schultz W. 2007. Behavioral dopamine signals. *Trends Neurosci* 30(5): 203–210.

52. Schultz W. 2016. Dopamine reward prediction-error signalling: A two-component response. *Nat Rev Neurosci* 17(3): 183–195.

53. Watabe-Uchida M, Eshel N, Uchida N. 2017. Neural circuitry of reward prediction error. *Annu Rev Neurosci* 40:373–394.

54. Soltani A, Wang XJ. 2010. Synaptic computation underlying probabilistic inference. *Nat Neurosci* 13(1): 112–119.

55. Gurney KN, Humphries MD, Redgrave P. 2015. A new framework for cortico-striatal plasticity: Behavioural theory meets in vitro data at the reinforcement-action interface. *PLoS Biol* 13(1): e1002034.

56. Morris G, Schmidt R, Bergman H. 2010. Striatal action-learning based on dopamine concentration. *Exp Brain Res* 200(3–4): 307–317.

57. Izhikevich EM. 2007. Solving the distal reward problem through linkage of STDP and dopamine signaling. *Cereb Cortex* 17(10): 2443–2452.

58. Schultz W. 1997. Dopamine neurons and their role in reward mechanisms. *Curr Opin Neurobiol* 7(2): 191–197.

59. Buonomano DV, Mauk MD. 1994. Neural network model of the cerebellum: Temporal discrimination and the timing of motor responses. *Neural Comput* 6: 38–55.

60. Medina JF, Garcia KS, Nores WL, Taylor NM, Mauk MD. 2000. Timing mechanisms in the cerebellum: Testing predictions of a large-scale computer simulation. *J Neurosci* 20(14): 5516–5525.

61. Marr D. 1969. A theory of cerebellar cortex. *J Physiol* 202(2): 437–470.

62. Albus JS. 1971. A theory of cerebellar function. *Math Biosci* 10: 25–61.

63. Ito M. 1989. Long-term depression. *Annu Rev Neurosci* 12: 85–102.

64. Laje R, Buonomano DV. 2013. Robust timing and motor patterns by taming chaos in recurrent neural networks. *Nat Neurosci* 16(7): 925–933.

65. Merchant H, Harrington DL, Meck WH. 2013. Neural basis of the perception and estimation of time. *Annu Rev Neurosci* 36: 313–336.

Chapter 9

1. Rigotti M, Barak O, Warden MR, et al. 2013. The importance of mixed selectivity in complex cognitive tasks. *Nature* 497(7451): 585–590.

2. Murray JD, Bernacchia A, Roy NA, Constantinidis C, Romo R, Wang XJ. 2017. Stable population coding for working memory coexists with heterogeneous neural dynamics in prefrontal cortex. *Proc Natl Acad Sci USA* 114(2): 394–399.

3. Jun JK, Miller P, Hernandez A, et al. 2010. Heterogeneous population coding of a short-term memory and decision task. *J Neurosci* 30(3): 916–929.

4. Vargas-Irwin CE, Shakhnarovich G, Yadollahpour P, Mislow JM, Black MJ, Donoghue JP. 2010. Decoding complete reach and grasp actions from local primary motor cortex populations. *J Neurosci* 30(29): 9659–9669.

5. Mante V, Sussillo D, Shenoy KV, Newsome WT. 2013. Context-dependent computation by recurrent dynamics in prefrontal cortex. *Nature* 503(7474): 78–84.

6. Morcos AS, Harvey CD. 2016. History-dependent variability in population dynamics during evidence accumulation in cortex. *Nat Neurosci* 19(12): 1672–1681.

7. Afshar A, Santhanam G, Yu BM, Ryu SI, Sahani M, Shenoy KV. 2011. Single-trial neural correlates of arm movement preparation. *Neuron* 71(3): 555–564.

8. Fusi S, Miller EK, Rigotti M. 2016. Why neurons mix: High dimensionality for higher cognition. *Curr Opin Neurobiol* 37: 66–74.

9. Lehky SR, Tanaka K. 2016. Neural representation for object recognition in inferotemporal cortex. *Curr Opin Neurobiol* 37: 23–35.

10. Sussillo D, Abbott LF. 2009. Generating coherent patterns of activity from chaotic neural networks. *Neuron* 63(4): 544–557.

11. Laje R, Buonomano DV. 2013. Robust timing and motor patterns by taming chaos in recurrent neural networks. *Nat Neurosci* 16(7): 925–933.

12. Miller P. 2013. Stimulus number, duration and intensity encoding in randomly connected attractor networks with synaptic depression. *Front Comput Neurosci* 7: 59.

13. Rigotti M, Ben Dayan Rubin D, Morrison SE, Salzman CD, Fusi S. 2010. Attractor concretion as a mechanism for the formation of context representations. *Neuroimage* 52(3): 833–847.

14. Rajan K, Harvey CD, Tank DW. 2016. Recurrent network models of sequence generation and memory. *Neuron* 90(1): 128–142.

15. Barak O. 2017. Recurrent neural networks as versatile tools of neuroscience research. *Curr Opin Neurobiol* 46: 1–6.

16. Chandrasekaran C. 2017. Computational principles and models of multisensory integration. *Curr Opin Neurobiol* 43: 25–34.

17. Buzsaki G. 2004. Large-scale recording of neuronal ensembles. *Nat Neurosci* 7: 446–451.

18. Churchland MM, Yu BM, Sahani M, Shenoy KV. 2007. Techniques for extracting single-trial activity patterns from large-scale neural recordings. *Curr Opin Neurobiol* 17(5): 609–618.

19. Shenoy KV, Sahani M, Churchland MM. 2013. Cortical control of arm movements: A dynamical systems perspective. *Annu Rev Neurosci* 36: 337–359.

20. Ganguli S, Sompolinsky H. 2012. Compressed sensing, sparsity, and dimensionality in neuronal information processing and data analysis. *Annu Rev Neurosci* 35: 485–508.

21. Maass W, Natschlager T, Markram H. 2002. Real-time computing without stable states: A new framework for neural computation based on perturbations. *Neural Comput* 14(11): 2531–2560.

22. Mazor O, Laurent G. 2005. Transient dynamics versus fixed points in odor representations by locust antennal lobe projection neurons. *Neuron* 48(4): 661–673.

23. Franke F, Jackel D, Dragas J, et al. 2012. High-density microelectrode array recordings and real-time spike sorting for closed-loop experiments: An emerging technology to study neural plasticity. *Front Neural Circuits* 6: 105.

24. Grienberger C, Konnerth A. 2012. Imaging calcium in neurons. *Neuron* 73(5): 862–885.

25. Stirman JN, Smith IT, Kudenov MW, Smith SL. 2016. Wide field-of-view, multi-region, two-photon imaging of neuronal activity in the mammalian brain. *Nat Biotechnol* 34(8): 857–862.

26. Ziv Y, Burns LD, Cocker ED, et al. 2013. Long-term dynamics of CA1 hippocampal place codes. *Nat Neurosci* 16(3): 264–266.

27. Welch LR. Hidden Markov models and the Baum-Welch algorithm. *IEEE Information Theory Society Newsletter* . 2003: 53(4).

28. Baum LE, Petrie T, Soules G, Weiss N. 1970. A maximization technique occurring in the statistical analysis of probabilistic functions of Markov chains. *Ann Math Stat* 41: 164–171.

29. Abeles M, Bergman H, Gat I, et al. 1995. Cortical activity flips among quasi-stationary states. *Proc Natl Acad Sci USA* 92(19): 8616–8620.

30. Seidemann E, Meilijson I, Abeles M, Bergman H, Vaadia E. 1996. Simultaneously recorded single units in the frontal cortex go through sequences of discrete and stable states in monkeys performing a delayed localization task. *J Neurosci* 16(2): 752–768.

31. Jones LM, Fontanini A, Sadacca BF, Miller P, Katz DB. 2007. Natural stimuli evoke dynamic sequences of states in sensory cortical ensembles. *Proc Natl Acad Sci USA* 104(47): 18772–18777.

32. Camproux A-C, Saunier F, Chouvet G, Thalabard J-C, Thomas G. 1996. A hidden Markov model approach to neuron firing patterns. *Biophys J* 71: 2404–2412.

33. Miller P, Katz DB. 2010. Stochastic transitions between neural states in taste processing and decision-making. *J Neurosci* 30(7): 2559–2570.

34. Miller P, Katz DB. 2011. Stochastic transitions between states of neural activity. In: Ding M, Glanzman DL, eds. *The Dynamic Brain: An Exploration of Neuronal Variability and Its Functional Significance* (29–46). New York: Oxford University Press.

35. Brown EN, Frank LM, Tang D, Quirk MC, Wilson MA. 1998. A statistical paradigm for neural spike train decoding applied to position prediction from ensemble firing patterns of rat hippocampal place cells. *J Neurosci* 18(18): 7411–7425.

36. Eden UT, Frank LM, Barbieri R, Solo V, Brown EN. 2004. Dynamic analysis of neural encoding by point process adaptive filtering. *Neural Comput* 16(5): 971–998.

37. Truccolo W, Eden UT, Fellows MR, Donoghue JP, Brown EN. 2005. A point process framework for relating neural spiking activity to spiking history, neural ensemble, and extrinsic covariate effects. *J Neurophysiol* 93(2): 1074–1089.

38. Deneve S, Duhamel JR, Pouget A. 2007. Optimal sensorimotor integration in recurrent cortical networks: A neural implementation of Kalman filters. *J Neurosci* 27(21): 5744–5756.

39. Faragher R. 2012. Understanding of the basis of the Kalman filter via a simple and intuitive derivation. *IEEE Signal Process Mag* 29(5): 128–132.

Index

Computational Neuroscience

Terrence J. Sejnowski and Tomaso A. Poggio, editors

Neural Nets in Electric Fish, Walter Heiligenberg, 1991

The Computational Brain, Patricia S. Churchland and Terrence J. Sejnowski, 1992

Dynamic Biological Networks: The Stomatogastic Nervous System, edited by Ronald M. Harris-Warrick, Eve Marder, Allen I. Selverston, and Maurice Moulins, 1992

The Neurobiology of Neural Networks, edited by Daniel Gardner, 1993

Large-Scale Neuronal Theories of the Brain, edited by Christof Koch and Joel L. Davis, 1994

The Theoretical Foundations of Dendritic Function: Selected Papers of Wilfrid Rall with Commentaries, edited by Idan Segev, John Rinzel, and Gordon M. Shepherd, 1995

Models of Information Processing in the Basal Ganglia, edited by James C. Houk, Joel L. Davis, and David G. Beiser, 1995

Spikes: Exploring the Neural Code, Fred Rieke, David Warland, Rob de Ruyter van Steveninck, and William Bialek, 1997

Neurons, Networks, and Motor Behavior, edited by Paul S. Stein, Sten Grillner, Allen I. Selverston, and Douglas G. Stuart, 1997

Methods in Neuronal Modeling: From Ions to Networks, second edition, edited by Christof Koch and Idan Segev, 1998

Fundamentals of Neural Network Modeling: Neuropsychology and Cognitive Neuroscience, edited by Randolph W. Parks, Daniel S. Levine, and Debra L. Long, 1998

Neural Codes and Distributed Representations: Foundations of Neural Computation, edited by Laurence Abbott and Terrence J. Sejnowski, 1999

Unsupervised Learning: Foundations of Neural Computation, edited by Geoffrey Hinton and Terrence J. Sejnowski, 1999

Fast Oscillations in Cortical Circuits, Roger D. Traub, John G. R. Jeffreys, and Miles A. Whittington, 1999

Computational Vision: Information Processing in Perception and Visual Behavior, Hanspeter A. Mallot, 2000

Graphical Models: Foundations of Neural Computation, edited by Michael I. Jordan and Terrence J. Sejnowski, 2001

Self-Organizing Map Formation: Foundations of Neural Computation, edited by Klaus Obermayer and Terrence J. Sejnowski, 2001

Neural Engineering: Computation, Representation, and Dynamics in Neurobiological Systems, Chris Eliasmith and Charles H. Anderson, 2003

The Computational Neurobiology of Reaching and Pointing, Reza Shadmehr and Steven P. Wise, 2005

Dynamical Systems in Neuroscience, Eugene M. Izhikevich, 2006

Bayesian Brain: Probabilistic Approaches to Neural Coding, edited by Kenji Doya, Shin Ishii, Alexandre Pouget, and Rajesh P. N. Rao, 2007

Computational Modeling Methods for Neuroscientists, edited by Erik De Schutter, 2009

Neural Control Engineering, Steven J. Schiff, 2011

Understanding Visual Population Codes: Toward a Common Multivariate Framework for Cell Recording and Functional Imaging, edited by Nikolaus Kriegeskorte and Gabriel Kreiman, 2011

Biological Learning and Control: How the Brain Builds Representations, Predicts Events, and Makes Decisions, Reza Shadmehr and Sandro Mussa-Ivaldi, 2012

Principles of Brain Dynamics: Global State Interactions, edited by Mikhail Rabinovich, Karl J. Friston, and Pablo Varona, 2012

Brain Computation as Hierarchical Abstraction, Dana H. Ballard, 2015

Visual Cortex and Deep Networks: Learning Invariant Representations, Tomaso A. Poggio and Fabio Anselmi, 2016

Case Studies in Neural Data Analysis: A Guide for the Practicing Neuroscientist, Mark A. Kramer, and Uri T. Eden, 2016

From Neuron to Cognition via Computational Neuroscience, edited by Michael A. Arbib and James J. Bonaiuto, 2016

The Computational Brain, 25th Anniversary Edition, Patricia S. Churchland and Terrence J. Sejnowski, 2017

An Introductory Course in Computational Neuroscience, Paul Miller, 2018